LIGHT VEHICLE MAINTENANCE AND REPAIR

LEVEL 2

Patrick Hamilton, John Rooke and Robert Sharman

Edited by Roy Brooks

DELMAR
CENGAGE Learning

Australia • Brazil • Japan • Korea • Mexico • Singapore • Spain • United Kingdom • United States

Light Vehicle Maintenance and Repair Level 2
Patrick Hamilton, John Rooke and Robert Sharman
Edited by Roy Brooks

Publishing Director: Linden Harris

Commissioning Editor: Lucy Mills

Editorial Assistant: Claire Napoli

Project Editor: Alison Cooke

Production Controller: Eyvett Davis

Marketing Executive: Lauren Redwood

Typesetter: MPS Limited, a Macmillan Company

Cover design: HCT Creative

Text design: Design Deluxe, Bath

For product information and technology assistance, contact **emea.info@cengage.com**.
For permission to use material from this text or product, and for permission queries, email **emea.permissions@cengage.com**.

British Library Cataloguing-in-Publication Data
A catalogue record for this book is available from the British Library.

ISBN: 978-1-4080-5749-0

Cengage Learning EMEA
Cheriton House, North Way,
Andover, Hampshire SP10 5BE
United Kingdom

Cengage Learning products are represented in Canada by Nelson Education Ltd.

For your lifelong learning solutions, visit **www.cengage.co.uk**

Purchase your next print book, e-book or e-chapter at **www.cengagebrain.com**

Printed in the United Kingdom by Ashford Colour Press Ltd
Print Number: 10   Print Year: 2023

MIX
Paper from
responsible sources
FSC
www.fsc.org   FSC® C011748

I would like to thank my wife Sharon, daughters Jessica and Kirsten, also my late step son Peter and step son Simon, for the patience and support whilst I have worked on this book

Patrick Hamilton

I would like to thank Jackie for supporting and encouraging me during the many hours writing the workbook

John Rooke

I would like to thank my wife Julie and my son Jonathan, for walking the dogs and supplying copious amounts of hot drink, whilst I have spent my time researching and writing.

Bob Sharman

CONTENTS

FOREWORD

Welcome to this very latest edition in the Light Vehicle Maintenance and Repair series, popularly know as the "Brooks Books". For more than 40 years these books have helped many tens of thousands of motor vehicle students to gain and conveniently record knowledge in the exciting world of automobile engineering.

Although the basic essentials, such as "suck, squeeze, bang, blow" must remain the same, automotive technology and the way in which it is taught is constantly evolving. Similarly so with the books which from the very beginning have been frequently subject to technical revision and variation of content to suit ever changing needs.

This latest full colour edition with a high proportion of specially drawn illustrations, maintains the original ease of use and understanding, but adds considerable value by including helpful practical tips, guidance to useful web sites, word puzzles and the like – all designed to stimulate interest and remain a valuable source of reference. With due deference to the long-established highly successful format of the books, the new team of authors, all practicing and experienced lecturers in motor vehicle work, clearly show that they understand and accurately cater for the needs of both students and teachers.

Roy Brooks, Editor

ACKNOWLEDGEMENTS

The Publisher wishes to thank the following companies for granting permission to use images, drawings and artworks:

A-C Delco	Ford (UK)	Tata (UK), Rover
Autodata Ltd.	Gates	The Tool Connection Ltd and Laser Tools
Betex	Goodyear, Dunlop	Schaeffler.com
BGI	Honda (UK)	Sealey Group
BMW	Lotus	Snap On Industrial
Bosch (UK)	Motor Industry Magazine	Talbot
Bridgestone	NRMA Motoring & Services	TRW, lucas
Champion	Peugeot	Vauxhall Motors
Chris Longhurst – www.carbibles.com	Pico Scope	Volkswagen
Delphi Automotive Systems, AP Ltd	Renault	
Draper Tools Ltd	Ridetech	

Photo Research and Permissions by James Clift, Jason Newman and Alex Goldberg of www.mediaselectors.com

ABOUT THE AUTHORS

Patrick Hamilton CertEd PGDE LCGI EngTech MIMechE AAE MIMI MIfL

Patrick has over 20 years light and heavy vehicle practical experience gained at leading vehicle manufacturers' main dealerships, including six years in the Royal Air Mechanical Transport Servicing Section. He is currently Head of the School of Engineering at West Suffolk College and teaches on automotive and engineering courses. Patrick has experience of teaching on a range of light vehicle and heavy vehicle programmes ranging from entry level to Level 4. He has also worked for the Sector Skills Council writing some of their QCF units and has been a technical consultant and author for City and Guilds and the IMI Awards, helping to develop and write their qualifications.

John Rooke CertEd, MIMI, MifI

John Rooke has over 30 years experience in the motor trade trained at a main dealer and he then took the opportunity to enter teaching as an instructor in a private training school, teaching practical skills. In 1999 he became a lecturer at Cambridge Regional College teaching Automotive Engineering to learners from entry level to Level 3. John presently has an additional role of supporting engineering staff to improve teaching and learning. During the past nine years he has been the colleges IMI centre coordinator enabling him to keep fully up to date with qualification and assessment requirements.

Robert Sharman B.Ed. I. Eng. MIMI. MIRTE. MSOE.

Robert has over 15 years of practical experience, working on a wide range of light/heavy vehicles and road construction plant. He also has 30 years experience of teaching full time in Further Education, on a full range of Motor Vehicle Craft and Technician programmes up to Level 4, during which time he was also a City and Guilds moderator, IMI co-ordinator and Programme Leader of a large Automobile section. Robert is now working on a consultancy basis for the Sector Skills Council and IMI Awards, assisting in the development of new qualifications.

QUALIFICATION MAPPING GRID

Light Vehicle Maintenance & Repair Level 2 QCF unit	Part 1 – Introduction	Section 1 Health & Safety	Section 2 Good Housekeeping	Part 2 – The Motor Industry	Section 1 Workplace structure and job roles	Section 2 Tools and Equipment	Part 3 – Engines	Section 1 Construction and operating principles	Section 2 Cooling	Section 3 Lubrication	Section 4 Fuel injection systems	Section 5 Ignition Systems	Section 6 Exhaust Systems	Part 4 – Chassis	Section 1 Vehicle Construction	Section 2 Wheels & Tyres	Section 3 Braking systems	Section 4 Suspension systems	Section 5 Steering systems	Part 5 – Transmission	Section 1 Vehicle layouts	Section 2 Clutch	Section 3 Gearbox	Section 4 Final drive	Section 5 Driveline	Part 6 – Electrical Systems	Section 1 Understanding electricity	Section 2 Circuits and operation	Section 3 Batteries	Section 4 starting system	Section 5 Charging system	Section 6 Auxiliary systems	Section 7 Diagnosis and testing
G01/02 Health, Safety and Good Housekeeping in the Automotive Environment		●	●																														
G03 Support working relationships in the Automotive Work Environment					●																												
G04 Health, Safety and Good Housekeeping in the Automotive Environment						●																											
LV02.1 Principles of Light Vehicle Engine Mechanical, lubrication and Cooling System Units and Components								●	●	●																							
LV02.2 Principles of Light Vehicle Fuel, Ignition, Air and Exhaust System Units and Components											●	●	●																				
LV04 Principles of Light Vehicle Chassis Units and Components																●	●	●	●														
LV012 Principles of Light Vehicle Transmission and Driveline Units and Components																						●	●	●	●								
LV03 Principles of Motor Vehicle Electrical Units and Components																											●	●	●	●	●	●	
LV03/AE06 (L3) Principles of Diagnosis and Rectification of Light Vehicle Auxiliary Electrical Faults																																	●

Note: Aspects of Units LV01 - Carry Out Routine Light Vehicle Maintenance and LV05 - Inspect Light Vehicles Using Prescribed Methods, as well as optional units are contained within relative sections of the workbook. Vehicle construction and layout, whilst not specific to any unit has been included with regard to basic principles. It should be remembered that whilst the main learning objectives of the qualification have been included, it is important to refer to the latest QCF structure and units to ensure full coverage. The workbook is not intended to be a complete syllabus.

ABOUT THE BOOK

Correctly set up a DTI and check the end float of a wheel hub bearing. Compare the readings with those of the manufacturer.

Activity boxes provide additional tasks for you to try out

www

For more information on alignment go to **www.alignmycar.co.uk**

Web link boxes suggest websites to further research and understanding of a topic

TIP

When cleaning seat belt webbing, only use a mild soap solution and water.

Tip boxes share author's experience in the automotive industry, with helpful suggestions for how you can improve your skills

If you need to use a hazardous cleaning material, read the label on the container. This will tell you how to use it safely, and what to do if you *do* have an accident – for example, if the cleaning material touches your eyes or skin.

Health and safety tip boxes draw your attention to important health and safety information

Multiple choice questions

Choose the correct answer from a), b) or c) and place a tick [✓] after your answer.

1 **Which one of the following is a type of file?**

   a) bastard [✓]
   b) morse [ ]
   c) vernier. [ ]

2 **Which one of the following would be used to measure the thickness of a brake disc?**

   a) DTI [ ]
   b) micrometer [✓]
   c) rule. [ ]

Multiple choice questions are provided at the end of each chapter. You can you use questions to test your learning and prepare for assessments

Learning objectives

After studying this section, you should be able to:

● **Identify light vehicle engine mechanical system components.**
● **Describe the construction and operation of light engines.**

Learning objectives at the start of each chapter explain key skills and knowledge you need to understand by the end of the chapter

Online lecturer resource Check answers to all of the student activity questions in this book by visiting: www.cengage.co.uk/auto2

Traditionally learners would have attended college full time to achieve a Vocationally Related Qualification (VRQ) or part time as part of an apprenticeship, gaining a National Vocational Qualification (NVQ) and a VRQ as well as the required key skills. To an extent this has not really changed.

What has changed is how the qualifications have been developed, allowing a more flexible approach to the selection of units and their delivery. This has been developed by the Sector Skills Council (SSC) for the industry area concerned. The SSC for the automotive industry is known as Automotive Skills, a division of the Institute of the Motor Industry (IMI). The SSC is employer led and acts in response to employer needs. It does this by producing National Occupational Standards (NOS), developed by employer partnerships and working parties. The NOS describe the different functions carried out by people working throughout the range of sectors in the industry. The Skills, Knowledge and Competency requirements are identified for all the levels of technical, parts, sales and operations management.

Qualification Credit Framework (QCF) units have been developed from the NOS. These are the common units which awarding bodies, such as IMI Awards, City and Guilds and Edexcel Btec use to develop their qualifications. This creates a method of standardisation across all awarding bodies.

One of the advantages of the QCF units is that a learner can complete individual units at one training provider using a specific awarding body. They may, for whatever reason, have to move to another training provider who uses a different awarding body, whereby the accredited units are transferable.

The QCF units cover Competency, Skills and Knowledge. Each unit is allocated a "Credit" value. A predetermined number of "Credits" are required to achieve a qualification. The QCF units and "Credits" are transferable across awarding bodies, allowing the learner to build and complete their VRQ and/or Vocational Competence Qualification (VCQ) even if they move around the country.

The Knowledge units cover the technical understanding of the subject area. The Skills units show that the learner is able to carry out practical tasks to a required standard. These units are designed to be delivered in a college and training provider environment. The Competency units show that a learner not only has the required skills but that they are now able to perform tasks independently within given timescales and with limited support and guidance. The Competency units can only be completed in the workplace.

The VRQ is specifically designed for a training environment based delivery on either a full or part-time basis. With the use of the QCF units it is divided into two areas, those of Skills and Knowledge. A variety of qualifications have been designed to meet the VRQ criteria which are constructed with a mixture of the QCF units. These can provide learners with practical skills and knowledge preparing them for work in the automotive industry.

The Vocational Competence Qualification (VCQ) (originally known as the NVQ) covers the Competency requirements needed for an individual to satisfactorily perform and function, for example, at Level 2.

Employers can select QCF units which meet their business and strategic needs.

This workbook has been designed to assist the learner in developing their knowledge and skills and to provide support for final competence, in Light Vehicle Maintenance and Repair at Level 2.

PART 1
INTRODUCTION

USE THIS SPACE FOR LEARNER NOTES

_____

_____

_____

_____

_____

_____

_____

_____

SECTION 1

Health and safety in the automotive environment

USE THIS SPACE FOR LEARNER NOTES

_____

_____

_____

_____

_____

Learning objectives

After studying this section you should be able to:

- List the main legislation relating to automotive environment health and safety and describe the general legal duties of employers and employees.
- Identify key hazards and risks and describe policies and procedures for reporting them.
- Identify key warning signs and their characteristics.
- Explain the importance of wearing Personal Protective Equipment (PPE).
- State the meaning of common product warning labels.
- Identify fire extinguishers in common use and which types of fire they should be used on.
- State procedures that need to be taken with tools, equipment and materials.
- Describe vehicle and personal safety considerations when working at the roadside.

Key terms

Match the definition using the terms at the bottom of the page.

Cancer forming. _____

Likelihood or chance of harm being caused. _____

Poisonous, likely to cause injury or death (often chemical). _____

Something likely to cause harm or loss – a source of danger. _____

Substance likely to catch fire. _____

An unplanned event that results in injury, ill-health or damage. _____

Substance that can destroy tissue, usually strong acid or alkali. _____

Substance that can cause ill-health or injury. _____

Evaporates readily – can cause fire or explosion, e.g. petrol. _____

Hazard, Risk, Accident, Carcinogenic, Flammable/inflammable, Volatile, Toxic, Harmful, Corrosive

Health and safety starter

Try your hand at this word search to become familiar with some of the terms in this chapter.

```
T N B A T W E W C W X B H N R
N L O K X H Q O K M B R D X E
A D W I R P R I O V E O M I G
T D Z N T R P K F H Y D I Z U
I P B N O C H C S H T J W Y L
R B R S W L U I H A Z A R D A
R G I O E Q U D Y L Q C D Z T
I V I N H G H J N V F T Z H I
E C O R N I S K R I J S Y M O
V Z F I R E B L A N K E T Z N
B R T L L E M I P W F Q O C L
P X I G C Y R O T A D N A M H
E R G S V T K H I I M F I B V
M O U D K P L I O N O K F R I
G X I T X R R X Z G A N Z R P
```

ACTS
CORROSIVE
EXTINGUISHER
FIREBLANKET
GOGGLES
HAZARD
INDUCTION
IRRITANT
MANDATORY
PROHIBITION
REGULATION
RISK

HEALTH AND SAFETY AT WORK ACT 1974 (HASAWA)

The Health and Safety at Work Act (HASAWA) covers all people at work whatever their occupation and sets out employer's and employees' duties regarding health and safety in the workplace. The workplace could include anywhere that people are employed such as garages, colleges or training providers, therefore it affects you!

The Act is enforced by the Health and Safety Executive (HSE), which has inspectors who give advice to employers, check workplaces, and investigate accidents.

Employers must display a Health and Safety Law poster or provide employees with a booklet, which is available from the HSE website.

New poster

Find your Health and Safety Law poster or visit the HSE website to view a Health and Safety Law pocket card and list at least four employer duties.

1 _____

_____

2 _____

_____

3 _____

4 _____

_____

TIP Health and safety regulations are updated frequently.

Check **www.hse.gov.uk** for the latest regulations.

Enforcement

A body called The Health and Safety Executive Inspectorate enforces the HASAWA. Its inspectors have various powers and penalties at their disposal.

Describe the two main types of enforcement that the HSE inspectorate may take.

1 _____

_____

2 _____

_____

TIP Look back at the Health and Safety poster for your responsibilities.

SPECIFIC ACTS AND REGULATIONS

Below is a list of various Regulations which mostly apply to special situations. The Regulations add further depth and detail to the Acts. They are made as the need arises and carry as much legal authority as the Act to which they relate. New or amended Regulations therefore keep the Acts up-to-date.

Electricity at Work Regulations 1989
Workplace (Health, Safety and Welfare) Regulations 1992
Personal Protective Equipment at Work Regulations 1992
Provision and Use of Work Equipment Regulations 1998 (PUWER)
Lifting Operations and Lifting Equipment Regulations 1998 (LOLER)
Management of Health and Safety at Work Regulations 1999
Control of Substances Hazardous to Health Regulations 2002 (COSHH)
Dangerous Substances and Explosive Atmospheres Regulations (DSEAR) 2002
Manual Handling Regulations 1992 (amended 2002)
Control of Noise at Work Regulations 2005
Health and Safety (Display Screen Equipment) Regulations 1992
Health and Safety (First Aid) Regulations 1981
Employers' Liability (Compulsory Insurance) Act 1969
Reporting of Injuries, Diseases and Dangerous Occurrences Regulations 1995 (RIDDOR)
Pressure Systems Safety Regulations 2000

State which Regulation would be relevant to each of the following situations.

1 Removing a road wheel from a heavy-duty four-wheel drive off-road car.

_____

2 Cleaning the dust from rear drum brakes.

_____

3 Draining petrol from a fuel tank.

_____

4 Using a bench grinder to sharpen a cold chisel.

_____

5 Lifting an engine/gearbox assembly from a vehicle using chains and an engine hoist.

_____

6 Using a mains electricity power washer to clean a vehicle exterior.

_____

7 Carrying out a major routine service on a car.

_____

8 Working on a rolling road with performance vehicles.

_____

9 Working using a computer for 5 hours a day.

_____

10 Removing a steering wheel with an air bag fitted.

_____

 TIP The HSE website now has a dedicated motor vehicle repair section dealing with our industry.

RISK ASSESSMENT

In 1992 The Management of Health and Safety at Work Regulations were created which gave more detail to the general requirements of the HASAWA.

In general, employers have to carry out risk assessments where specific hazards have been identified.

This is the suggested method from the HSE:

- **Identify the hazards.**
- **Decide who might be harmed and how.**
- **Evaluate the risks and decide on precautions – change process, modify or PPE.**
- **Record findings and implement (communicate to employees – training may be required).**
- **Review assessments and update if necessary.**

Who would be likely to carry out a risk assessment in the workplace?

_____

Here are some examples of garage hazards that require a risk assessment:

- **draining vehicle fuel tanks.**
- **handling vehicle air bags.**
- **using mains electrical equipment.**

Suggest three hazards in your workshop that would require a risk assessment.

1 _____

2 _____

3 _____

 TIP The motor vehicle repair section of the HSE website gives good examples of hazards.

INITIAL INDUCTION TO HEALTH AND SAFETY REQUIREMENTS

When new members of staff start at the garage they will need to receive an in-company health and safety induction. This is important to ensure new staff members are aware of safe procedures and processes.

You are the supervisor in a large dealership running the light vehicle maintenance and repair workshop. You have a new trainee starting next week.

In your group discuss what should be covered in a good company initial induction for a 16-year-old trainee.

- _____
- _____
- _____
- _____
- _____
- _____

- _____
- _____
- _____
- _____
- _____

How is it reported?

- _____
- _____
- _____
- _____

REPORTING TO THE HEALTH AND SAFETY EXECUTIVE

The Reporting of Injuries, Diseases and Dangerous Occurrences Regulations 1995 (RIDDOR)

Certain situations at work require employers to report by various methods to the HSE, to identify where and how risks arise, and to investigate serious accidents.

Find information on the Reporting of Injuries, Diseases and Dangerous Occurences Regulations (RIDDOR) and state what should be reported.

- _____
- _____

Dangerous occurrence

Even if no one was injured, incidents such as hydraulic failure of lifting equipment including jacks or engine hoists are reportable.

ACCIDENT RECORDING

If you have an accident at work, which types of injuries have to be reported to your employer?

_____

Injuries are recorded and records held for 3 years.

Most companies use an accident book, which must contain the following:

- **Injured person's personal details – Name, age (DOB), job description.**

- **Details of the injury** – Nature of the injury and how it occurred.
- **When the accident happened** – Date, time and location.

Who can fill out the accident book?

_____

REPORTING SAFETY CONCERNS

You have a legal responsibility to protect yourself and others; therefore, if you find a potential hazard then it is your duty to report it to your supervisor.

Why should you report a hydraulic jack that is leaking?

_____

_____

In your organization there may be a procedure of labelling faulty equipment to prevent use until action is taken.

Pictures supplied by Draper Tools Limited

CAUSES OF ACCIDENTS

Generally speaking, accidents are caused by:

Human	Environmental
1 Ignorance of the dangers involved.	1 Unguarded or faulty machinery.
2 Failure to take adequate precautions.	2 Incorrect or faulty tools.
3 Tiredness, causing lack of concentration.	3 Inadequate ventilation.
4 Fooling about.	4 Badly-lit workshops.

State six other causes of accidents in the workplace:

1 _____

2 _____

3 _____

4 _____

5 _____

6 _____

SAFETY SIGNS

There are four types of safety signs, identify each of the following:

_____   _____   _____   _____

Contrasting colours make the sign more conspicuous.

Black is used with Yellow. White is used with Red, Blue or Green.

Prohibition signs

State the meaning of prohibition.

_____

State what prohibition is indicated in the following signs:

HSE

_____   _____   _____   _____   _____

_____   _____   _____   _____   _____

Warning signs

State the risk that each sign indicates:

_____ _____ _____ _____

_____ _____ _____ _____

Mandatory signs

State the meaning of mandatory – _____

State the type of protection that must be worn:

_____ _____ _____ _____

 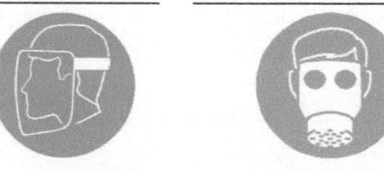  

_____ _____ _____ _____

Safe condition

State what each sign indicates:

_____ _____ _____ _____ _____

PERSONAL PROTECTIVE EQUIPMENT (PPE)

All Personal Protective Equipment (PPE) in use at work should carry the CE mark and where appropriate should comply with a European Norm (EN) standard.

Which parts of the body do the following items of PPE protect?

Overalls – _____

Steel toecapped boots or shoes – _____

Safety goggles – _____

Welding mask – _____

Hard hat – _____

Gloves – _____

Ear defenders and plugs – _____

Dust mask or respirator – _____

Eye protection

Give examples of where the following eye protection may be worn in a garage environment.

Face mask _____
_____

Darkened face mask _____
_____

Clear goggles or safety glasses _____
_____
_____
_____
_____
_____

Darkened goggles _____
_____

Hearing protection

Ear defenders (ear muffs)

Disposable ear plugs

Give **two** examples of where hearing protection may be needed in a garage environment.

1 _____

2 _____

Head protection

Three types of head protection may be used in a garage workshop.

Describe what the following head wear protects.

1 Bump cap

_____

2 Hair net cap

_____

3 Hard hat

_____

Skin protection

Thoroughly cleaning the skin, particularly hands, face and neck, is extremely important. How should hands be protected?

● _____

● _____

● _____

_____

_____

Hand protection

Give examples of where these gloves could be used to protect your hands.

Heavy-duty leather gloves

_____

_____

_____

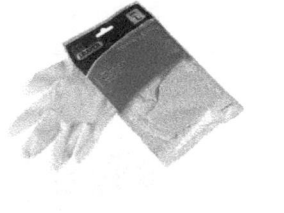

Latex gloves

_____

_____

_____

Latex gloves can cause an allergic reaction on those sensitive to latex. Vinyl or nitrile may be an alternative.

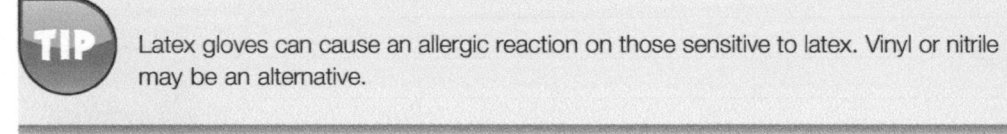

Nitrile gloves

_____

_____

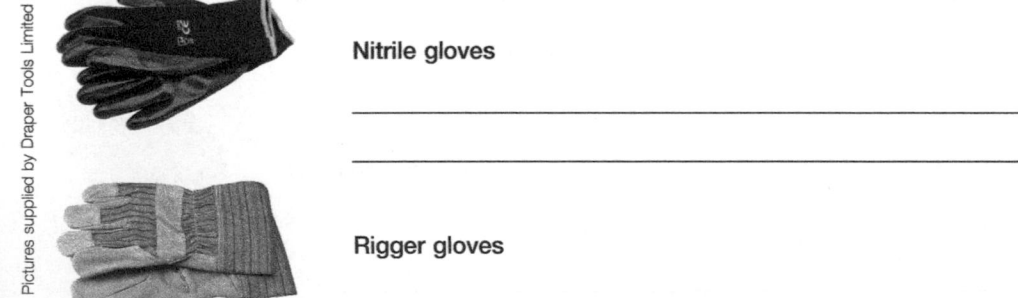

Rigger gloves

_____

_____

_____

_____

TIP If during servicing or repair activities hypodermic needles are found, inform your supervisor to organise for specialist removal.

Special gloves, which cannot be penetrated, are required for this procedure.

Protection of lungs

Dust masks may be worn to protect your lungs from dust.

Give examples where dust masks would be used.

_____

_____

_____

Care of Personal Protective Equipment

Report any faults with PPE to your supervisor immediately.

State what you would check for on each of following items of PPE:

Safety goggles – _____

Overalls – _____

Safety boots – _____

Gloves – _____

Ear defenders – _____

Dust masks – _____

HAZARDOUS SUBSTANCES

Under COSHH, or the Control of Substances Hazardous to Health Regulations 2002, all persons at work need to know the safety precautions to take so as not to endanger themselves or others through exposure to substances hazardous to health.

Using the table below state the hazard each symbol represents and explain the likely effects caused by a substance that is labelled with each symbol.

Symbol	Hazard	Effect

A new system of marking hazardous substances is to be fully in place by 1 June 2015. The table below gives the new symbols under **C**lassification **L**abelling and **P**ackaging Regulations, abbreviated to **CLP.**

You will see more use of these symbols before this date.

	Example of hazard statement	Example of precautionary statement
	Heating may cause an explosion	Keep away from heat/ sparks/open flames/ hot surfaces – no smoking
	Heating may cause a fire	Keep only in original container
	May intensify fire; oxidizer	Take any precaution to avoid mixing with combustibles
	Causes serious eye damage	Wear eye protection
	Toxic if swallowed	Do not eat, drink or smoke when using this product
	Toxic to the aquatic life, with long-lasting effects	Avoid release to the environment
	New pictogram, reflects serious longer-term health hazards such as carcinogenicity and respiratory sensitization e.g. may cause allergy or asthma symptoms or breathing difficulties if inhaled	In case of inadequate ventilation, wear respiratory protection
	New pictogram, refers to less serious health hazards such as skin irritancy/sensitisation and replaces the CHIP X symbol e.g. may cause an allergic skin reaction	Contaminated work clothing should not be allowed out of the workplace
	New pictogram, used when the containers hold gas under pressure e.g. may explode when heated	None

HSE

The Control of Substances Hazardous to Health Regulations 2002 (COSHH) data sheets

All hazardous substances supplied must have either a paper-based or web-based **M**aterial **S**afety **D**ata **S**heet (MSDS) available.

Harmful substances

Certain activities in a motor vehicle repair premises present particular health hazards. The hazards may, for example, be due to breathing in polluted air or coming into contact with harmful substances. (See COSHH Regulations.)

 List three toxic gases or substances likely to be present in a motor vehicle repair workshop:

- _____
- _____
- _____

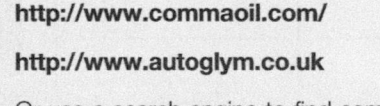 **http://www.commaoil.com/**

http://www.autoglym.co.uk

Or use a search engine to find companies' MSDS

Group activity

Discuss the hazardous actions in the table opposite. Identify the potential hazards and agree on precautions which can be taken to reduce the risk of each hazard occurring.

Hazardous actions	Potential hazard	Suitable precautions
Engine tuning	exhaust fumes: _____	pipe gases outside, adequate ventilation, use of extractor fans, gas not aimed into confined space
Welding	_____	_____
Degreasing parts washer	_____	_____
Battery charging	_____	_____

ELECTRICAL SAFETY

Two dangers arising as a result of using electricity in a workshop are:

- *fire* **possibly caused by an electric circuit overheating or a burst bulb igniting fuel**
- *electric shock* **as a result of someone coming into contact with a live circuit.**

For safety reasons hand-held electrically operated equipment and hand lamps should use reduced voltage.

To further reduce the risk of fire and electrocution many garages use battery-operated hand-held equipment.

What could cause a circuit to overheat?

_____

_____

_____

Pictures supplied by Draper Tools Limited

Checks on mains electrical equipment

Checks that should be made on the body of the power tool are given below. Complete the table stating the checks that should be completed on the cable and the plug.

Checks on the body of the power tool	
Pictures supplied by Draper Tools Limited	● make sure the tool is free from dirt or grease and vents are clear ● check for cracks ● check the tool has a label to show that it passed a portable appliance test (PAT) and is in date ● check for correct operation of the switch.
Checks on the cable *Pictures supplied by Draper Tools Limited*	● _____ ● _____ ● _____ ● _____ ● _____
Checks on the plug	● _____ ● _____ _____ _____ ● _____ ● _____

FLAMMABLE LIQUIDS AND GASES

Many flammable substances are used in garages. List five other examples below.

1 petrol

2 _____

3 _____

4 _____

5 _____

6 _____

Some liquids are *volatile*. What is meant by this and what particular hazards can this present during the normal course of repair work?

_____

_____

_____

Draining a fuel tank

You are about to drain the petrol from a vehicle's tank.

Describe how to safely complete the procedure in the form of a bullet point list. Use the image to help you.

● _____

_____

● _____

Sealey Group

13

- _____
- _____
- _____
- _____
- _____
- _____

WWW Visit **http://www.hartleige.com/product/fuel-tank-drainers** to find out more about fuel retrievers.

Look at the scenarios below and answer these questions relating to garage situations where flammable substances or gases are involved.

1 Why should petrol not be drained near this vehicle inspection pit?

_____

_____

_____

2 What could happen if you disconnected the battery charger from the battery terminals while the charger is still operating?

_____

_____

_____

_____

3 What could happen if the fuel pressure was not relieved?

_____

FIREFIGHTING EQUIPMENT

Liquids or chemicals that are highly combustible are commonly found in motor vehicle workshops such as petrol, cleaning solvents, paints, etc. It is therefore important that everyone tries to prevent a fire and has a working knowledge of how to use the correct type of fire extinguisher required to eliminate a fire.

Complete the triangle to show:

- **The three elements needed to start a fire.**
- **One action needed to stop each element and stop the fire.**

Fires are classified by their type and a letter is used to identify which fire extinguisher will put out each type of fire.

Fire extinguishers

There are four main types of fire extinguishers available in garages which are suitable for stopping different types of fire.

 Look at the **TYPES OF FIRE** under the Classification of Fire Risk, and **TICK** in the grid which type of FIRE extinguisher would be suitable.

Classification of Fire Risk	WATER	FOAM	CO$_2$ GAS	POWDER
Class A Paper, wood, textiles				
Class B Flammable liquids				
Class C Flammable gases, liquids				
Electrical hazards ⚡				

Which extinguishers must not be used on mains electrical fires and why?

_____

Colour coding for fire extinguishers

Since January 1997 the British standard for fire extinguishers has been **BS EN 3.** This standard states that all fire extinguishers must be red, although 5 per cent may be colour coded using the former colours.

Fire extinguishers sold before January 1997 which are painted in the old colours are still allowed providing they are in good condition and recharged correctly.

State which of the three elements the extinguisher removes and briefly describe how to use each extinguisher.

Foam

Also known as AFFF (Aqueous Film-Forming Foam).

Removes – _____

How to use – _____

_____

_____

Foam fire extinguisher

Water

Removes – _____

How to use – _____

_____

Water fire extinguisher

TIP Do not use a water fire extinguisher on burning liquids as it spreads the fire.

CO₂ gas (carbon dioxide)

Removes – _____

How to use – _____

_____

_____

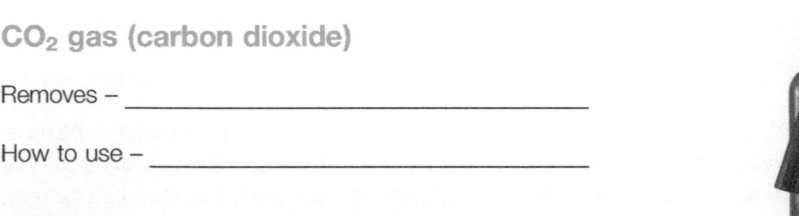

CO₂ fire extinguisher

All pictures are supplied by Draper Tools Limited

16

TIP
- Only hold the insulated horn and handle of a CO_2 gas extinguisher as the brass fittings get extremely cold and could cause freeze burns to the skin.
- Only use where there is adequate ventilation.
- There is little cooling effect, therefore a fire may reignite once the gas is no longer present.

Dry powder (multi-purpose)

Removes – _____

How to use – _____

_____

_____

Dry powder fire extinguisher

TIP Dry powder fire extinguishers are effective but do leave a powder residue which may damage equipment and cause corrosion of electrical connections.

Fire blankets can be used to smother a fire or wrap around a person if their clothes are on fire.

Fire blanket

FIRE PREVENTION AND CONTROL

Doors and passages must be kept clear and a positive routine established, to be followed in the event of a fire.

Briefly describe the procedure to be followed in the event of a fire in the workshop:

1 _____

2 _____

3 _____

4 _____

5 _____

_____

6 _____

FIRST AID

Employers have to assess the first aid requirements of their business. The garage industry is classed as a higher hazard and therefore arrangements for first aid need to be put in place.

Complete the statements using the word bank below:

telephone   remove   assessment   move   danger   call

What to do if you are first on the scene of an accident at work:

1 Make an _____ of the situation; ensure that you are not endangering yourself. (Two casualties are not helpful!)

2 If possible _____ the cause of injury – e.g., turn off electricity, turn off machines.

3 If you are not a trained first aider, _____ for a first aider or _____ for an ambulance.

4 Stay with the person to assist the first aider, or if no first aider to reassure the person that help is on the way. Do not attempt to _____ the person unless they are in immediate _____.

Basic first aid

First aid is best left to personnel trained to carry out this role. If you wish to become first aid trained then talk to your employer about enrolling on a local course.

Useful contacts are:

https://www.sja.org.uk

http://www.hse.org.uk

http://www.redcross.org.uk

GARAGE WORKSHOP HAZARDS

Examine the drawing of the garage workshop below. Circle the hazards you can find on the drawing and list them in the table opposite. You should be able to circle and list at least 20 hazards.

1 _____
2 _____
3 _____
4 _____
5 _____
6 _____
7 _____
8 _____
9 _____
10 _____
11 _____
12 _____
13 _____
14 _____
15 _____
16 _____
17 _____
18 _____
19 _____
20 _____
21 _____
22 _____
23 _____
24 _____

WORKSHOP PLAN

Workshop safety familiarization

Draw a plan of your college, training centre or company workshop, identifying the following items on your drawing:

- Fire exits.
- Fire alarm points (if fitted).
- Position of fire extinguishers and types.
- Vehicle hoists.
- Bench grinder.
- Bench or pillar drill.
- Statutory notices (Acts or Regulations e.g., Health and Safety Law or Electricity at Work Regulations).
- First aid box.
- Hydraulic or mechanical press.
- Location of accident book.
- Power isolation points.
- Exhaust extraction.

Safe use of machinery and equipment

Many accidents in garages are caused either by the employee not taking adequate precautions or by faulty equipment.

If you were asked to work on the vehicle shown below, what TWO precautions would you take before starting?

1 _____

2 _____

State TWO other precautions that should be observed:

1 _____

2 _____

State precautions, other than those shown, that are necessary when working on a vehicle raised by a ramp (hoist):

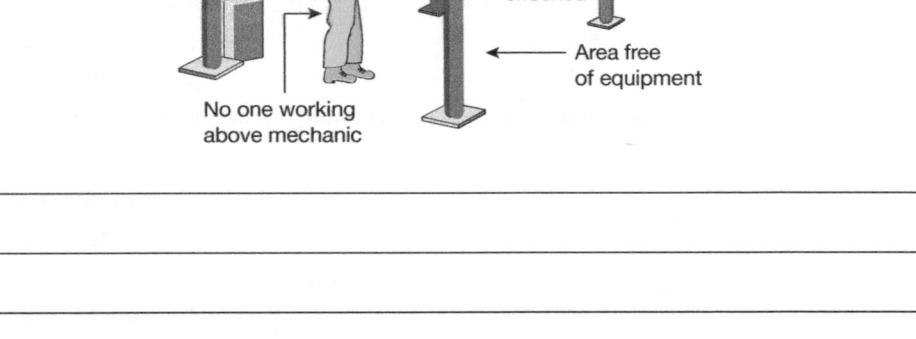

Vehicle central on hoist

Wheel chocked

Area free of equipment

No one working above mechanic

1 _____

2 _____

_____

3 _____

TIP

Only trained staff deemed competent can change the grindstone.

Good posters can still be found in workshops under Abrasive Wheels Regulation (now superseded by PUWER).

Compressed air

Most garages have a compressor to enable air tools and equipment to be used. The storage tank will have the safe working pressure (SWP) marked on the tank. Compressors can often work at 150 PSI (approximately 10 bar).

Pictures supplied by Draper Tools Limited

COMPRESSED AIR CAN KILL!

Serious, sometimes fatal, injuries can be caused by compressed air being injected into the body through the skin or into a body opening, such as your mouth, ear or rectum.

What precautions should be taken when using compressed air equipment?

● _____

● _____

● _____

State what PPE should be worn when using a compressed air blow gun.

1 _____

2 _____

3 _____

MOVEMENT OF LOADS

Any heavy object which requires moving manually or by mechanical lifting equipment is considered to be a load. In a large garage or parts department, heavy loads may be transported in the manners shown. Name each method of transport.

Which of the above units is loaded correctly? _____

 TIP

When using a sack barrow:

- ensure you walk on level ground
- lean forward slightly
- put your foot on the axle when lowering.

Below is a wheeled bench. When would this bench be good for moving loads in a workshop?

_____

_____

Sealey Group

Manual handling of loads

More than one-third of all over-three-day injuries (an injury which causes the injured person to be away from work for more than three days) reported each year to the HSE and local authorities are caused by manual handling – the transporting or supporting of loads by hand or by bodily force. The pie chart shows the pattern for over-three-day injuries reported in 2001/02.

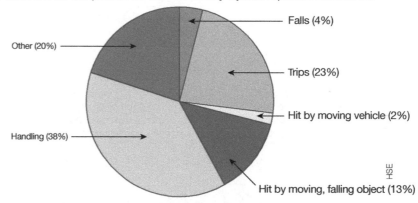

Falls (4%)

Other (20%)

Trips (23%)

Hit by moving vehicle (2%)

Handling (38%)

HSE

Hit by moving, falling object (13%)

Under the Manual Handling Regulations 1992 (amended 2002) employers must:

- **AVOID the need for hazardous manual handling, so far as is reasonably practicable.**
- **ASSESS the risk of injury from any hazardous manual handling that can't be avoided.**
- **REDUCE the risk of injury from hazardous manual handling, so far as is reasonably practicable.**

LIFTING

One person lift (squat lift)

TIP When turning while holding a load move your feet, do **not** twist your body.

TIP When the lift is awkward always ask for assistance.

Whenever possible use lifting equipment.

Team lifts

Now we turn

Starting with 1, arrange the following statements in order to give a good team lift procedure.

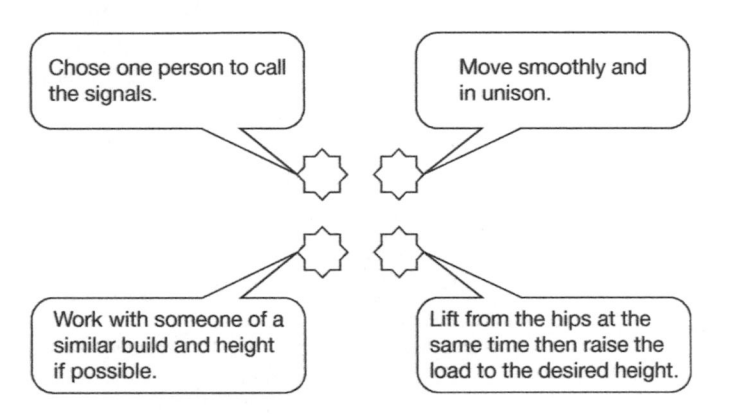

Chose one person to call the signals.

Move smoothly and in unison.

Work with someone of a similar build and height if possible.

Lift from the hips at the same time then raise the load to the desired height.

Pushing and pulling safely

TIP When pushing a car it is better to start the push with your back to the car.

Keep the strain off your back and let your body weight and leg muscles do the work.

The Lifting Operations and Lifting Equipment Regulations 1998 (LOLER)

Generally, the Regulations require that lifting equipment provided for use at work is:

- **Strong and stable enough for the particular use and marked to indicate safe working loads.**
- **Equipment is checked every 6 months by a competent person.**

In your garage, college or training workshop there is likely to be lifting equipment which is covered by these Regulations.

Lifting devices

Vehicle hoists (ramp)

State three checks that should be made before using a 4-post ramp to raise a vehicle.

1 _____

2 _____

3 _____

© Chris Howes/Wild Places Photography/Alamy

© Paul Rapson/Alamy

What would you check before raising a vehicle on a wheel-free or 2-post hoist?

- _____
- _____
  _____
- _____

TIP Rear engine vehicles usually need to be reversed on to a hoist or ramp to keep the centre of weight in-line with the posts.

Take care when removing front engine power units as the rear may overbalance.

Trolley jack

Before using a jack check for:

- hydraulic fluid leaks
- wheels move freely and are not damaged.

Give two more checks that should be carried out:

1 _____

2 _____

Pictures supplied by Draper Tools Limited

 **TIP** The SWL does not need to be above the vehicle weight as only one end or side of the vehicle is lifted by the jack.

Engine hoist

State four checks to be made before using an engine hoist:

1 _____

2 _____

3 _____

4 _____

Pictures supplied by Draper Tools Limited

As a general rule any load over _____ requires some form of powered lifting gear to support or move it.

Position of chains and slings

Which is the correct way to lift an engine using a chain sling as shown below?

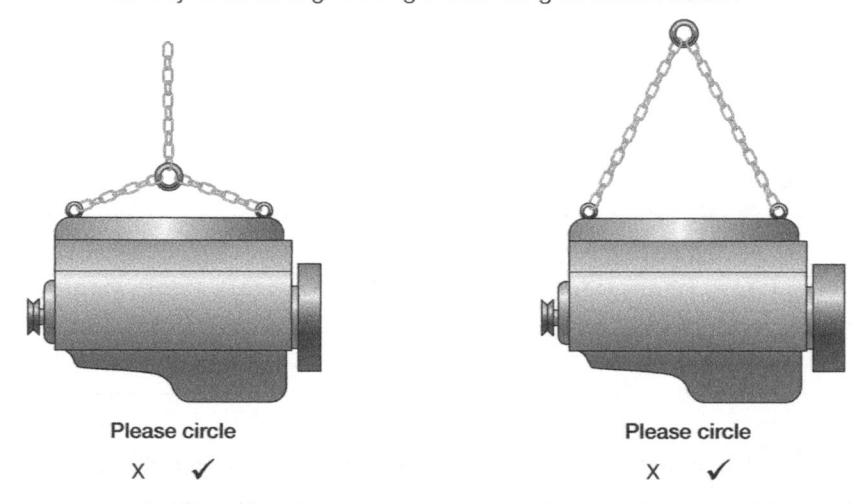

Please circle   **Please circle**

X   ✓   X   ✓

Slings and chains should be checked for wear at least once every _____

The angle made by the slings is very important.

What is the maximum recommended angle between the slings? _____

If the angle were to be substantially increased, what would be the effect of the 'pull' on the slings?

_____

ROADSIDE RECOVERY

Before recovering a vehicle from the roadside you need to have suitable training. Here is a suggested procedure.

During a vehicle recovery or breakdown follow a safe system of working:

- **On arrival switch on hazard warning lights and flashing amber beacon.**
- **Keep back from the broken-down vehicle by at least 10 m.**
- **Turn wheels of the recovery vehicle to the kerb once stationary.**

- Wear hi-visibility clothing (Hi-Viz).
- If the light is poor then carry a switched on torch.
- Stay close to the kerb.
- Make sure everyone is out of the broken-down vehicle and are safely positioned well back from the kerb on the verge.

What position should the wheels on this breakdown vehicle be pointing?

Read the procedure above and answer the following questions.

1 What would be required at the roadside if the light is poor?

_____

2 How far should you park from the broken-down vehicle?

_____

http://www.avrouk.com

Abbreviation buster

Using your knowledge, complete the following health and safety related abbreviations. They are all contained in this chapter.

HASAWA – _____

PPE – _____

COSHH – _____

PUWER – _____

LOLER – _____

RIDDOR – _____

HSE – _____

SWL – _____

PAT – _____

SWP – _____

MSDS – _____

Multiple choice questions

Choose the correct answer from a), b) or c) and place a tick [✓] after your answer.

1 **The Health and Safety at Work Act applies to:**

a) employees only [ ]

b) employers only [ ]

c) all people at work. [ ]

2 **What type of safety sign is shown:**

a) prohibition [ ]

b) mandatory [ ]

c) warning. [ ]

3 **The best type of fire extinguisher to be used on an electrical fire is:**

a) water [ ]

b) foam [ ]

c) carbon dioxide. [ ]

4 **What does the abbreviation COSHH stand for?**

a) Control of Substances Harmless to Health [ ]

b) Control of Substances Hazardous to Health [ ]

c) Carrying of Substances Hazardous to Health [ ]

5 **What should be done when hazardous products are being used in a workplace?**

a) manufacturer's data sheets obtained from supplier [ ]

b) risk assessments carried out for products used [ ]

c) all the above. [ ]

6 **What will happen if compressed air is forced through the skin?**

a) death if air is forced into the bloodstream [ ]

b) skin irritation [ ]

c) come out in a rash. [ ]

7 **If you discover someone who has suffered an electric shock, what is the first thing you would do?**

a) rush over and drag them out of the workshop [ ]

b) turn off the power [ ]

c) hit the fire alarm button. [ ]

8 **When attending a broken-down vehicle on the motorway when it is dusk you should:**

a) wear light-coloured overalls and carry a torch [ ]

b) wear Hi-Viz clothing and carry a switched on torch [ ]

c) use a flashing red torch. [ ]

9 **All employers must have a written health and safety policy. True or false?**

a) true [ ]

b) false. [ ]

10 **On inspection of a mains electric drill, you notice the coloured wires showing as the flex leaves the plug. You should:**

a) carry on using it as the coloured wires are insulated [ ]

b) not use the drill and report to your supervisor [ ]

c) wrap some insulation tape around it. [ ]

SECTION 2

Good housekeeping

After studying this section you should be able to:

- Identify vehicle protective equipment for a range of repair activities.
- Describe why the automotive environment should be properly cleaned and maintained.
- Describe requirements and systems which may be put in place to ensure a clean automotive environment using appropriate procedures and precautions.
- Describe procedures for starting and ending the working day which ensure effective housekeeping practices are followed.
- Describe how to minimize waste when using utilities and consumables.
- Describe the selection and use of cleaning equipment when dealing with general cleaning, spillages and leaks.
- Describe procedures for correct disposal of waste materials.

USE THIS SPACE FOR LEARNER NOTES

_____

_____

_____

_____

_____

_____

_____

Key terms

EPA Environmental Protection Act.

Detergent Chemical used for cleaning, usually diluted with water.

Solvents Chemicals used to clean and remove oil or grease that are often highly flammable.

VPE Vehicle Protective Equipment.

COSHH Control of Substances Hazardous to Health.

VEHICLE PROTECTION

To keep the vehicle clean during servicing and repair work, several types of protection are available.

Seat covers

State why seat covers should always be used in service and repair work.

_____

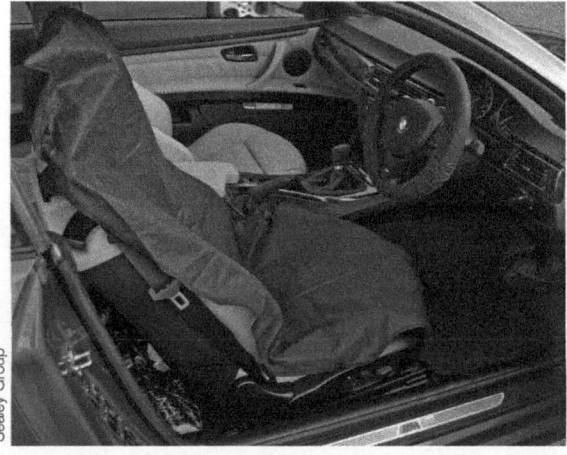

What types of seat cover are available and give advantages/disadvantages for each.

● _____

_____

● _____

_____

Give three more types of vehicle protection.

1 _____

2 _____

3 _____

State four consequences of not using appropriate vehicle protection when carrying out a major routine service.

1 _____

2 _____

3 _____

_____

4 _____

GOOD HOUSEKEEPING

Maintaining a clean and tidy work area

We are all impressed when we see a clean and shiny car, even though we know it would work just as well dirty. In the same way customers will be impressed if you keep your workshop clean and tidy.

Identify **ten** housekeeping issues in this garage workshop scenario.

1 _____

2 _____

3 _____

4 _____

5 _____

6 _____

7 _____

8 _____

9 _____

10 _____

Slips, trips and falls

Poor housekeeping accounts for many of these types of accidents in garages.

Which of the following scenarios are a possible cause of slips, trips and falls?

Complete the true or false table.

Garage scenario	True	False
Air lines lying across the floor		
Using a bench grinder		
Using an impact gun to remove wheel nuts		
Oil left on the floor following a vehicle routine service		
Using an angle grinder with a long extension lead trailing across the floor		
Cleaning a trolley jack		
Removing a cooling system pressure cap		
Using a ladder to gain access to tyres on a high shelf		
Working near an open vehicle pit		

Housekeeping routines

Match the list of housekeeping routines to the part of the working day when they should be completed.

Opening

During the day

Closing

- Keep the lift and the floor clean.
- Switch off power to equipment.
- Check that all tools and equipment are clean and tidy.
- After use, put special tools back where they are normally kept.
- Check that any customers' vehicles are secure.
- Check the special workshop tools and equipment such as the air line, the tyre remover and the wheel balancer are working.
- Put away neatly all special tools.
- Make sure that items such as wheels and removed tyres do not obstruct work areas or pathways.
- Check that the lifts and floor area are clean and free from obstructions.
- Clean and tidy the work area.
- Move parked vehicles away from the work areas.
- Lock up your personal toolbox.

Cleaning

Cleaning equipment should be kept in a separate store, as many chemicals are highly concentrated.

Always read the instructions on the labels before using them (see COSHH Regulations in Section 1). If specialized cleaning is required, your employer will provide protective clothing.

When you have finished cleaning, put the cleaning equipment and unused chemicals back in the store.

 If you need to use a hazardous cleaning material, read the label on the container. This will tell you how to use it safely, and what to do if you *do* have an accident – for example, if the cleaning material touches your eyes or skin.

Different materials are used for cleaning and are usually classified as:

- solvents
- detergents.

Solvents

What are the dangers when using solvents to clean workshop equipment?

_____

Detergents

These are often mixed with water to dilute before use.

Find a detergent that is used in your workshop and suggest the PPE that should be used with the product.

_____

TIP Barrier cream used on your hands can reduce the risk of industrial dermatitis.

HEALTH AND SAFETY

What information does the supplier of the cleaning material have to provide when requested?

_____

Why can this be useful?

_____

_____

This sign is typically found on cleaning products.

Suggest at least **three** items of garage equipment that require regular cleaning:

● _____

● _____

● _____

⚡ While cleaning, place cones and notices to warn others. Section off areas that could be dangerous, such as slippery floors.

State a number of areas of the workshop that require regular cleaning:

● _____

● _____

● _____

EMERGENCY CLEANING

Breakages and spillages must be cleaned up immediately.

If this is not done, someone may be injured. The firm could also be in breach of the Health and Safety at Work Act (see Section 1 – Health and Safety). If an accident happens, the firm may be fined.

© Lyroky/Alamy

Oil spillages are best dealt with using absorbent granules that soak up the oil and can be swept away.

Give **two** dangers of leaving oil on the floor:

1 _____

2 _____

_____

DISPOSING OF DANGEROUS WASTE MATERIAL

All workshops produce dangerous waste materials, which must be disposed of correctly by a licensed contractor.

Waste management is covered by the **Environmental Protection Act 1990.**

Items must be disposed of in different ways. Usually this is decided by the local council who pass by-laws. Refuse disposal requirements differ from place to place.

Some types of dangerous material must be kept separate. They will be collected by specialist agencies, or taken to the local refuse collection point.

State **six** products which are encountered in a service and repair workshop that have to be disposed of correctly:

1 _____

2 _____

3 _____

4 _____

5 _____

6 _____

USING RESOURCES ECONOMICALLY

How can the following resources be used economically?

Utilities

Electricity

_____

Heating

_____

Water

_____

Telephone

_____

Consumables

Paper towelling

_____

_____

Lubricants

_____

Cleaning materials

_____

Fasteners

_____

RECYCLING WASTE MATERIALS

Vehicle manufacturers are producing vehicles that have an increasing amount of components that can be reused or recycled and are working towards the use of sustainable materials during manufacture.

What are the advantages of recycling?

_____

List six items that are designed to be recycled:

1 _____

2 _____

3 _____

4 _____

5 _____

6 _____

Multiple choice questions

Choose the correct answer from a), b) or c) and place a tick [✓] after your answer.

1 **You have been asked to collect a car from a customer's home while wearing overalls and boots. What vehicle protection must you take with you?**

   a) seat cover and wing cover [ ]

   b) seat cover and floor mat [ ]

   c) floor mat and steering wheel cover. [ ]

2 **How could efficiency and productivity in a vehicle workshop be improved?**

   a) ensure tools are put back straight away on a shadow board [ ]

   b) buy extra tools in case one cannot be found [ ]

   c) leave any cleaning up of spills to the end of the day. [ ]

3 **Which of the following must be taken away by a licensed contractor?**

   a) paper air filters [ ]

   b) coolant hoses [ ]

   c) used anti-freeze. [ ]

4 **How could electricity be saved in a garage environment?**

   a) use rechargeable battery operated equipment [ ]

   b) remove some of the light bulbs in the workshop [ ]

   c) turn off lights and other electrical equipment when not required. [ ]

5 **What housekeeping tasks should be carried out during the working day?**

   a) remove all vehicles from the workshop [ ]

   b) after use, put special tools back where they are normally kept [ ]

   c) switch off power to equipment. [ ]

PART 2
THE MOTOR INDUSTRY

USE THIS SPACE FOR LEARNER NOTES

SECTION 1
Workplace structure and job roles 35

SECTION 2
Tools and equipment 43

Workplace structure and job roles

Learning objectives

After studying this section you should be able to:

- Identify organizational structures, their purpose and the roles of people within the structures.
- Understand the importance of obtaining, interpreting and using information in order to support job roles within the work environment.
- Understand different types of communication and their requirements when carrying out vehicle repairs.
- Explain how to develop good working relationships with colleagues and customers in the automotive workplace.

Key terms

After sales   The section of a business which deals with the repair, maintenance and fitment of auxiliary components once a new vehicle has been sold to the customer.

Approved repairer   A company/business which has been approved to undertake repairs on behalf of specific manufacturers.

Franchised dealer   A firm selling and servicing a particular make of vehicle, appointed by the manufacturer.

Job description   States the duties and responsibilities of a particular job role.

Mission statement   An explanation of the main objectives of the company, its aims and why it is trading.

Organizational chart   A chart with vertical lines of authority and horizontal lines linking people with equal status.

Organizational structure   Who does what in a company.

Working relationships   The interaction between colleagues when working together towards a common goal.

ORGANIZATIONAL STRUCTURES AND FUNCTIONS

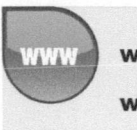

www.ukmot.com/mot_check.asp

www.direct.gov.uk/en/TravelAndTransport/Highwaycode/index.htm

The automotive industry is very diverse. Companies vary in size and in what they actually do. Many specialize in certain types of vehicle and their sales or repair. Within the industry there are recognized subsections. The main subsections that could be incorporated into a dealership are listed below:

| Reception | Vehicle repair workshop | Vehicle parts | Main office |
| Vehicle sales | Body shop | Paint shop | Valeting |

Describe the function of the following sections:

Reception – _____

_____

Vehicle repair workshop – _____

_____

Vehicle parts department – _____

_____

Main office – _____

_____

Vehicle sales – _____

_____

Body and paint shops – _____

_____

_____

Valeting – _____

_____

_____

Some of the sections above may need to communicate directly with each other during the course of their work. Draw coloured lines between the sections which would communicate with each other directly. Show the direction of communication on the diagram.

Discuss the reasons for your choices with a colleague or within your class group.

The diagram below a typical organizational structure for a medium-sized business.

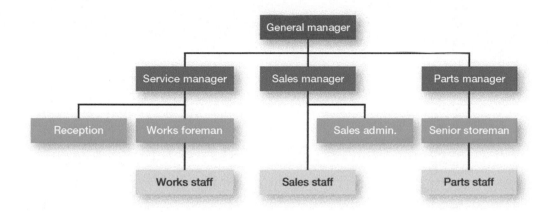

Each employee in a company should be aware of their own role and the role of others. Disputes may arise if anyone is uncertain about who does what.

Identify from the diagram above who the following people would be in direct communication with:

Workshop team leader – _____

Parts manager – _____

Workshop staff – _____

General manager – _____

Complete the following statement.

The Service manager is responsible to the _____ for _____

_____.

Note: Another term for 'team leader' which may be used could be 'foreman' or 'forewoman'.

In the modern automotive repair workshop, information and specifications on vehicles, components, oils and fluids need to be used constantly. This can bo obtained from a number of sources.

Give an example of what type of information you would expect to get from the sources listed:

Garage staff – _____

Manuals – _____

Data books – _____

Parts lists – _____

Computer software and the Internet – _____

Manufacturer's data – _____

Diagnostic equipment – _____

Wall charts – _____

It is important that the information from these sources is correct and that it is properly used.

Why is this? _____

_____

It is also important that any replacement parts used on vehicles are of good quality and meet the manufacturer's original equipment specification.

Why is this? _____

_____

When sourcing parts from suppliers, many will work from the vehicle registration number. However, every modern vehicle has an individual 'Vehicle Identification Number' (VIN). It is important to locate this and record it correctly, as it can be very useful when trying to find information about a vehicle.

Give examples of what the VIN can tell a technician about a vehicle.

_____

_____

The VIN can be found in different places on various vehicles. Name three popular locations.

1 _____

2 _____

3 _____

TIP For an interactive website which will decode a VIN number, go to
http://www.analogx.com/contents/vinview.htm

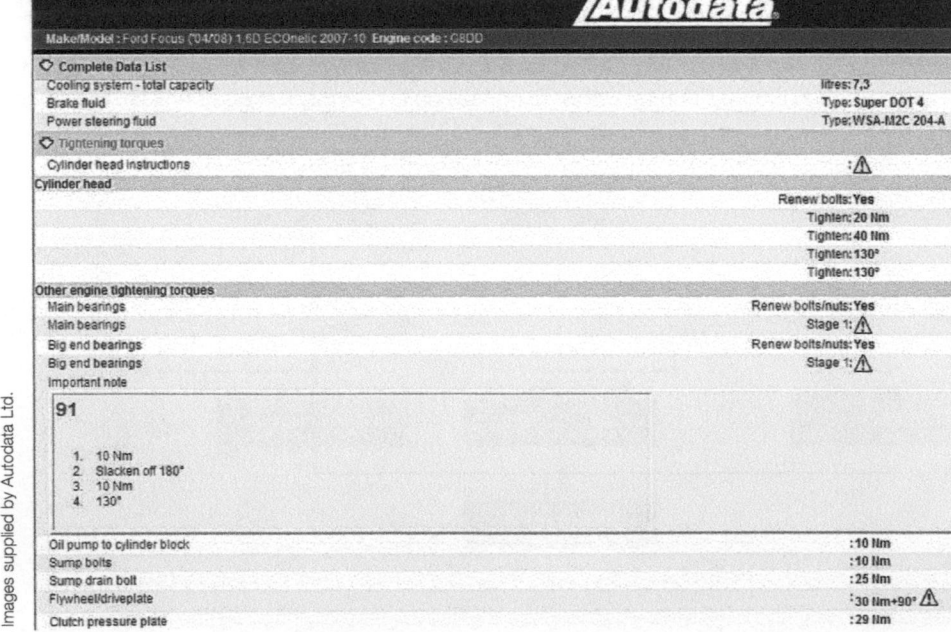

Images supplied by Autodata Ltd.

Select data from the Autodata chart and answer the following:

1 What make and model of vehicle do the data refer to? _____

2 What is the capacity of the cooling system? _____

3 What is the recommended torque for the sump drain bolt? _____

Not all information is related to the actual repair of a vehicle. In some cases vehicles are inspected and not repaired. Give two examples of when this may happen.

1 _____

2 _____

General inspection and testing sheets provided by various motoring organizations may be used, others are produced by specific manufacturer's for their own models.

List five systems or components which are checked when carrying out an inspection on a vehicle:

1 _____

2 _____

3 _____

4 _____

5 _____

TIP Go to **www.motester.co.uk** for information on the MOT test, to see what is checked and if your answers above are correct.

Remember to respect the customer's vehicle – no unnecessary high speed cornering or wheel spinning!

What are the MOT requirements for VIN numbers?

_____

When carrying out vehicle tests and inspections, it is important to work to the correct logical procedures. Why is this important?

_____

_____

Vehicles that are diagnosed for faults may also need to be road tested. When else may a vehicle be driven by a tester?

_____

_____

When road testing a vehicle, what code must the driver adhere to at all times?

_____

COMMUNICATION WITHIN THE AUTOMOTIVE WORK ENVIRONMENT

When problems and misunderstandings do arise, it is often because of poor communication. Good communication is an essential part of every working day, for employees at all levels.

In a typical garage, there will be daily communication between colleagues within a department, between departments, between management, supervisors and shop-floor staff, and between customers and reception staff. The company will also communicate with suppliers, subcontractors, vehicle manufacturers, advertising agencies, banks, accountants, lawyers, the council, and so on.

What might happen if a customer is not dealt with using good, clear communication and there are misunderstandings?

_____

_____

Non-verbal communication

It is not only your words that communicate – so too does your facial expression and the way you stand and move. Customers will notice your body language: for example whether you smile and look directly at them or slouch and avoid their eyes.

Problems in communication

Some things can cause problems in communication.

Lack of communication

Problems can arise if people are not told things they need to know. Changes to systems, safety issues, shop-floor problems and the like must be passed on to the right people without delay.

Delays in getting information to staff may be caused by pressure of work or different hours of work (such as part-time, full-time, or shift work).

Incorrect communication

Problems can also arise if the information given is wrong. An example of this is shown below.

If a message is passed orally via several people, it may become totally changed!

How can the problem shown be avoided?

_____

_____

Methods of communication

This section is particularly concerned with the effects of communication on working relationships. The method of communication chosen will depend on the situation. List where different types of communication would best be used.

Direct discussion – face-to-face conversation

1 _____

2 _____

3 _____

Writing

1 _____

2 _____

3 _____

Telephone, text or email

1 _____

2 _____

3 _____

TIP Remember that communication of information can change with the target audience, depending on how much they know about the subject concerned.

Where written reports are required they should be factual and contain sufficient detail to avoid confusion.

Records of work carried out are an important part of the communication requirements of the automotive industry. These may take the form of inspection sheets as well as manufacturer's fleet and company records. They may also be customer records and job cards.

Give **five** items of information that would be expected to be completed on a finished job card:

1 _____

2 _____

3 _____

4 _____

5 _____

Explain why records are important.

_____

_____

_____

_____

GOOD WORKING RELATIONSHIPS WITH COLLEAGUES AND CUSTOMERS

Working relationships develop with the people who work alongside you. As you interact in the workshop with the manager, other mechanics and fitters, or your supervisor, you build working relationships with them. Good working relationships are very important to the success of every business. You may not be friends with all your colleagues; occasionally you may even dislike some of them. But to help the business run smoothly, you must get on with all of them professionally.

Working as a team

To make the company successful, all of its employees must work together. They must cooperate, like members of a football team: this is teamwork.

If the firm does well, and employees get on with each other and trust each other, there will be a good feeling in the workplace, and people will be enthusiastic about their jobs. This feeling is called good morale.

When everyone works hard, and no one wastes time or resources, the firm will be efficient. By being efficient, employees will get a lot done – they will be productive.

If a team with good morale works efficiently and productively, customers will be satisfied and pleased to come again. They are also likely to recommend the firm to others, so it will gain a good reputation. And this in turn will bring more business, and the firm will become even more successful, and will grow. It will gain a good company image.

What are the key points of teamwork?

- _____
- _____
- _____
- _____
- _____

Building good relationships

Here are a few examples of the kind of things that can upset good working relationships:

- **In everyday conversation you may discuss your social life, sports, films, and so on. You may find that a workmate has opinions on some topics which are different from yours. Never allow these differences to spoil your working relationship.**

- **Sometimes you may find colleagues who lack interest or enthusiasm, who are lazy or incompetent, who keep being absent, and so on. If people do not 'pull their weight' in the workplace, this can cause anger and frustration among other members of the team.**

- **Beware of anyone who ignores company rules and regulations, or safe working practices. This behaviour can create problems or even dangers for everyone else in the company.**

- **Personal appearance and hygiene are important. Employees who do not bother with these may upset colleagues, and this may affect the performance of a workforce.**

- **Managers and supervisors should not show favouritism and should pay everyone fairly.**

List ways in which you can build good working relationships with your colleagues:

- _____
- _____
- _____
- _____
- _____
- _____

WORKING RELATIONSHIPS

Find the words which make up the following statement hidden in the word search below:

'It is important to make and honour realistic commitments to colleagues and customers.'

```
I T I S I M P R O R T A N T
T O M A K E A N E N D H O N S
C O U R E A L A I S T I T C
C U O M M I T T L T M E N T
S T S O C O L L I E R E D N A
A G U T E S A N S D M O C U C
S H O N O U R T T O M P O E
R S D M T M C I X T S L M Y
X C U S I X E M C Õ Y L F N I
K T J N J K M R F E B Z E J
K M F X X O E B S A R D W Õ U
U H L L C Õ F H G C U X P E
J G M P V T Õ U N I A X W C J
D Õ I N W E D O A E A T E M
W E F X Õ S C A Z Õ B E V Y
```

AND
COLLEAGUES
COMMITMENTS
CUSTOMERS
HONOUR
IMPORTANT
IS
IT
MAKE
REALISTIC
TO
TO
TO

Multiple choice questions

Choose the correct answer from a), b) or c) and place a tick [✓] after your answer.

1 The valeting section in a large dealership would support the:
a) repair workshops and car sales section [ ]
b) body shop and parts department [ ]
c) reception and office administration. [ ]

2 The receptionist in a large organization will have direct links with the:
a) parts department [ ]
b) customer [ ]
c) cleaners. [ ]

3 In an automotive environment, a trainee should expect to:
a) be supervised by a technician [ ]
b) deal directly with customers [ ]
c) work on vehicles unsupervised. [ ]

4 In an automotive environment, a general manager would be responsible for:
a) supervising the workshop cleaning rota [ ]
b) selling cars to customers at weekends [ ]
c) all aspects of the business. [ ]

5 When changing the camshaft drive on an engine, the correct procedures would be found in:
a) an unleaded petrol information manual [ ]
b) timing belts and chains manual [ ]
c) service data book. [ ]

6 If recognized procedures and processes are not carried out when working on a vehicle:
a) the job can be completed much quicker and more cheaply [ ]
b) the job may take longer or not be completed properly [ ]
c) the foreman will be pleased that you have taken a short cut. [ ]

7 A 'VIN' number allocated to a vehicle is important when:
a) renewing road tax [ ]
b) completing an invoice [ ]
c) ordering spare parts. [ ]

8 Important messages for colleagues, taken over the telephone, should be:
a) passed on to colleagues immediately, in writing [ ]
b) passed on to colleagues verbally, when next seen [ ]
c) recorded and played back at lunch-time. [ ]

SECTION 2

Tools and equipment

USE THIS SPACE FOR LEARNER NOTES

_____

_____

_____

_____

_____

_____

_____

Learning objectives

After studying this section you should be able to:

● Explain how to use specialist measuring equipment.
● State the different materials and their properties as used in the construction of the motor vehicle.
● Explain the different methods of joining materials.

Key terms

Ductility Ability for a cold metal to be stretched and formed without breaking.

Ferrous A metal which contains iron, making it magnetic.

Laminated glass Mainly used in windscreens. Can break and puncture but will not shatter.

Malleability Ability of a metal to be shaped by compression loads.

Yield strength or yield point The stress at which a material begins to deform.

TOOLS USED WHEN SERVICING VEHICLES

www.draper.co.uk

www.britool.com

Taps and dies

Taps and dies are used to cut internal and external threads.

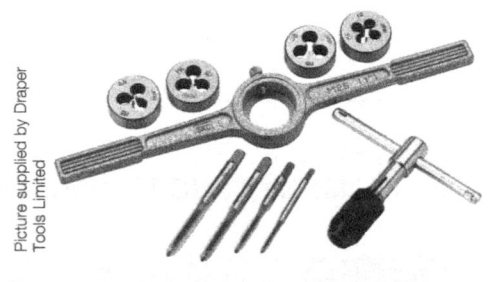

Picture supplied by Draper Tools Limited

Correctly label the tap, die, die holder and tap holder in the diagram above.

A tap is used to cut an internal thread for a bolt or stud hole. A die is used to cut an external thread on a piece of rod.

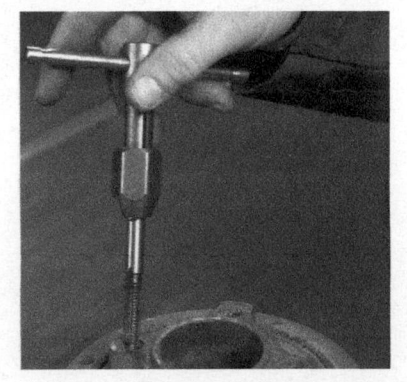

A tap being used to repair a damaged thread

Thread restorers and extractors

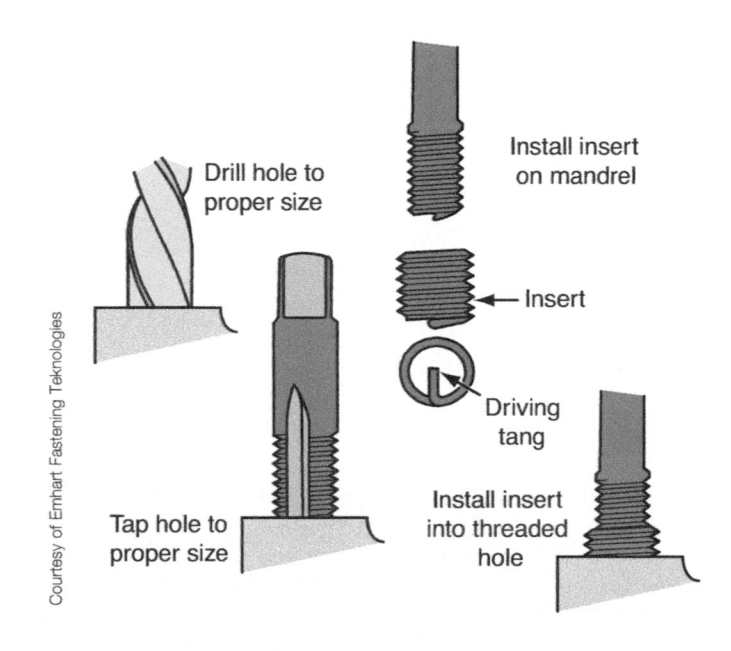

Courtesy of Emhart Fastening Teknologies

Drill hole to proper size

Install insert on mandrel

Insert

Driving tang

Tap hole to proper size

Install insert into threaded hole

The pictures above show the use of a heli-coil® threaded insert to repair damaged threads.

TIP Allow a cylinder head to cool before changing spark plugs. If the cylinder head is not allowed to cool, the bores for the plugs can become oval as the head cools without spark plugs in the bores.

The thread file shown is used for restoring and cleaning internal or external threads of varying pitches.

A die nut can be used to clean external threads on studs and bolts.

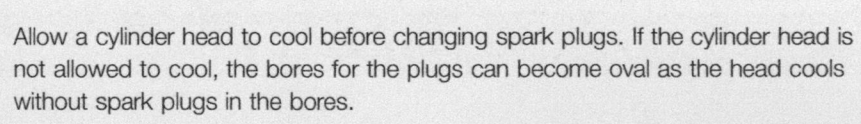

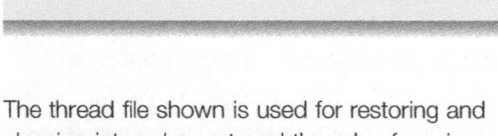

Picture supplied by Draper Tools Limited

Give two examples of when a thread file or die nut could be used:

1 _____

2 _____

The pictures below show a stud extractor and a stud remover. Identify each of them:

_____   _____

Special-purpose tools

These are made to do a specific job on a vehicle and are numerous and varied. They range from a spark plug spanner to inner door handle removal tools.

Give four examples of such special tools:

1 _____

2 _____

3 _____

4 _____

Vernier caliper

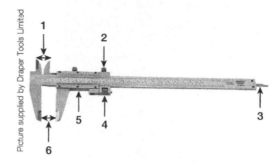

Vernier caliper

These can measure external and internal diameters as well as depth and have a degree of accuracy of 0.02 mm. They come with either metric or imperial scales. It can take some practice to read them, but when understood it is straightforward and accurate for measuring. These can be purchased with a digital readout, which makes them even easier to use.

Identify, and briefly describe the purpose of the numbered parts on the vernier caliper:

1 _____

2 _____

3 _____

4 _____

5 _____

6 _____

Give examples of where this Vernier caliper would be used.

_____

_____

Principle of reading scales

The vernier caliper consists of two slightly different scales, one fixed and one moving.

These gauges may be found calibrated in both English and metric scales.

There are two types of metric scales, in both cases the reading on the sliding scale is multiplied by 0.02.

(a) 25–division vernier (sliding) scale. The main scale is graduated in 0.5 mm divisions.

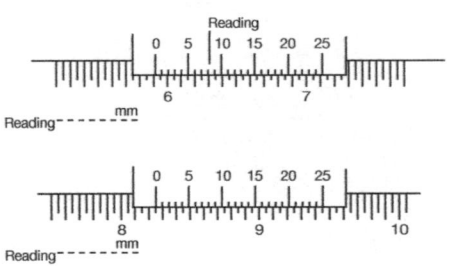

Reading at 1	60.00 mm
from 1 to 2	6.00 mm
from 2 to 3	0.36 mm (reading as shown x 0.02)
Reading	66.36 mm

(b) 50–division vernier (sliding) scale. The main scale is graduated in 1 mm divisions.

Reading at 1	30.00 mm
from 1 to 2	3.00 mm
from 2 to 3	0.30 mm (reading as shown x 0.02)
Reading	33.30 mm

State the readings on the scales below:

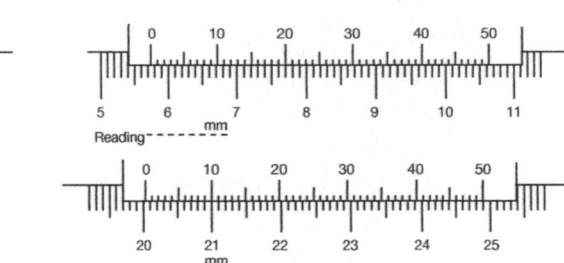

Reading - - - - - - - mm

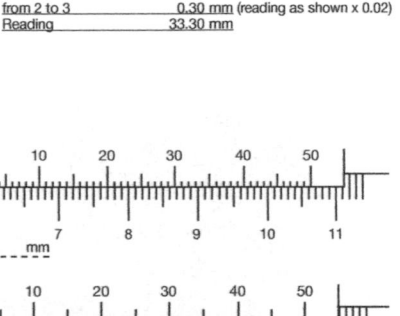

Sketch vernier readings of 17.86 mm and 53.24 mm:

Micrometer

The micrometer caliper is used to measure the diameter of components to an accuracy of 0.01 mm (1/100 mm).

Micrometers are made in size ranges of 0–25 mm, 25–50 mm, 50–75 mm, etc. A big end bearing journal of 45.50 mm diameter would be measured by using a 25–50 mm micrometer.

Examine a micrometer. Correctly label the arrowed parts on the micrometers shown from the list:

Sleeve
Lock nut
Anvil
Frame
Thimble
Ratchet stop

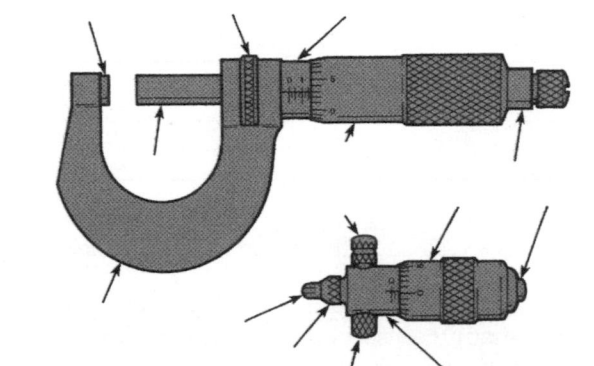

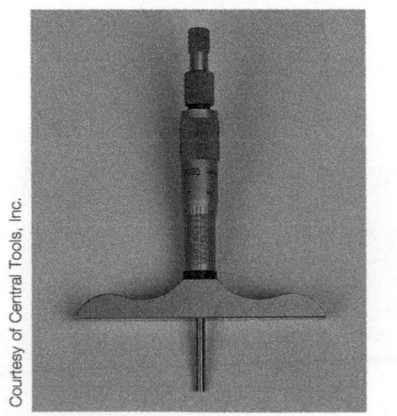

Courtesy of Central Tools, Inc.

What type of micrometer is shown above? _____

Reading the micrometer scale.

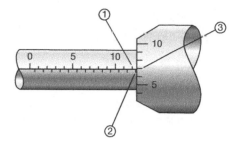

Reading at 1 = 12.00 mm
Reading at 2 = 0.50 mm
Reading at 3 = 0.07 mm

Micrometer
    reading = 12.57 mm

State the readings of the scales of the metric micrometers below.

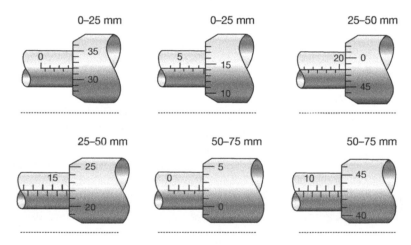

0–25 mm   0–25 mm   25–50 mm

25–50 mm   50–75 mm   50–75 mm

Sketch scales on the micrometers to give the readings indicated:

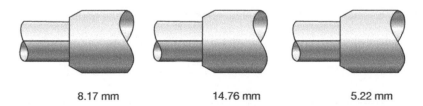

8.17 mm   14.76 mm   5.22 mm

Dial indicators

A dial test indicator (DTI) is an instrument which may be used to give comparative measurements from an item of specific or standard size. A typical clock gauge with a measuring range of 10 mm would have an accuracy of 0.01 mm.

Picture supplied by Draper Tools Limited

A DTI set up to show the amount of hub bearing end float

 Correctly set up a DTI and check the end float of a wheel hub bearing. Compare the readings with those of the manufacturer.

Give two other examples of the uses of a DTI in motor vehicle repair:

1 _____

2 _____

State the procedure for setting up and using a DTI:

1 _____

2 _____

3 _____

4 _____

Which gauge would be used to check the run out of a brake disc?

_____

TOOLS USED FOR MEASURING

Identify the measuring tools shown in the following figures.

Used to take internal measurements and when offered up to a rule, or a similar calibrated measuring device, readings can be taken.

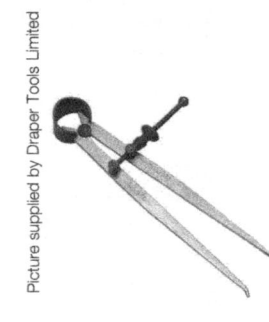

Picture supplied by Draper Tools Limited

_____

Used to take external measurements and when offered up to a rule, or a similar calibrated measuring device, readings can be taken.

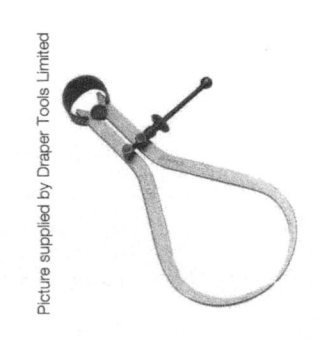

Picture supplied by Draper Tools Limited

_____

These are a selection of thin strips of metal of specific thicknesses and have an accuracy of 0.025 mm. They are used to measure the size of clearance gaps. They can be obtained in metric, imperial or a combination of both measurements.

Picture kindly provided by permission of Snap-On Industrial

_____

Give two examples where these would be used:

1 _____

2 _____

MARKING OUT

This is when a simple component is to be made, lines are reproduced on the material's surface from an engineering drawing.

Tools used for marking out

Tool/Equipment	Purpose
Bevel protractor	
Dividers	
Odd leg calipers	
Scribe	
Trammels	

(Continued)

Tool/Equipment	Purpose
Combination set	_____
	_____
Parallels and parallel blocks	_____
Surface plate and surface table	_____
Set square	_____
Angle plate and surface gauge	_____
Vee blocks and clamps	_____
Template	_____
	_____

Characteristics	Description	Typical MV component
Ductility	_____	_____
	_____	_____
Malleability	_____	_____
	_____	_____
Elasticity	_____	_____
	_____	_____
Plasticity	_____	_____
	_____	_____
Softness	_____	_____
Thermal conductivity	_____	_____
Electrical conductivity	_____	_____

MATERIALS USED IN VEHICLE CONSTRUCTION

Characteristics (or properties) of materials

A material may possess one or more different characteristics.

Define each characteristic and give an example of a motor vehicle component which demonstrates this characteristic.

Characteristics	Description	Typical MV component
Hardness	_____	_____
	_____	_____
Strength	_____	_____
Brittleness	_____	_____
	_____	_____
Toughness	_____	_____
	_____	_____

Strength

Materials in the motor vehicles are subjected to a number of different forces.

Briefly describe the following terms:

Yield strength or yield point – _____

Shear strength – _____

Compressive strength – _____

Tensile strength or tenacity – _____

State the type of stress to which the following motor vehicle components are mainly subjected.

Component	Types of stress
Cylinder head bolt	_____
Connecting rod	_____

(Continued)

Component	Types of stress
Clutch disc rivets	
Handbrake rods	
Coil spring	
Cylinder head gasket	
Gudgeon pin	
Propeller shaft	

Ferrous and non-ferrous metals

Metals may be split into two main groups: Ferrous and non-ferrous metals.

What does ferrous mean?

_____

_____

_____

What does non-ferrous mean?

_____

_____

_____

Most pure non-ferrous metals are not used separately but are alloyed with other materials when used to produce motor vehicle components.

Ferrous metals	Non-ferrous metals
Iron	Aluminium
Low carbon steel	Copper
Medium carbon steel	Brass

Ferrous metals	Non-ferrous metals
High carbon steel	Bronze
Cast iron	Zinc
Stainless steel	Lead
Tool steels	Tin

Iron in its pure state is soft and ductile. This makes it not very practical in the pure state for use in the motor vehicle. By adding carbon a useful set of alloys are produced, known as carbon steel. The hardness of the steel is determined by the amount of carbon added, ranging from 0.1 per cent to 4 per cent.

The following chart shows the percentage of carbon contained in the metals.

Material	% Carbon
Wrought iron	0.01
Mild steel	0.25
Medium carbon steel	0.50
High carbon steel	1.20
Cast iron	3.00

Alloy steels

Alloy steels are used for most of the highly stressed components in the modern car. What is meant by an alloy steel?

_____

_____

_____

Most of the tools in a technician's tool box are made from alloy steel, for example, spanners, pliers and hammers.

Aluminium and its alloys

Pure aluminium is not commonly used in a vehicle because it is too ductile and malleable. But when small amounts of other materials are added, alloys can be produced that are much stronger, harder, able to retain strength at high temperatures and corrosion resistant. Describe two types of aluminium alloy.

1 _____

_____

2 _____

_____

Name typical motor vehicle components made from the materials below.

Ferrous metals	Components	Non-ferrous metals	Components
Low-carbon (mild) steel	_____	Aluminium alloy	_____
	_____		_____
High-carbon steel	_____	Copper	_____
	_____		_____
Cast iron	_____	Copper-based alloys, brass and bronze	_____
	_____		_____
	_____		_____
	_____		_____
Alloy steels	_____	Lead-based alloys	_____
	_____		_____
	_____		_____
	_____		_____
	_____	Zinc-based alloys (die casting)	_____
	_____		_____

Copper

Name four important properties that make copper particularly useful in cooling systems and electrical cables:

1 _____

2 _____

3 _____

4 _____

Cast iron

Cast iron (iron which contains two per cent carbon or more) is hot cast as a liquid into moulds. Cast iron has good wear characteristics but is brittle.

Tin

When alloyed with lead it is used as solder in the joining of electrical circuits, components and wiring.

Give reasons why the following non-ferrous metals are considered very suitable materials for the following components:

Component	Material	Reason for choice
Piston Some cylinder heads	Aluminium	_____ _____
Radiator core (or stack)	Copper	_____ _____
Electrical cables	Copper	_____ _____
Fuel pumps	Zinc-based aluminium alloy	_____ _____

(Continued)

Component	Material	Reason for choice
Small plain bearings	Bronze	_____
Radiator header tanks	Brass	_____
Bearings	Aluminium, tin,	_____
(thin shell)	copper, lead	_____

The main alloys of copper are brass and bronze.

Brass is an alloy of copper and _____.

Bronze is an alloy of copper and _____.

HARDENING, TEMPERING, ANNEALING, NORMALIZING AND CASE HARDENING

Read through the following descriptions (Hardening, Tempering, Annealing and Normalizing). Correctly fill in the missing words with the following:

indentations	880°C	200°C	naturally
water	reheated	ductility	brittle
hardness	soften	sand	oil
impact	stresses	restore	cherry
ductile	slowly	temperature	worked
resistant	properties	coarse	internal
hardened	heating	heating	deformed

Hardening

So that steel can be _____ to scratching, abrasions, wear and _____ it needs to be _____. This is achieved by heating the steel to _____, which is when it turns red in colour. It is then quenched by immersing it in _____.

The downside of hardening is that it makes the steel _____ and considerably reduces its resistance to _____ shocks.

Tempering

This is used to remove some of the _____ from steel, making it tough and _____. To achieve this, the brittle steel needs to be _____ to a certain temperature. This temperature depends upon what _____ are required. It is then quenched in either water or _____.

Tempering at _____ reduces brittleness and at 300°C reduces hardness. Most hand tools are tempered within 200°C to 300°C.

Annealing

This process is used to _____ the material, which increases its _____ and relieves some of the internal _____. To anneal a metal it needs to be heated until it is _____ red then _____ cooled. This is achieved by covering with ashes, _____ or lime.

Normalizing

The purpose of this is to _____ the grain structure when steel has been _____, either hot or cold. When steel is kept at a red hot _____ for a long period of time (such as welding), the _____ structure becomes long and _____. When it is cold worked, for example, being bent without _____, the internal structure will become _____ and stressed.

Normalizing is achieved by _____ the steel to cherry red and allowing to cool _____ in the surrounding air.

Research 'case hardening'. The following websites will be of use:

www.heat-treat.co.uk

http://www.technologystudent.com/equip1/heat2.htm

PLASTICS

Plastics are a large group of man-made materials. They may be formed into any required shape under the application of heat and pressure. There are two groups of plastics. State their properties and give a typical use.

1 Thermosetting

_____

_____

_____

_____

Formica and epoxy resins are thermosetting materials.

2 Thermoplastic

_____

_____

_____

_____

Polystyrene, PTFE (Teflon), terylene, polythene, PVC and nylon are all thermoplastic materials.

Glass reinforced plastic (GRP)

A material made from a plastic matrix (can be epoxy or thermosetting plastic) which is reinforced with fine glass fibres. The most common name given to this material is fibreglass. It is a lightweight strong material that can be relatively easily formed into different shapes.

Where is GRP likely to be used on a vehicle?

_____

CERAMICS

A ceramic material is one that has been produced through a heating process and forms a pot-like substance that is very hard and brittle.

One of the few ceramic substances used on a vehicle is for:

_____

_____

KEVLAR®

This is a man-made fibre, which is a high tensile strength material, belonging to the family of nylons. It is also tough, with high cut resistance as well as being flame proof and self-extinguishing. It is an aramid, an abbreviation for aromatic polyamide, synthetic fibre. It is more commonly known as a para-aramid. It is claimed to be five times stronger than the same weight of steel. Some motor vehicle components, especially in high performance cars, can be made from Kevlar.

List four motor vehicle components which can include Kevlar® in their construction:

1 _____

2 _____

3 _____

4 _____

GLASS

There are two different types of glass used in motor vehicle windows. Both types are known as safety glass.

One type is called laminated glass (to laminate means to layer). This is used for the windscreen. The other type is known as tempered glass. This is used for the side and rear windows.

Laminated glass

How is laminated glass made?

_____

_____

_____

What is the purpose of the PVB layer?

_____

_____

_____

The laminated glass also deflects up to 95 per cent of the sun's ultraviolet rays (UV).

Tempered glass

The process of making this glass is by heating and cooling it rapidly at room temperature by using a system of blowers. Because the surface of the glass cools much faster than the centre of the glass the surface has good compressive strength and the centre has good tensile strength. The differences between these two gives the laminated glass five to ten times its original strength.

The strength of the glass makes it suitable for daily use in the motor vehicle.

What happens when a large impact hits the laminated glass?

_____

RUBBER

There are two types of rubber, natural and synthetic rubber (man-made). Natural rubber is made from the sap of a rubber tree. It has to be vulcanized (heated), having sulphur added to improve its elasticity and prevent it perishing, all of which make it a useful material. This is when it takes on elastic properties.

Synthetic rubber is more suited to vehicle applications. It is an elastic polymer and chemical engineers can change its structure, making it resistant to oils and brake/clutch fluids.

Where is rubber used in the construction of a motor vehicle?

_____

_____

_____

_____

_____

JOINING OF MATERIALS AND COMPONENTS

Mechanical joining devices

These provide the means of joining one component to another and are traditionally nuts, bolts, screws, keys and pins; plus adhesives on modern vehicles.

ISO Metric screw thread

Screw threads may be external (male) – for example, bolts, studs and screws, or internal (female) – for example, nuts and threaded holes.

The metric thread is now extensively used in motor vehicle engineering. A coarse or fine pitch is used to suit different applications.

The diagram below shows a typical screw thread. Correctly label with the following:

Crest   Pitch   Minor diameter   Major diameter   Root   Thread angle

A thread designation such as M10 × 1.5 × 40 describes a metric where:

M = _____   1.5 = _____

10 = _____   40 = _____

ISO Metric is an abbreviation for _____.

Define the term lead used in connection with screw threads.

_____

Screw threads can be known as fine or coarse pitch.

For one turn of a fine thread it will move a _____ distance than would a nut on the _____ thread.

Its _____ is less.

Coarse threads are often used in soft alloys because they are less likely to strip during dismantling and assembly. They also give a greater depth of thread that is stronger in soft materials.

Fine threads are very commonly used for motor vehicle applications because they are less likely to work loose when subjected to vibrations.

Locking devices

Most of the nuts and bolts used on motor vehicles are fitted with locking devices. This is to counteract the loosening effect caused by strain and vibration.

Name the following numbered locking devices:

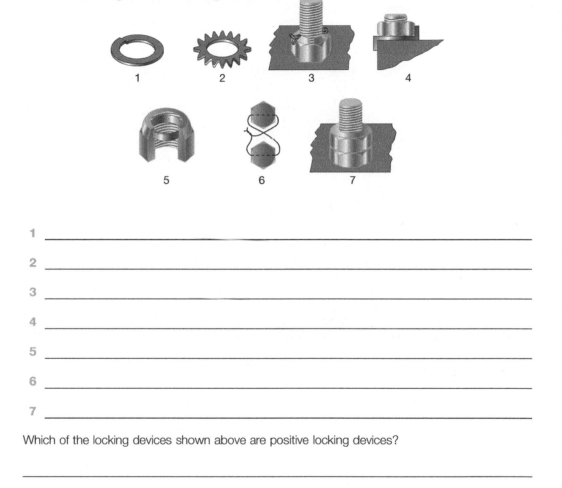

1 _____

2 _____

3 _____

4 _____

5 _____

6 _____

7 _____

Which of the locking devices shown above are positive locking devices?

_____

Which are frictional locking devices?

_____

Describe an alternative method of locking nuts and bolts that is very often used in engine assembly.

_____

_____

Both left- and right-hand threads are used on motor vehicles. However, in most cases it is the right-hand thread which is used.

When tightening nuts and bolts, care must be taken to ensure that they are not over-tightened or left slack. What tool should be used to ensure the correct tightness is applied?

_____

What problems are associated with over-tightening nuts or bolts?

_____

How is the nylock nut prevented from working loose?

_____

_____

List other common methods of joining materials:

_____

_____

_____

_____

Types of fit

Briefly explain the following terms:

Clearance fit – _____

_____

Interference fit – _____

Push fit – _____

Multiple choice questions

Choose the correct answer from a), b) or c) and place a tick [✓] after your answer.

1 **Which one of the following is a type of file?**

   a) bastard [ ]

   b) morse [ ]

   c) vernier. [ ]

2 **A ferrous metal contains:**

   a) Kevlar [ ]

   b) iron [ ]

   c) adhesive. [ ]

3 **Which one of the following would be used to measure the thickness of a brake disc?**

   a) DTI [ ]

   b) micrometer [ ]

   c) rule. [ ]

PART 3
ENGINES

USE THIS SPACE FOR LEARNER NOTES

SECTION 1
Constructing and operating principles 59

SECTION 2
Engine cooling systems 87

SECTION 1

Constructing and operating principles

USE THIS SPACE FOR LEARNER NOTES

_____

_____

_____

_____

_____

_____

After studying this section you should be able to:

- Identify light vehicle engine mechanical system components.
- Describe the construction and operation of light vehicle engines.
- State common terms used in light vehicle engines.
- Identify the key engineering principles that are related to light vehicle engine mechanical systems.
- Identify alternative types of engine power systems.

Key terms

Four-stroke The number of cycles which take place during full combustion.

Spark ignition A term used when an engine uses a spark to ignite the fuel and air mixture.

Compression ignition A term used when an engine compresses the air mixture to generate ignition.

Hybrid A vehicle powered by at least two power sources, e.g. petrol engine and electric motor.

Compression ratio (CR) The ratio between the maximum cylinder volume and the minimum volume it is compressed into.

Torque A turning force measured in Newton metres (Nm).

Firing order The sequence in a multi-cylinder engine in which ignition takes place in each of the cylinders.

Crankshaft Rotates in the cylinder block. Converts reciprocating motion to rotational motion.

Camshaft Rotates at half engine speed and opens the valves at the correct time and duration.

FOUR-STROKE CYCLES

The basic function of an engine is to convert chemical energy to mechanical energy and to produce usable power and torque.

It might be useful to refer to the figure opposite as an aid to understanding the engine cycles when working through the following.

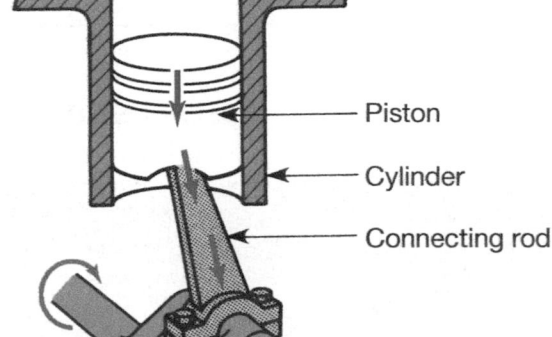
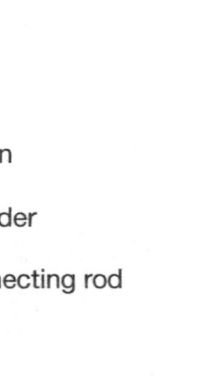

- Piston
- Cylinder
- Connecting rod
- Crankshaft

Four basic engine parts

How many degrees rotation of the crankshaft takes place in one complete four-stroke cycle?

Describe, emphasizing their basic differences, the four-stroke petrol (spark ignition) and diesel (compression ignition) engine operating cycles. Mention typical timings and compression ratios. Indicate the gas flow into and out of the cylinders. Show the positions of the valves at the commencement of each stroke and label them inlet or exhaust. Indicate the direction in which the piston is moving in each case. Label the spark plug and injector for each of the two systems.

Petrol (spark ignition)

Induction:
_____
_____
_____
_____
_____
_____
_____

Compression:
_____
_____
_____
_____
_____
_____
_____

Power:
_____
_____
_____
_____
_____
_____
_____

Exhaust:
_____
_____
_____
_____
_____
_____

Diesel (compression ignition)

Induction:
_____
_____
_____
_____
_____
_____
_____

Compression:
_____
_____
_____
_____
_____
_____
_____

Power:
_____
_____
_____
_____
_____
_____
_____

Exhaust:
_____
_____
_____
_____
_____
_____

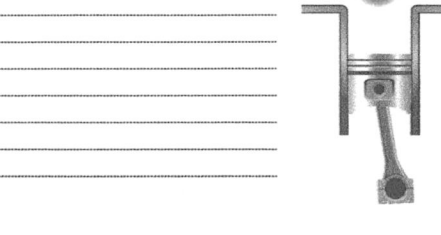

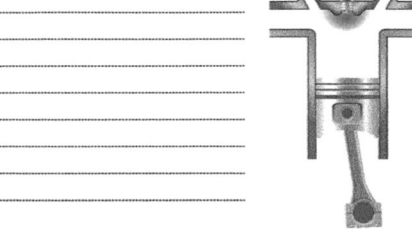

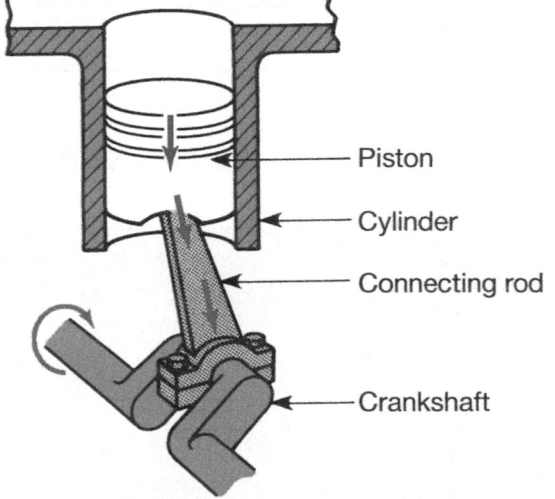

ENGINE TERMINOLOGY

Explain the meaning of the following terms:

tdc _____

bdc _____

Bore _____

Stroke _____

Cylinder capacity _____

_____

Engine capacity _____

Cylinder swept volume

The volume of an engine cylinder is found by multiplying the area of the cylinder by the distance moved by the piston (stroke).

Area of cylinder = πr^2

where
r = cylinder bore ÷ 2
Swept volume = $\pi r^2 h$
where h = stroke

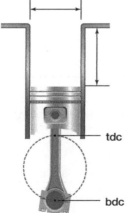

Complete the lettered drawings below to show: on (a) cylinder bore; on (b) tdc and (by shading) the clearance volume; and on (c) the stroke and (by shading) the swept volume.

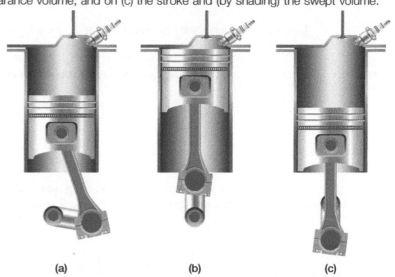

(a)   (b)   (c)

The swept volume of a cylinder in a four-cylinder engine is 249 cm³. Calculate the total volume of the engine.

_____

_____

This engine would be known as a _____ litre engine.

Compression ratio

The engine cylinders opposite have a compression ratio of

_____.

The clearance volume is _____

_____

The compression ratio is the proportion by which _____

_____

This may be expressed as:

$$\text{Compression ratio (CR)} = \frac{\text{Total volume}}{\text{Clearance volume}}$$

$$= \frac{\text{Swept volume (SV)} + \text{Clearance volume (CV)}}{\text{Clearance volume}}$$

$$CR = \frac{SV + CV}{CV}$$

Calculate the CR of a cylinder when the SV is 720 cm^3 and the CV is 90 cm^3.

_____

_____

_____

Torque

When a spanner is placed on a bolt and an effort is made to turn the bolt, the applied force is said to have created a turning moment or torque.

To calculate torque, the following formula is used:

Torque = _____

The force is usually expressed in _____ and the radius in _____

The SI (system international) unit for torque is _____

Work

The two factors which govern the amount of work done are:

_____

Work done is defined as _____

_____.

Force is expressed in _____.

Distance is expressed in _____.

These two values would produce _____.

But the unit used to measure work is a _____

and 1 _____ = 1 _____.

Consider the situation below.

Force required to push vehicle 320 N

20 m

The man in the picture above is pushing the car with a force of 320 N over a distance of 20 m. Therefore the work done by the man moving the vehicle will be: 320 × 20 = 6400 Nm.

This is the same as _____

Power

This is the rate of doing work. In other words, power = work done (in Joules) divided by time taken (in seconds). Power is measured in Watts (W).

The power of a vehicle engine is normally quoted in kilowatts (kW).

Brake horsepower

Engines were originally compared with the horse which they were replacing. The original calculation was that a powerful horse could move 33 000 lb 1 foot in 1 minute. This was soon adopted as the standard in the UK and US.

1 BHP is equivalent to _____.

ENGINE COMPONENTS – THEIR BASIC FORMS AND LOCATIONS

Purpose and functional requirements of engine components

Name the parts numbered in the figures opposite. Briefly state their basic function.

	Part	Function
1		
2		
3		
4		
5		
6		
7		
8		

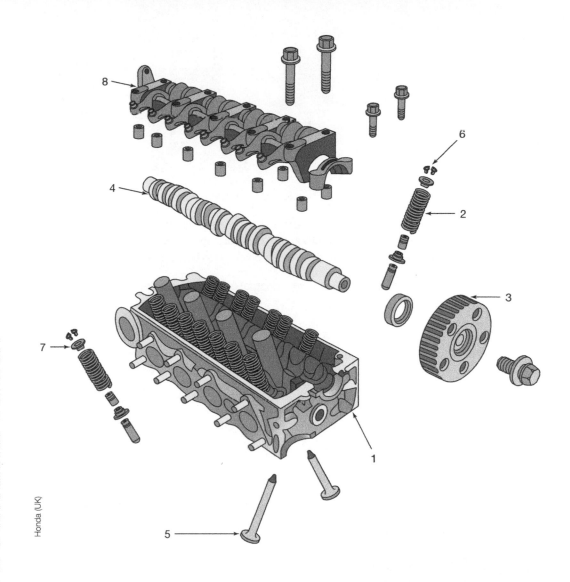

Honda (UK)

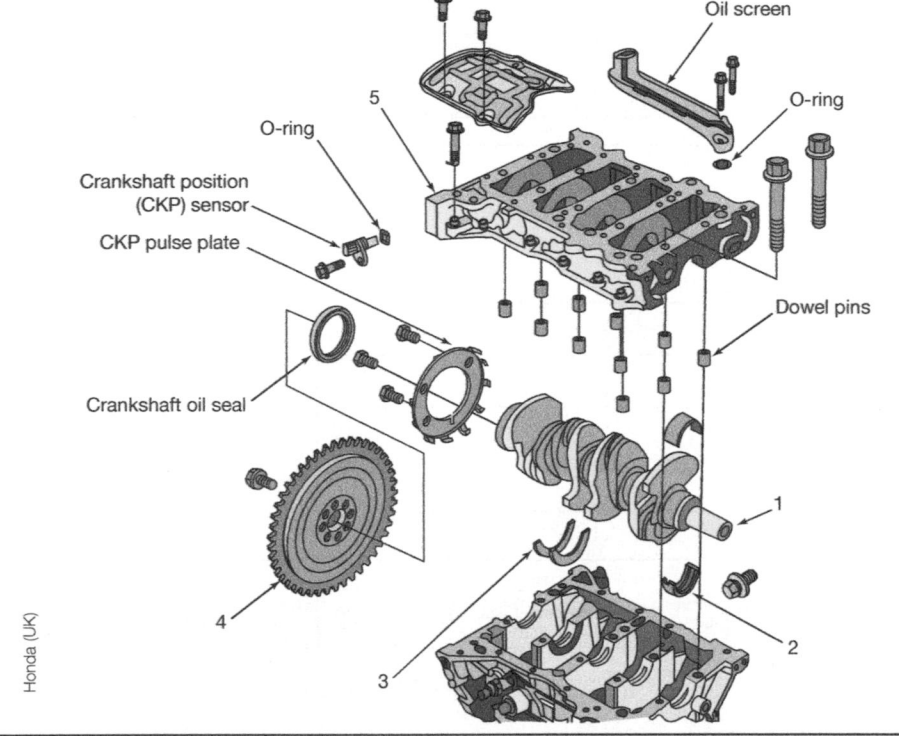

Honda (UK)

	Part	Function
1	_____	_____ _____
2	_____	_____ _____ _____
3	_____	_____
4	_____	_____ _____
5	_____	_____ _____

Camshafts

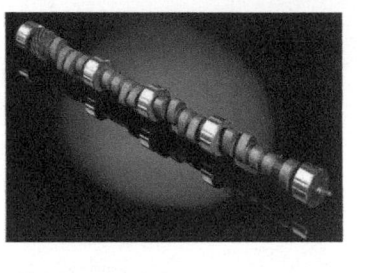

The function of the camshaft is to operate the valves, and frequently it forms a convenient mounting point for various auxiliary drives. Its position on the engine and the actual shape of the cams themselves vary according to individual manufacturers' requirements.

How does the shape of the lobes on the camshaft affect engine performance?

_____

Suggest methods to accurately locate the cam drive gear on the shaft:

_____

_____

_____

The cam is generally quite simple in purpose. It can however be manufactured to engine requirements. For example, the cam determines when and how long the valve will open.

Valve operation

One type of valve arrangement is the overhead valve (ohv) with side camshaft.

The most common valve arrangement used is the overhead camshaft (ohc) where the camshaft is above the piston. With the ohc design, the most common arrangement is to allow the cam to operate directly on the tappet block which is in direct contact with the valve.

The sketches below show three different types of ohc valve layout. Name the types of design:

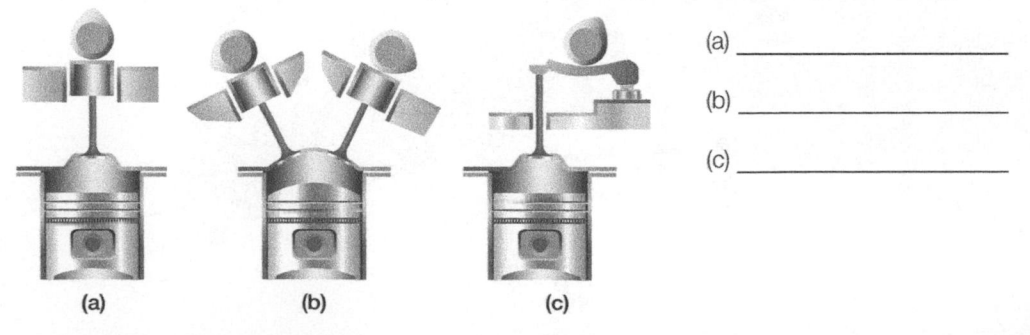

(a) _____

(b) _____

(c) _____

(a)   (b)   (c)

When twin ohcs are fitted they need to be timed together. In the figure to the right the punch mark on the exhaust camshaft gear has to be aligned in the centre of the two punch marks on the intake camshaft gear.

The camshafts will also need to be correctly timed to the crankshaft.

Briefly describe the reason why twin ohcs need to be timed together.

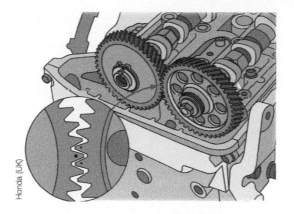

Honda (UK)

_____

_____

Valve timing

The inlet valve, as well as being open on the induction stroke, is also partly open on two other strokes:

1 The inlet valve, opens on the _____ and closes on the _____.

2 The exhaust valve opens on the _____ and closes on the _____.

The reasons for these early openings and late closings are _____

_____.

 TIP The valve timing on a diesel four stroke engine is similar to the spark ignition four stroke engine timing.

The valve-open period can be represented on a valve timing diagram. This shows the number of crankshaft degrees during which the valves are open. The diagram opposite shows a valve timing diagram and indicates the open and close points. Measure and state the angles before and after tdc.

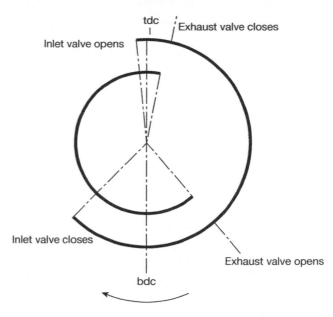

tdc
Exhaust valve closes
Inlet valve opens
Inlet valve closes
Exhaust valve opens
bdc

Valves

Valves allow the fuel/air mixture to enter into the combustion chamber. What do they also allow?

_____

What is the name given to the most common type of valve used in the four-stroke engine?

_____

Generally the inlet valves have a larger head (creating a larger surface area). Why is this?

_____

_____

_____

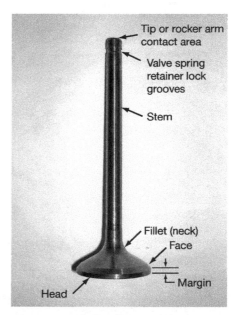

Tip or rocker arm contact area
Valve spring retainer lock grooves
Stem
Fillet (neck)
Face
Margin
Head

The parts of a typical valve

Valve clearance

The valves in an engine require clearance, there are two reasons for this:

1 _____

2 _____

Self-adjusting tappets

This is the most commonly used system on the modern motor car. Describe the operation of the hydraulic tappet shown below.

_____

_____

_____

Add the following labels to the diagram:

Pressure chamber
Non-return ball valve
Feed chamber
Plunger

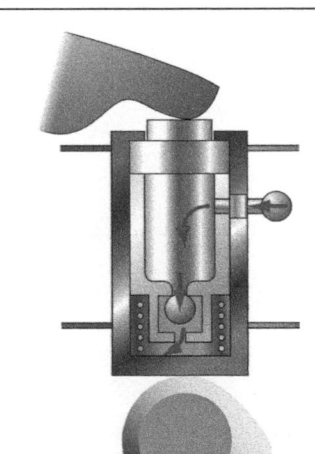

List common faults found during service associated with:

1 **camshafts and tappets**

   a _____

   b _____

   c _____

2 **timing drives**

   a _____

   b _____

   c _____

Valve guides

The valve stem is located in a guide which may be integral with the head or in the form of a sleeve which is an interference fit in the head or block.

Name the types of valve guides shown below and state the most probable material used in each case.

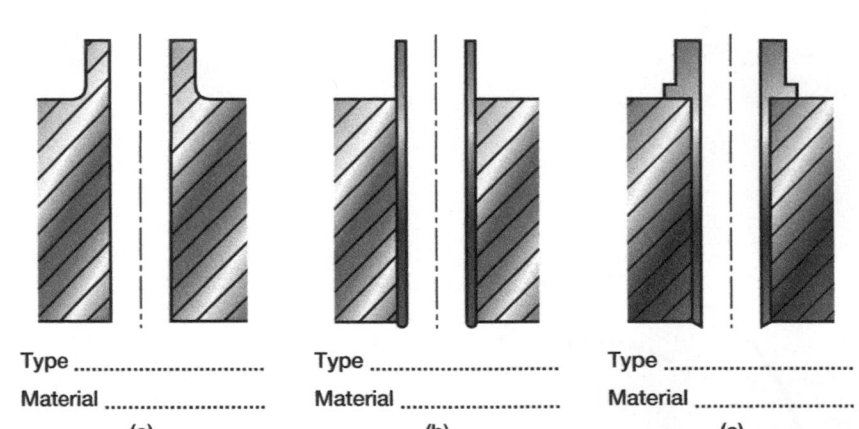

Type ....   Type ....   Type ....
Material ....   Material ....   Material ....
(a)   (b)   (c)

What would be the symptom of excessively worn valve guides?

_____

_____

_____

Valve springs

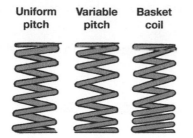 

| Uniform pitch | Variable pitch | Basket coil | Mechanical vibration dampeners |

Common valve spring designs

The valve is opened, either directly or indirectly, by a cam on the camshaft and is usually closed by a coil spring.

The valve spring may be a single spring with either:

● uniform coils
● the coils wound closer together at the cylinder head end
● two springs, one inside the other.

What is meant by valve bounce?

_____

Valve spring retention

The most common arrangement is by split collets.

 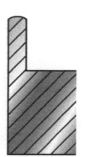

 Remove a valve from a cylinder head, examine it and complete the sketch shown above. Clearly show the method of spring attachment and stem oil-sealing arrangement.

What does the oil seal prevent?

_____

_____

_____

Two alternative types of valve spring retention are:

1 _____

2 _____

USE OF MORE THAN ONE CYLINDER

Two of the main reasons why a conventional vehicle uses an engine with more than one cylinder are:

1 A multi-cylinder engine has a higher power-to-weight ratio than a single-cylinder engine.

2 With multi-cylinder engines there are more power strokes for the same number of engine revolutions. This gives fewer fluctuations in torque and smoother power output.

CYLINDER ARRANGEMENTS – FIRING ORDERS

The most common way of arranging the position of cylinders for multi-cylinder engines is *in-line*.

State the typical firing orders for the 2, 4, 5 and 6 cylinder in-line engines:

2 cylinder _____

4 cylinder _____

5 cylinder _____

6 cylinder _____

Twin-cylinder

The cranks on a twin-cylinder in-line engine may be arranged in two ways:

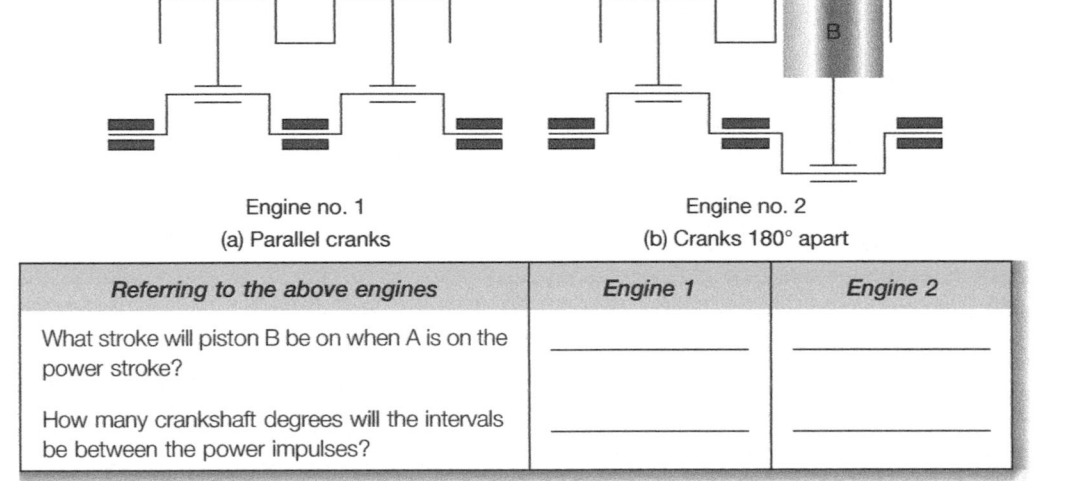

Engine no. 1
(a) Parallel cranks

Engine no. 2
(b) Cranks 180° apart

Referring to the above engines	Engine 1	Engine 2
What stroke will piston B be on when A is on the power stroke?		
How many crankshaft degrees will the intervals be between the power impulses?		

Four-cylinder

Complete the line diagram to show a four-cylinder in-line engine:

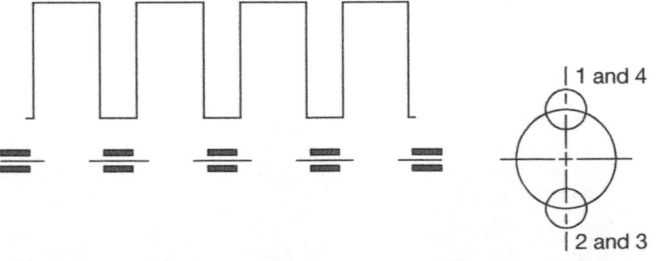

| 1 and 4

| 2 and 3

Considering a four-cylinder four-stroke in-line engine, if the firing order is 1342 what stroke would the following be on:

1   When number 2 cylinder is on the power stroke, number 4 cylinder is on _____

2   When number 4 cylinder is on the exhaust stroke, number 1 cylinder is on _____

Six-cylinder

The line diagram below shows a six-cylinder in-line engine:

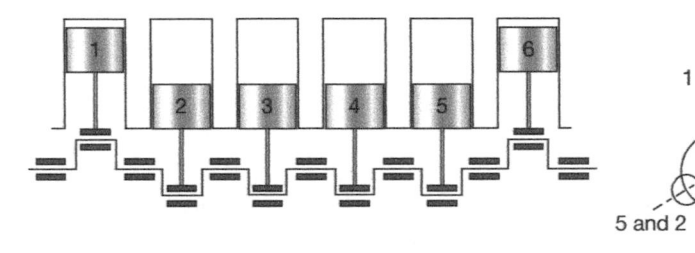

1 and 6

5 and 2   3 and 4

Determine the firing orders on a four- and six-cylinder engine.

Remove the valve-rocker covers and chalk all the inlet valves. Turn the engine in the correct direction of rotation. The sequence in which the inlet valves open is the same as the engine firing order.

Vehicle checked_____ Firing order_____

Vehicle checked_____ Firing order_____

On each engine below, number the cylinders as quoted by engine manufacturers, and state their firing orders.

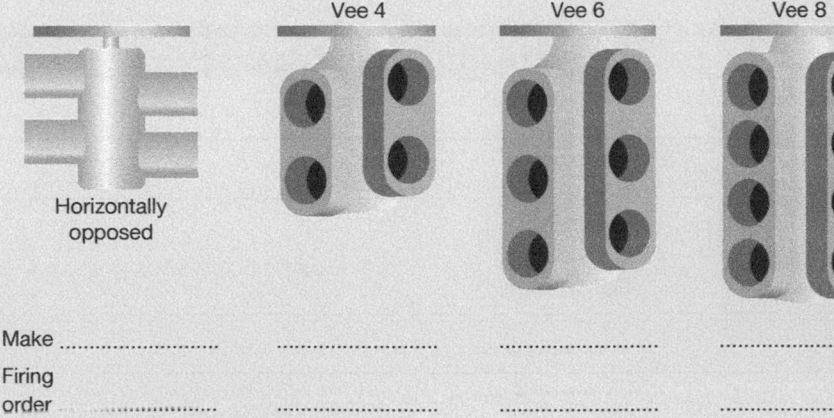

Vee 4   Vee 6   Vee 8

Horizontally opposed

Make ....

Firing order ....

COMBUSTION CHAMBER SHAPES (SPARK IGNITION)

Describe the three main types of combustion chambers shown, making reference to the position of the valves and spark plug.

_____
_____
_____
_____
_____
_____

Wedge

Hemispherical

_____
_____
_____
_____
_____
_____
_____
_____

Bowl in piston

_____
_____
_____
_____
_____
_____

Spark-Ignition Design–Petrol

COMBUSTION CHAMBERS USED IN DIESEL (CI) ENGINES

Direct injection

On the first diagram draw a sectioned view of the type of piston used in a direct injection engine.

Show on both diagrams the direction of air and fuel swirl as the piston is nearing tdc. Name the main parts.

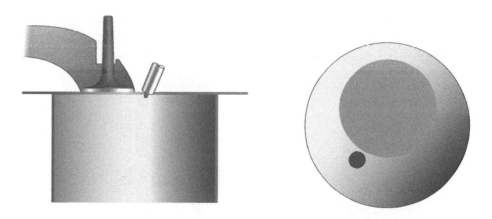

Describe the important features of the basic layout shown above and explain how turbulence and swirl are induced and controlled.

_____
_____
_____
_____

Indirect injection

The type of indirect injection shown below is known as the Ricardo design.

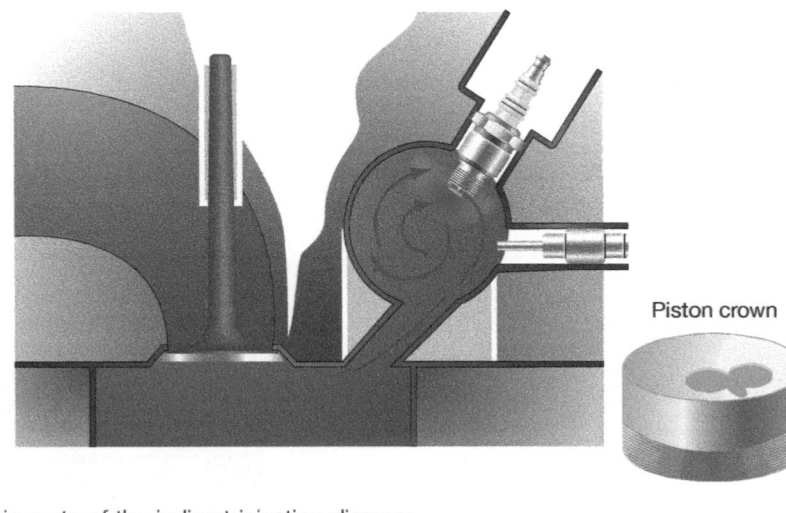

Piston crown

Label the main parts of the indirect injection diagram.

Explain how turbulence is created in the pre-combustion chamber shown.

_____

_____

_____

_____

CYLINDER HEADS

The cylinder head is mounted on top of the cylinders. It confines the pressure of combustion and directs it down on to the piston. The head also provides water-cooling passages, inlet and exhaust passages and supports the valve gear.

The most common cylinder head arrangement used in today's engines is the pentroof combustion chamber (see figure opposite). It is most commonly found in engines with four valves per cylinder. The inlet and exhaust valves are on opposite sides of the chamber, with the spark plug located in the centre. This cylinder head arrangement has a squished area around the entire cylinder.

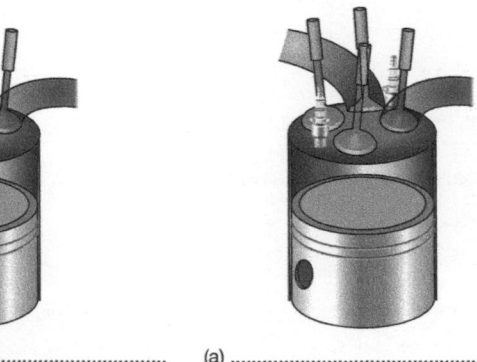

Cylinder head with pentroof combustion chambers

Label the important features of the following cylinder heads. For each cylinder head state a) the number of valves, b) the port arrangements and c) the spark plug location.

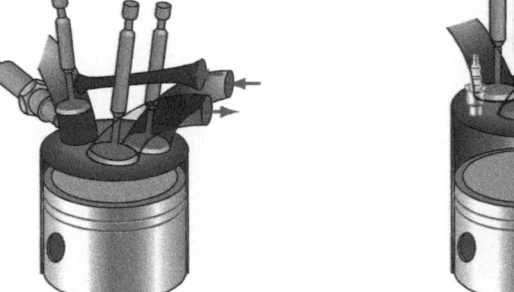

(a) ....
(b) ....
(c) ....

(a) ....
(b) ....
(c) ....

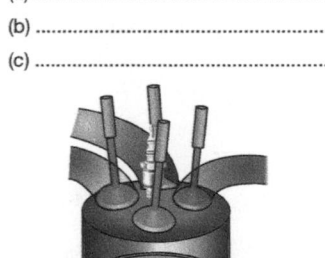

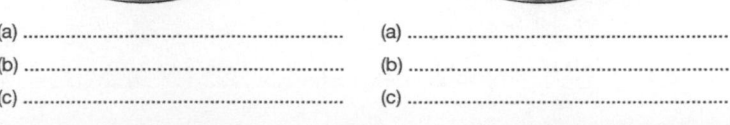

(a) ....
(b) ....
(c) ....

(a) ....
(b) ....
(c) ....

Nut-tightening sequences

Cylinder head bolts (or nuts) must be tightened in the correct sequence and to the correct torque. What might be the results if this procedure is not followed?

Light vehicle cylinder head

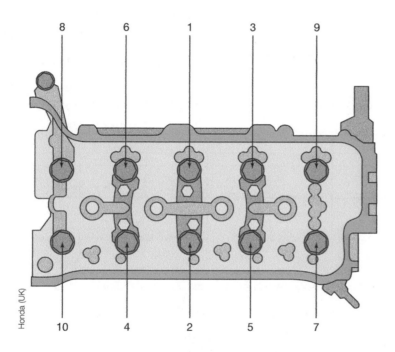

Honda (UK)

When tightening the cylinder head bolts on older type engines it is usually necessary to tighten the bolt in the correct sequence according to their size or Newton metre range.

A more common modern method is to tighten the cylinder head bolts, first by using a torque wrench and then by using an angle measuring tool, and follow a sequence pattern similar to that in the diagram above.

In the example shown opposite, the cylinder head bolts need to be tightened in stages. Initially they are torqued up in the correct order and then further tightened according to the figures below. The procedure for this example is different for new bolts and re-used bolts.

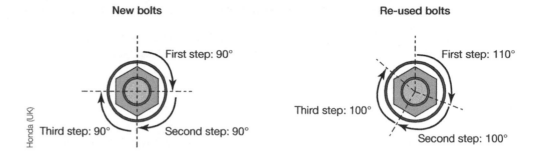

Honda (UK)

Cylinder head faults

List five common service faults associated with cylinder heads:

1 _____

2 _____

3 _____

4 _____

5 _____

CRANKSHAFT-TO-CAMSHAFT MOVEMENT RATIO

On simple camshaft drive arrangements the crankshaft-to-camshaft movement ratio can be proved mathematically by either counting the number of teeth on each gear or by measuring the gears' diameters and dividing the **driven** gear value by the **driver** gear value.

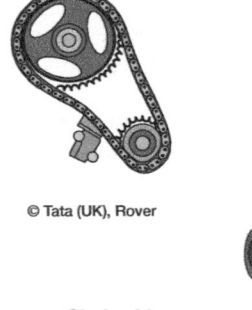

Chain drive   Belt drive   Gear drive

© Tata (UK), Rover   © Ford

Examine engines with drive arrangements similar to those shown above and complete the table.

Engines examined	No. of teeth camshaft gear	No. of teeth crankshaft gear	Gear ratio

A chain drive is normally tensioned by a hydraulic tensioner. This is spring loaded and pressured engine oil applies the required hydraulic tension.

Most modern belt drives are tensioned by a _____

It is important to refer to the manufacturer's fitting instructions when installing a new belt. This is to check the correct procedure for tensioning and fitting the belt.

What must be checked and lined up on all of these drive systems?

_____

_____

Camshaft drive arrangements

Camshafts may be driven by gears, chains or belts.

Engines which have camshafts positioned in the crankcase may be driven by:

_____

The rotational speed of the crankshaft is _____ that of the camshaft.

The speed ratio of the crankshaft to camshaft is _____

How does this ratio affect the size of the gears?

_____

In certain cases the camshaft gear is made of a fibre material.

The reason for this is _____

⚡ Keep fingers away from all moving parts when the engine is running. Extra care and attention needs be taken with the timing gears and belts on a running engine.

Overhead camshaft belt drives

Why should the crankshaft not be rotated when the belt is removed?

_____

What damage can occur if the timing belt fails when the engine is running?

_____

_____

If the timing belt fails, the damage can be expensive to repair, therefore manufacturers specify intervals for timing belts to be changed.

Some engines are known as interference engines. What does this mean?

_____

_____

Now look at some technical data for three vehicles. State their recommended belt change mileages.

1 _____

2 _____

3 _____

How are the timing marks lined up?

_____

_____

_____

What check needs to be done after fitting a new timing belt?

_____

_____

What routine service checks need to be carried out on the timing belt?

_____

CYLINDER BLOCKS

On modern engines the 'cylinder block' is the term usually given to the combination of the block itself which carries the cylinders and the crankcase which supports the crankshaft. The cylinder block is the basic 'framework' of the engine.

Crankcases

The engine crankcase may be a separate unit supporting the crankshaft or may be an integral part of the cylinder block.

Describe the constructional details of:

1 Separate crankcases:

_____

2 A combined cylinder block and crankcase, where the crankcase lower face is in-line with the centre line of the main bearings, or is extended below the centre line of the main bearings as shown in the picture of a cylinder block of a typical in-line engine below:

_____

_____

Courtesy of Chrysler LLC

The cylinder block for a typical in-line engine

How are the effects of distortion and vibration created in the crankcase minimized?

_____

_____

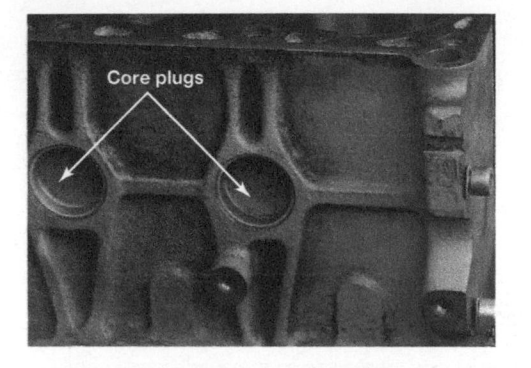

Core plugs in an engine block

Why are there always core (welch) plugs fitted to the side of the cylinder block of a multi-cylinder water-cooled engine?

_____

_____

Cylinder liners

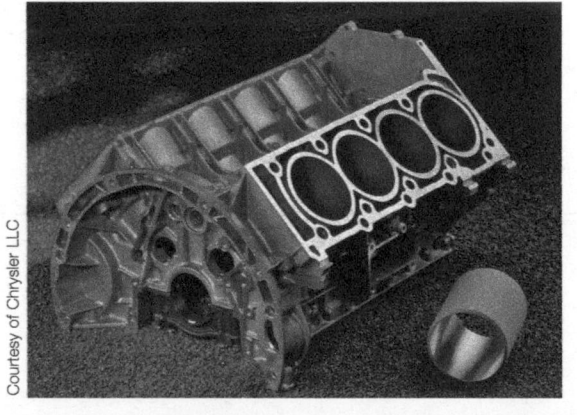

Courtesy of Chrysler LLC

A cylinder block and cylinder liner for a late model aluminium V8 engine

Cylinder liners are fitted into some engine blocks as shown in the figure above. What purpose do they serve when compared with integral type cylinders?

_____

_____

State two types of cylinder liner material:

1 _____

2 _____

List the main advantages of fitting cylinder liners:

_____

_____

What are the main causes of rapid cylinder bore wear?

1 _____

2 _____

3 _____

4 _____

What reasons may cause substantial variations in wear between different cylinders on the same engine?

1 _____

2 _____

3 _____

4 _____

Dry cylinder liners

These are pressed directly into the cylinder bore. Two types are used.

Describe their fitting procedures.

1 Interference fit.

_____

_____

2 Slip fit.

_____

_____

_____

Wet cylinder liners

This type of liner has its outer surface in direct contact with the cooling water.

There are two types of liner in common use:

a The liner's top shoulder is located in a recess in the cylinder block faces while the lower end is a push fit into the lower part of the block.

b This type of block has an 'open deck' layout with the liners inserted into spigots rising from the lower part of the block. These liners may be 'cast-in' on some engines.

On what type of engine is this design most common?

_____

_____

Name the items indicated on two diagrams of wet cylinder liners opposite.

Piston construction

The piston forms a sliding gas-tight (and almost oil-tight) seal in the cylinder bore and transmits the force of the gas pressure to the small end of the connecting rod. In achieving this the piston must form a bearing support for the gudgeon pin and take side thrust loads created by the angular displacement of the connecting rod.

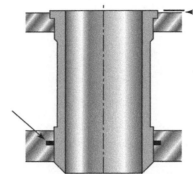
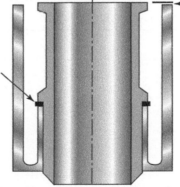

Type (a) commonly fitted in CI engines

Type (b) commonly fitted in ohc engines

On the piston diagram below indicate, using arrows, the main side thrust loads created by the connecting rod. Correctly label the piston diagram with the following:

Ring zone   Crown   Gas pressure   Skirt
Gudgeon pin boss   Top land   Thrust side (engine rotating clockwise)

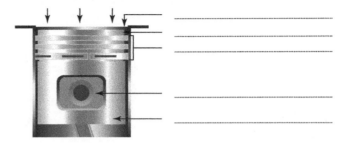

The shape, size, strength and material of a piston are largely dependent on the type of engine in which it is to be used, for example, spark ignition, compression ignition, type of combustion chamber, etc.

List some of the major points that must be considered when designing pistons.

_____

_____

_____

_____

_____

It is necessary to design pistons with operating clearances that vary according to their position on the piston. This is because of the temperature variation within the piston when it is operating.

The figure opposite shows how the temperature gradients vary on the crown and sides of a typical CI engine piston.

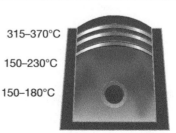

315–370°C

150–230°C

150–180°C

What design factors may be incorporated to control piston expansion?

_____

_____

_____

_____

Piston rings

What are the three basic functions of a piston ring?

1 _____

2 _____

_____

3 _____

Petrol engines usually have pistons with two compression rings and one oil-control ring above the gudgeon pin.

Correctly label the diagram of the piston with the following:

Compression ring – taper faced
Oil control ring – steel rail, multi-piece
Compression ring – plain

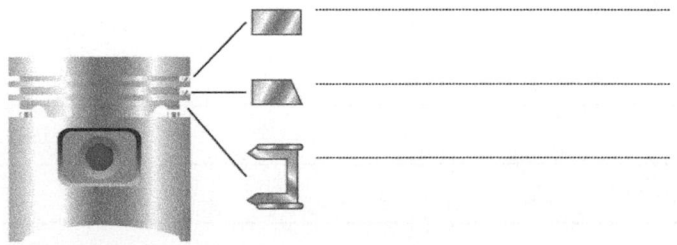

Compression rings

The top ring is usually a plain rectangular section with the outer edge plated in either chromium or molybdenum.

Look at a selection of compression ring designs.

Oil-control rings

Insert the missing words into the following descriptions.

Note that there are two extra distracter words.

    gudgeon   ring   film   slotted   oil   scrape   centre   scraper   iron   edge

This type of _____ should glide over the oil film as the piston moves up the cylinder yet _____ off all but a thin film of oil when descending.

The oil _____ ring above the _____ pin on older types of engines is usually of a _____ design.

The ring is made from cast _____. The _____ can freely drain through the _____ of the ring. They are suitable for use on long stroke engines.

Oil scraper rings

These are designed to exert a greater radial force on the cylinder wall than the above types of rings. Describe the fabricated construction shown.

_____

_____

_____

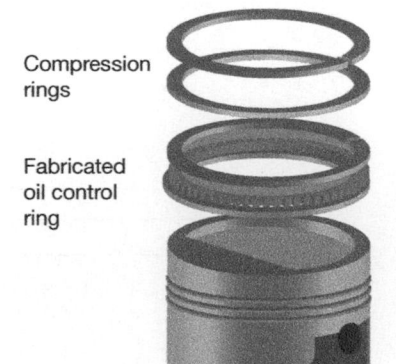

Compression rings

Fabricated oil control ring

The sketch below shows a section of a steel rail multi-piece ring.

Hepolite S E ring

What would be a symptom of excessively worn piston rings?

_____

Gudgeon pins

The gudgeon pin connects the piston to the connecting rod and has to withstand the shock loads created by the forces of combustion.

The pin must be free to rotate in either the piston or the connecting rod or both, in order to allow these two components to move relative to one another.

The pin must also be prevented from moving sideways and scoring the cylinder walls. There are a number of ways in which this can be done:

1   The gudgeon pin is fully floating held with circlips fitting into the gudgeon-pin bosses. The gudgeon pin is free-floating in the connecting rod small end. The small end is usually fitted with a bush.

   The pin is free to turn in the connecting rod and becomes free in the piston as the engine warms up due to the greater expansion rate of the piston.

2   The gudgeon pin is held in the small end with a clamp-bolt (the pin is a thermal fit in piston). This is shown in the diagram on the right.

3   The gudgeon pin which is an interference fit in small end, free in piston. (**Note**: A small end bush is not fitted.)

Describe the method of fitting the gudgeon pin and piston to the connecting rod on the type shown.

_____

_____

_____

Gudgeon pin offset

Solid skirt pistons require greater running clearances than other types and so can produce noisy 'piston slap' when the engine starts up from cold. This effect tends to be more pronounced on the modern, shorter stroke type engines.

What causes piston slap?

_____

_____

To reduce piston slap, the gudgeon pin may be offset about 1.5 mm from the piston centre line towards the maximum thrust side of the cylinder. This allows the piston to tilt during the compression stroke so that the skirt is touching the maximum thrust side of the cylinder wall and as the piston rocks when it moves onto the power stroke the slapping effect is controlled.

Show the position of the gudgeon pin offset on the diagram on the right, by drawing centre lines and the gudgeon pin.

Why are gudgeon pins normally hollow?

_____

Maximum thrust side

d.o.r.

Connecting rod

Small end

The small end of the connecting rod locates the gudgeon pin.

Where the gudgeon pin is fully floating and is fitted with a bush, lubrication is required. For the clamp type and interference fit types, the pin does not move in the rod and so no lubrication is required.

The figure to the right shows the little end of the connecting rod.

Describe how it is lubricated.

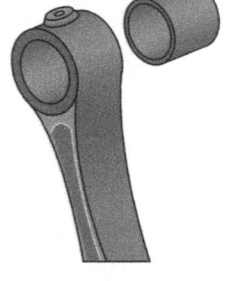

_____

_____

_____

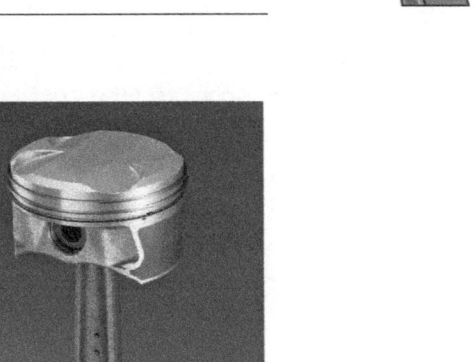

A piston and connecting rod assembly

Courtesy of BMW of North America, LLC

Correctly label the figure above with: **securing bolts, big end cap, thin shell bearings, connecting rod, gudgeon pin, piston skirt, piston crown, piston ring area.**

Steel backing _____

Cast copper/lead _____

Babbitt _____

Courtesy of Dana Corporation

The basic construction of a bearing composed of three materials

Describe how the bearings are:

1 Located (refer to the figure below). _____

_____

_____

Locating lug

Bearing

Slot

Housing bore

Housing

2 Prevented from rotating (refer to the diagram below). _____

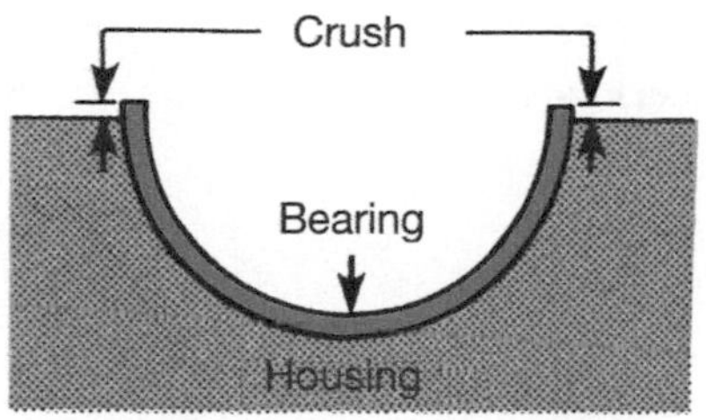

_____

_____

What other type of bearings may be used for big ends, particularly for motorcycles?

_____

The big end attaches on the crankshaft and via the connecting rod enables the reciprocating motion of the piston to be transferred into the rotary motion of the crankshaft.

What is the reason for designing the big end so that the cap face split line is at an oblique angle?

_____

_____

What is the disadvantage of splitting the big end at an angle?

_____

Four-cylinder in-line crankshafts

State the meaning of the following terms which apply to crankshafts and label each term on the photo on the next column:

Main bearing journal _____

Crankpin journal _____

_____

Throw _____

_____

Web _____

Journal radius _____

_____

Balance weight _____

_____

A crankshaft

Crankshafts are commonly made from drop forgings. What does this mean?

_____

_____

An alternative method to produce crankshafts is by casting. These can have large webs which may be hollow.

Why is it considered desirable to use 5 main bearings instead of 3?

_____

Attachment of components to the crankshaft

The flywheel is one of the main components attached to the crankshaft.

State the main functions of the flywheel:

1 _____

2 _____

3 _____

The figure below shows two alternative methods of attaching the flywheel to the crankshaft.

Both diagrams indicate dowel and locking washer arrangements and the first diagram shows a method of oil sealing. Name the main parts.

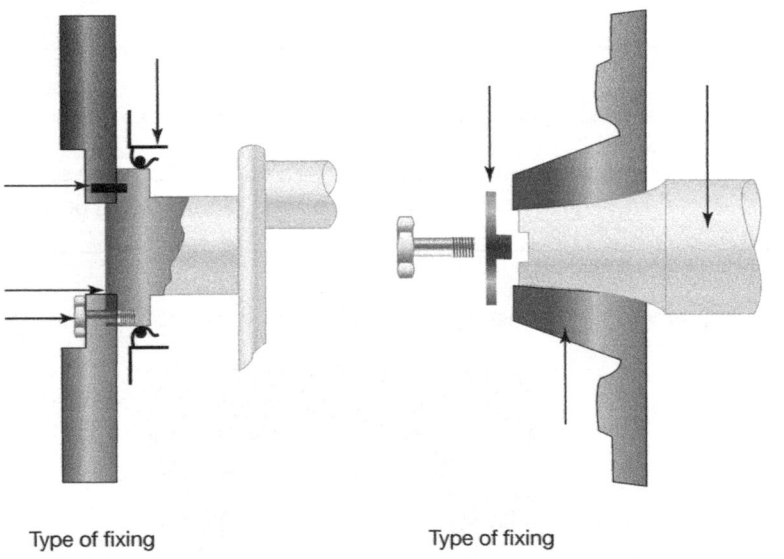

Type of fixing

_____

Type of fixing

_____

What is the object of fitting dowels as well as studs to the flange type flywheel mounting?

_____

_____

List common faults associated with:

1 Crankshafts.

  a _____

  b _____

  c _____

  d _____

  e _____

  f _____

  g _____

2 Main and big end bearings.

  _____

3 Flywheel.

  _____

  _____

TIP When stripping an engine down for repair it is important to label the pistons, connecting rods, bearing caps and valves to ensure that they are refitted where they came from. This is because components are machined and wear to each other.

End-float – crankshaft

Name the parts labelled 'A' on the diagram on the right and explain their function.

_____

_____

_____

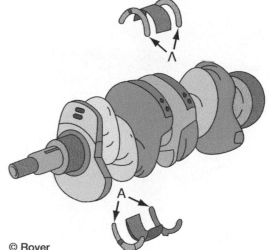

© Rover

The end-float clearance is checked using feeler gauges or by using _____

How can the end-float be altered?

_____

_____

_____

The end-float clearance when measured in the diagram opposite was 0.45 mm but the correct end-float should be 0.15 mm. The thickness of the thrust washers when removed was found to be 2.85 mm each. What should be the thickness of the new thrust washers?

_____

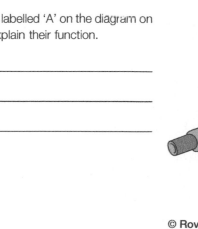

© Rover

Engine balance

Balance weights are added to the crankshaft to balance the main inertia forces that would affect main bearing loading and engine smoothness. Examination of any crankshaft will show the positions of such balance weights; they are usually cast as part of the crankshaft.

What factors create the out-of-balance forces that require balancing?

_____

_____

In some engines, balance can only be achieved by fitting a reverse rotating balance shaft together with extra weights in the flywheel and crankshaft pulley.

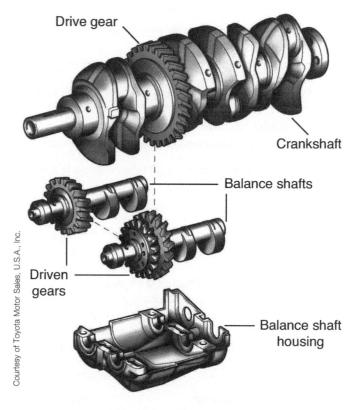

Drive gear

Crankshaft

Balance shafts

Driven gears

Balance shaft housing

Courtesy of Toyota Motor Sales, U.S.A., Inc.

Balance shaft assemblies for a four-cylinder engine

Manufacturers of certain high-quality engines consider perfect balance to be essential and fit two balance shafts which rotate at double engine speed (see figure above). These balance out the reciprocating forces which are known as secondary forces.

Crankshaft torsional vibration

It is easy to realize that metal components vibrate with a certain natural frequency. Many factors contribute to the precise vibratory frequency, one of which is length.

This can easily be demonstrated by clamping one end of a short steel rule and flicking the free end. Do the same with a longer steel rule and note the result:

The short steel rule had a _____ natural vibratory frequency.

The long steel rule had a _____ natural vibratory frequency.

Fill in the missing words in the following paragraph:

| cylinder | longer | shaft | power |
| vibration | twist | torsional | rotating |

Torsional _____ in a crankshaft is caused by the _____ impulses of the engine which gives the _____ crankshaft a series of sharp torsional impulses. The impulse from each _____ tends to momentarily _____ its particular crank ahead of the rest of the shaft, which untwists as soon as the impulse has passed. This sets up a _____ vibration in the _____, the _____the shaft the worse the effect.

Crankshaft vibration dampers

Identify the damper shown in the figure opposite.

_____

_____

Timing seal surface

Inner hub

Insert the missing words into the following descriptions.

Note that there are two extra distracter words.

inertia	camshaft	energy	higher
hub	vibration	damping	crankshaft
viscous	oscillate	torsion	lower
force	ring	silicone	

The operation of this type of damper is as follows.

It has a cast iron _____ and an outer cast iron _____ ring that is connected to the hub with an elastomer (rubber) sleeve which acts as a _____ spring. The hub is securely located onto the _____. At a dangerous vibration speed the excitation _____ becomes devoted to making this inertia ring _____. This eliminates the severe crankshaft _____ but sets up frequencies of _____ amplitude that are not dangerous.

Another type of damper is the _____ damper. As vibration occurs, the inertia _____ moves inside the casing shearing the _____ fluid. This requires a considerable _____ and it gives it the _____ effect.

 **TIP** Such dampers are only suitable for the engine which they are intended for.

List typical symptoms and causes of faulty engine vibration dampers:

Symptom	Cause

ENGINE MOUNTINGS

The engine/transmission unit is attached to the chassis or body structure through rubber blocks. Their main function is:

_____

What are the *basic* causes of engine vibration?

1 _____

2 _____

3 _____

4 _____

The mountings are considered to be fail safe. What is meant by this expression?

_____

_____

_____

Engine block

Engine mount bracket

Engine mount

Frame

A typical engine mount appearance

PRESSURE CHARGING

The performance of an internal combustion petrol engine depends to a very large extent on the density or weight of the charge in the cylinder at the beginning of the compression stroke, that is, on the volumetric efficiency.

The volumetric efficiency of an engine, and hence the power output, can be improved by using a supercharger (or pressure charger) to force the charge into the cylinders during induction.

Pressure charging systems used on spark ignition and compression ignition engines can be of various designs. Briefly describe the TWO most common methods:

1 Exhaust gas-driven turbochargers _____

_____

2 Mechanically driven superchargers _____

_____

Why is the pressure charging of CI engines much more popular than the pressure charging of petrol engines?

_____

_____

_____

Superchargers

One common type of supercharger is the roots blower.

How is a supercharger driven?

_____

_____

_____

Describe the basic operation of the supercharger in this figure:

_____

_____

_____

_____

A drawing showing the drive set up of a supercharger

How is the supercharger shown engaged and disengaged?

_____

Turbocharger

This type of pressure charger is driven by the flow of exhaust gas as it leaves the engine.

Identify and label the major parts on the diagram:

Compressor housing
Turbine housing
Turbine wheel
Fully floating shaft bearings

Describe the basic operation of a turbocharger:

_____

_____

_____

_____

_____

_____

© Chris Longhurst-www.carbibles.com

Cutaway of a typical turbo charger

Unless controlled, the turbo chargers are capable of 'blowing up' engines at maximum speeds.

State common types of control:

1 _____

2 _____

3 _____

The most common method of control, particularly with turbochargers, is to redirect the air/gas through a waste gate.

Intercoolers

In a basic turbocharged system, the compressed air is forced directly into the engine as shown in the figure below.

On many engines an intercooler or charge cooler is used to cool the air after passing through the turbocharger.

Why is an intercooler considered necessary?

_____

_____

_____

_____

_____

_____

ROTARY ENGINE

The rotary engine, or Wankel engine, has four cycles similar to the Otto cycle, except that the cycles occur in the space between a three-sided symmetric rotor and the inside of the housing.

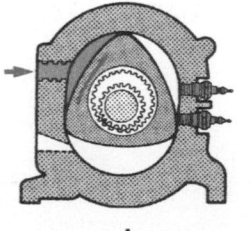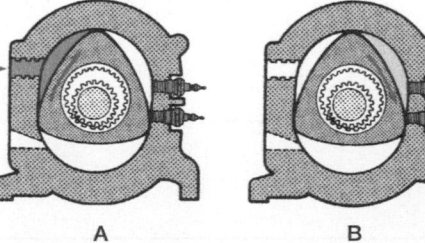

A   B   C   D

The rotor is triangular in shape and each face (side) has a recess which increases the effective displacement of the engine. The tips of the rotor are always in contact with the walls of the housing. The housing which the rotor rotates in is oval in shape.

Describe what is happening in the diagrams on page 84.

A _____

_____

B _____

C _____

_____

D _____

There are three separate chambers and at one given moment in time each chamber will be at a different stage of the cycle.

The advantages of a Wankel engine are: _____

_____

The disadvantages of the Wankel engine compared to a conventional petrol piston engine are:

_____

_____

HYBRID ENGINE

 When working on hybrid vehicles extra care must be taken, due to the high voltages used. Depending on the manufacturer and the system, these voltages can be up to 330 volts. Always refer to the manufacturer's safety procedures before carrying out any service or repair work.

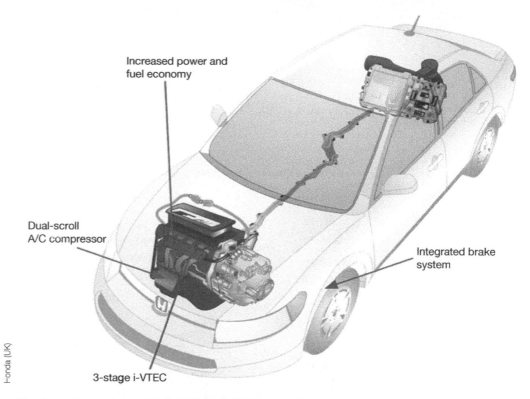

Increased power and fuel economy

Dual-scroll A/C compressor

Integrated brake system

3-stage i-VTEC

Honda (UK)

The figure above shows a typical Honda hybrid layout with the battery pack in the rear of the vehicle. **Note** the orange coloured high-voltage cables.

Honda (UK)

This Honda Hybrid i-DSi (intelligent Dual Sequential Ignition) engine is 1339cc and has a 10Kw brushless electric motor

With the need for greener, cleaner exhaust-emitting vehicles a number of manufacturers are producing hybrid engines. A hybrid vehicle will have at least two different power sources.

How is a hybrid vehicle powered?

_____

_____

_____

Find out what you can about the Atkins cycle engine. Write down the main differences between the Otto four-stroke cycle and the Atkins cycle.

_____

_____

_____

How is the electric motor powered? _____

_____

_____

Hybrid vehicles are classified as either being series hybrid or parallel hybrid. In a series hybrid the engine powers a generator which in turn charges the batteries. The engine never directly powers the vehicle.

In a parallel hybrid either the _____ powers the vehicle.

Most hybrid vehicles currently in use are known as _____ because_____

_____

Multiple choice questions

Choose the correct answer from a), b) or c) and place a tick [✓] after your answer.

1 **Which of the following is the correct sequence of events for a four-stroke cycle?**

   a) induction, power, exhaust, compression [   ]

   b) induction, compression, power, exhaust [   ]

   c) induction, exhaust, power, compression. [   ]

2 **A correct firing order of a four-cylinder engine is:**

   a) 1342 [   ]

   b) 1234 [   ]

   c) 4321. [   ]

3 **To help reduce piston slap:**

   a) the piston crown is oval [   ]

   b) a wet liner maybe used [   ]

   c) the gudgeon pin is offset. [   ]

4 **The component which attaches the piston to the crankshaft is the:**

   a) cylinder liner [   ]

   b) connecting rod [   ]

   c) gudgeon pin. [   ]

5 **The crankshaft to camshaft ratio is 3:1:**

   a) True [   ]

   b) False. [   ]

6 **Wedge, hemispherical and bowl are all types of:**

   a) camshaft lobe profiles [   ]

   b) SI combustion chambers [   ]

   c) wet type cylinder liners. [   ]

7 **Which one of the following terms relates to the crankshaft?**

   a) lobe [   ]

   b) skirt [   ]

   c) throw. [   ]

SECTION 2

Engine cooling systems

USE THIS SPACE FOR LEARNER NOTES

_____

_____

_____

_____

_____

_____

Learning objectives

After studying this section you should be able to:

- Identify engine cooling, heating and ventilation system components.
- Describe the construction and operation of engine cooling, heating and ventilation systems.
- Compare key engine cooling, heating and ventilation system components and assemblies against alternatives to identify differences in construction and operation.
- Identify the key engineering principles that are related to engine cooling, heating and ventilation systems.
- State common terms used in key engine cooling, heating and ventilation system design.

Key terms

Cross-flow radiator   Coolant flows horizontally and transfers heat from the engine to the surrounding air.
Thermostat   A temperature controlled valve which assists engine warm-up.
Pressure cap   Maintains the correct operating pressure for the cooling system.
Impeller   A water pump which circulates coolant around the engine.
Antifreeze   Is added to water in the correct percentage to reduce freezing and act as a corrosion inhibitor.
Hydrometer   Tests equipment to check the specific gravity of a liquid.

www   www.gm-radiator.com

www.commaoil.com

COOLING SYSTEMS

The principle of the internal combustion engine, which gets its power from the combustion of fuel and air and the resultant expansion of gases, necessitates the use of some type of cooling system.

During combustion, the temperature in the cylinder can momentarily be as high as 1800°C. Even when the gases expand and the temperature falls, it may still be higher than the melting point of aluminium.

What problems may occur if this heat is not dissipated at the correct rate?

____

____

The two main types of cooling systems are **Air** and **Liquid**. List TWO subdivisions of each type:

Air cooling

1 ____

2 ____

Liquid cooling

1 ____

2 ____

State THREE functional requirements of a cooling system:

1 ____

2 ____

3 ____

Heat Transmission

Heat is one type of energy. A piston engine relies on heat for its principle of operation. Heat is transmitted in one (or more) of three ways, what are they?

1 **Conduction –** ________________________________

____

2 **Convection –** ________________________________

3 **Radiation –** ________________________________

____

Heat flow through metal is by ________________________________.

Air Cooling System

With a few notable exceptions, this system is not very popular for multi-cylinder engines as difficulties are encountered trying to cool all the cylinders equally and maintain a constant temperature.

Water-Cooled Pressurized System

Water is a better cooling medium than air. It has a high specific heat and is able to transfer heat more efficiently.

Unfortunately, for water-cooling systems, the engine gives its best thermal efficiency when the cooling water is close to 100°C (that is, normal boiling point). To overcome this problem water-cooling systems are normally pressurized.

A cooling system is pressurized to __________ the __________ temperature of the coolant.

Why is such a design feature considered necessary?

____

What advantages are gained by pressurizing the system?

_____

_____

_____

What limits the pressure that can be imposed on a water-cooling system?

_____

For approximately every 5 kPa of pressure rise, the water boiling temperature increases by 1°C.

What would the approximate boiling temperature of water in a cooling system be, using:

1 A 4 lbf/in² (30 kPa) pressure cap? _____

2 A 15 lbf/in² (105 kPa) pressure cap? _____

TIP   14.7 lbf/in² (PSI) = 1 Bar

Advantages of Air and Liquid-Cooling Systems

State the relative advantages of **liquid** cooling compared with **air** cooling:

_____

_____

_____

_____

State the relative advantages of **air** cooling compared with **liquid** cooling:

_____

_____

_____

_____

The liquid-cooling system basically consists of water jackets surrounding the cylinders with provision for the heated water to pass into a radiator and cooled water from the radiator to flow back into the cylinder block.

Thermostat

This is a temperature-sensitive valve that controls water flow to the radiator. State the two main reasons why it is fitted:

1 _____

2 _____

Wax element type

This type of thermostat employs a special wax contained in a strong steel cylinder into which passes the thrust pin. This is surrounded by a rubber sleeve which also seals the upper end of the cylinder.

The wax element type thermostat is shown fully closed on the left and fully open on the right. Show the water flow direction on the right-hand diagram. Name the main parts on both diagrams.

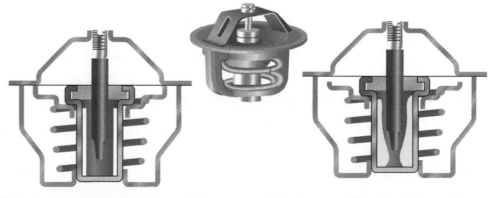

Fully closed wax element type thermostat   Fully open wax element type thermostat

Describe the wax element type thermostat's basic operation.

_____

_____

_____

Some thermostats have a small hole and pin in the valve disc. Why is this?

_____

All substances exist in one of three states: solid, gas or liquid. Most can change from one to the other. State a common example.

_____

What change of state causes the wax type thermostat to operate?

_____

Thermostat operation test

Check the serviceability of a thermostat when it has been removed from the vehicle.

Thermostat make _____ Type _____

Specified opening temperature _____

Actual temperatures: opening _____ closing _____

Visual defects (if any) _____

_____

Serviceability _____

TIP When replacing a thermostat, check the replacement has the same temperature rating as the original.

Pressure cap

This consists of a spring-loaded valve which resists the pressure of the expanding coolant, air and steam, in the header tank unit.

Name the essential parts of this typical of radiator cap.

How is the pressure controlled?

_____

_____

Explain the purpose of the small valve fitted in the centre of the pressure cap:

_____

_____

Pressure cap testing

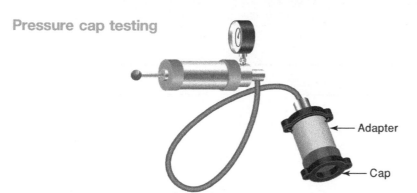

Adapter

Cap

Connect the cap to the adapter and the tester. Pressurize to check when the valve opens. The obtained pressure should be:

_____

A modern type of radiator cap

Examine various caps and state at what pressure they should operate. Use a pressure tester to check their condition.

Make of vehicle	Stated pressure of cap tested	Actual release pressure	State if cap maintained pressure for over 10 seconds

Matrix construction-tube radiator construction

This consists of thin, almost flat or oval, copper or brass tubes arranged in rows. Why are oval or almost flat tubes used?

_____

_____

Why are fins used? _____

RADIATOR TESTING

Pressure testing

© The Tool Connection Ltd and Laser Tools

Coolant pressure tester

A pressure test is carried out to determine external or internal leaks. Describe how to pressure test a system.

_____

_____

_____

Using a coolant pressure tester

Flow testing

If a radiator is suspected of being partially blocked how should it be checked when:

1 It is on the vehicle?

_____

_____

2 It has been removed from the vehicle?

_____

_____

_____

CROSS-FLOW RADIATORS

In the conventional radiator the coolant flows _____ through the core from _____ to _____.

In cross-flow radiators the coolant flows _____ through the core from the _____ of one side tank across to the _____ of the other side tank.

Outlet tank
(Plastic)

Bending tangs

Inlet tank
(Plastic)

Transmission
oil cooler

O-ring
gasket

Radiator core
(Aluminium)

O-ring
gasket

Draincock

Why is it considered necessary to fit such a radiator in preference to the vertical flow type?

_____

_____

What are the cross-flow radiator's basic disadvantages?

_____

_____

_____

With most cross-flow systems it is necessary to fit a remote header or expansion tank as shown below. Correctly number the following parts with their corresponding number on the diagram:

Header tank _____ Bottom hose _____ Pressure cap _____

Heater matrix _____ Drain tap _____ Radiator (cross-flow) _____

Top hose _____ Thermostat switch _____

Thermostat housing _____ Water pump _____

Why is the remote header tank necessary?

_____

_____

Most cooling systems (water or air) are fitted with fans. In the simplest arrangement the fan is permanently driven from the crankshaft via the fan belt.

The fan's function is _____

_____

Why is the type of fan described above rarely used on modern vehicles?

_____

Viscous coupling fans

These units operate on the viscous shear principle. There are TWO types of couplings:

1 **Torque limiting.** Here the thickness of the fluid determines the slipping effect and therefore the maximum speed at which the fan will rotate (3000 rev/min).

2 **Air temperature sensitive.** In this type warm air acts on a bi-metal strip sensor; this sensor opens a valve which allows fluid into the clutch and so provides maximum drive.

A viscous type fan clutch

State the function of the fins on the viscous coupling:

_____

Electrically driven fan

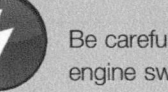

Be careful when working on an engine with an electric cooling fan. Even with the engine switched off the fan can still automatically operate to cool the engine.

The construction of this type includes an electric motor complete with fan, a temperature-sensitive control unit and a warning light. Name the parts indicated from the following word bank:

Thermostat housing
Fan motor
Temperature sensor
Pressure cap
Bleed screw

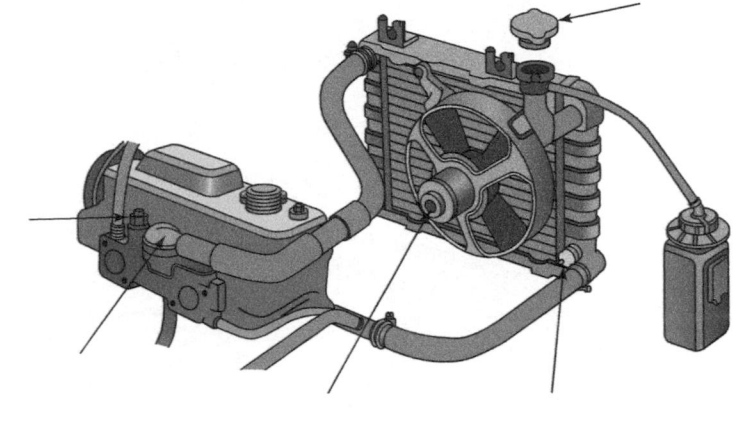

Tata (UK), Rover

Describe the operation:

_____

_____

_____

_____

Describe how a thermostatically-controlled cooling fan should be checked for correct operation (this could be done as a practical activity):

_____

_____

_____

_____

WATER PUMPS

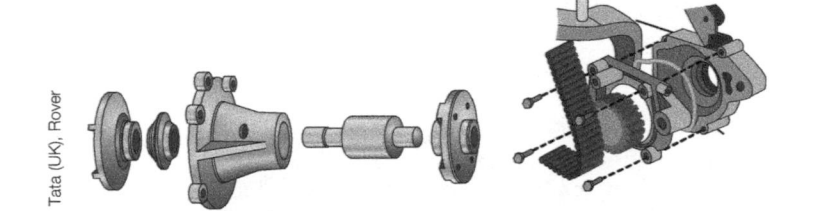

Tata (UK), Rover

Modern water pumps, properly called impellers, are usually bolted to the front of the cylinder block and are driven from the crankshaft by the timing belt.

Explain the provision normally made to circulate the water expelled from the water pump when the thermostat is closed:

_____

_____

A lot of hybrid vehicles use an electrically operated water pump.

In small groups discuss why you think these are used on hybrid vehicles. You will need to think about how a hybrid vehicle operates and how it is designed to be economical in and around a town or city.

SEALED COOLING SYSTEMS

A disadvantage of the ordinary cooling system is that small losses of coolant occur through the radiator overflow pipe. If the level of water in the header tank is not frequently checked it is possible for the water level to fall sufficiently to prevent circulation. A method of overcoming this is simply to immerse the lower end of the overflow pipe in coolant contained in an expansion chamber.

A sealed system ensures that the cooling system is kept completely full at all times. What design feature does this require?

_____

_____

Why is a sealed system considered essential on some heavy vehicles?

_____

WATER HEATERS

To give adequate comfort to the occupants of a vehicle in varying weather conditions, all modern vehicles are equipped with some form of heating and ventilation system. On vehicles fitted with water-cooled engines, the heat source is usually the hot water from the engine. Heater units which utilize the engine cooling water are normally situated on the bulkhead. These are normally a heater matrix type, similar in construction to that of a radiator matrix.

Describe the basic action of the heaters shown below:

_____

_____

_____

The diagram below shows a fresh-air cab heater with its operational flaps in the off position.

Show the flaps positioned to give:

1 maximum heat to interior and screen
2 warm air to screen only.

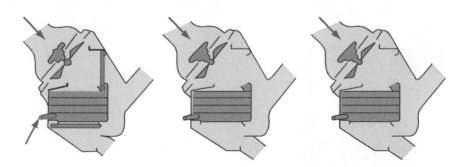

The drawings below show the cool air entries to the heater system. Using arrows show the flow of cool/heated air through each vehicle. Label where the vents can be found on these vehicles.

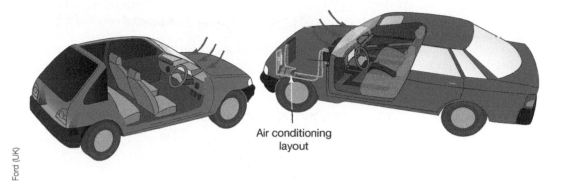

Ford (UK)

Air conditioning layout

HOSES

The cooling system components are connected by flexible hoses, that is, engine to cooling radiator, engine to heater radiator and possibly between cylinder head and water pump. This allows for the natural movement of the engine when running and the hoses can flex without fracturing.

Examine a hose and describe its construction.

_____

⚡ Always release the cooling system pressure by removing the radiator/header tank cap before attempting to remove hoses or cooling system components.

State common faults, and possible causes, related to hoses:

_____

_____

_____

_____

_____

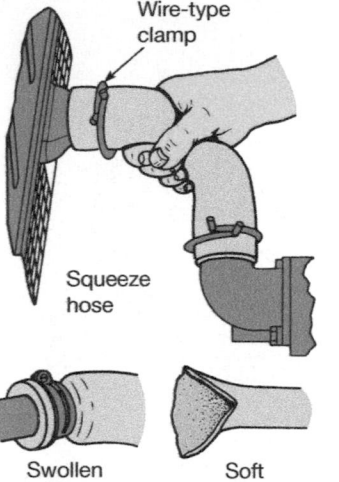

Wire-type clamp

Squeeze hose

Swollen   Soft

Chafed   Hardened

Cooling hose defects

Hoses need to be securely attached to the engine and radiator. Various types of clamps can be used. They fit around the outside of the hose and exert a clamping pressure around the hose where it connects to the radiator, water pump, engine block or heater matrix. It is the pressure of the clamp around the hose which creates the seal at that point.

 **TIP** During a routine vehicle maintenance service, visually check all coolant hoses for leakage, splits, swelling and chaffing.

Another type of hose clamp which is sometimes used is the **thermoplastic** type.

How is this clamp fitted to the hose?

_____

_____

_____

EFFECTS OF FREEZING

When water freezes it increases in volume and this can cause a cylinder block to crack. State at what temperature cooling water increases its volume.

_____

_____

Antifreeze

Fortunately, the freezing point of water can be lowered considerably by the addition of certain liquids. An example of such a substance is ethylene glycol.

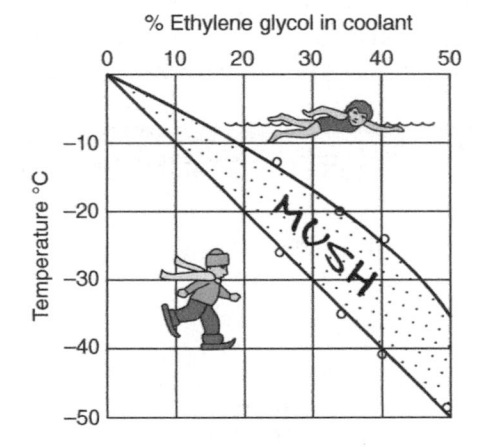

% Ethylene glycol in coolant

MUSH

The recommended percentage antifreeze mixture in the UK is:

_____

The protection depends on the percentage of antifreeze to water mix. It is important to check manufacturer's specific data for the vehicle being worked on.

The graph above shows two of the most important properties in an ethylene glycol-based antifreeze. These are:

1 _____

2 _____

What other properties must antifreeze possess?

_____

_____

_____

What are the disadvantages of using a methanol-based antifreeze?

_____

_____

_____

The proportion of an ethylene glycol-based antifreeze present in a cooling system can be determined by checking the specific gravity of the coolant and by reference to its temperature.

What instrument is used to measure coolant antifreeze content?

_____

_____

Name the parts indicated on the antifreeze hydrometer shown below:

State THREE operational tasks required when changing antifreeze:

1 _____

2 _____

_____

3 _____

_____

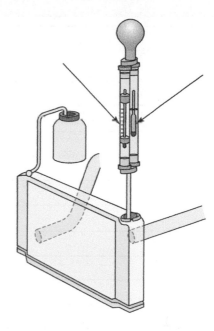

The following describes the procedure for measuring the strength of a cooling system's antifreeze. Insert the missing words into the following paragraphs. **Note** that there are two extra distracter words.

screen	scale	coolant	percentage
hydrometer	light	thermometer	shadow
multimeter	grid	mirror	amount

The engine is warmed up and the _____ is used to draw a quantity of _____ from the system. The readings on the float level and _____ are noted and these values are transferred to a reference _____ which indicates the _____ of antifreeze in the system and the _____ required to top up to the recommended proportions.

When using a refractometer tester a small quantity of coolant is placed on the _____ and the clear cover is lowered. The eyepiece is held to the eye and the tester is aimed towards a _____ source (not direct sunlight). A _____ will be cast and can be read off the _____ which shows the antifreeze content.

 Always clear up spilt antifreeze immediately. Antifreeze can be extremely slippery (it can also damage paintwork). Antifreeze can be TOXIC.

EFFECTS OF CORROSION: ACIDS AND ALKALIS

The main causes of corrosion in the cooling system are:

1 **Acids** which become very potent if there is any cylinder head blow-by. They accelerate rusting of the ferrous metal.

2 **Salts** contained in the water give excessive alkalinity and corrode alloy metals, in many cases causing a pinhole effect.

3 **Dissimilar metals in contact** create electrolytic (also referred to as galvanic) corrosion and this rapidly eats one of the metals away. This can affect soldered joints and gasket sealing.

Continual heating and circulation transfers these deposits around the circuit allowing them to settle in the slower flowing sections, eventually to cause blockages and overheating. Alloy engines are more likely to be affected than cast iron types.

These corrosive elements in the cooling system can be neutralized by keeping antifreeze in the system all year, or using a good quality corrosion inhibitor when antifreeze is not necessary.

Cooling system protection during use

Describe how the cooling and heater systems should be protected during use or repair by avoiding or preventing the following hazards:

1 Corrosion

_____

2 Freezing

_____

3 Excess pressure/vacuum

_____

_____

State any special tools that are required for carrying out cooling system maintenance:

_____

_____

_____

_____

_____

DIAGNOSTICS: COOLING/HEATER SYSTEM – SYMPTOMS, FAULTS AND CAUSES

State a likely cause for each symptom/system fault listed opposite. Each cause should suggest any corrective action required.

Symptom	System fault	Likely cause
System boiling, steam coming out from under bonnet	Overheating	
Engine is misfiring when accelerating and heater output is cool	Overcooling	
Engine whines when revving and there is a rumbling sound when idling	Noisy water pump	
Pool of liquid under vehicle and/or radiator empty	Coolant leakage	
Coolant discoloured and possible leakage at hose joints.	Corrosion	
Coolant discoloured and appears to be oily	Contamination	
Coolant level at header tank is correct, heater fan works but there is no warm air coming into car	Heater not operating	
Air Cooling The engine overheats when idling in traffic	Air-cooling system overheating	
Air Cooling In cold weather the engine runs cold when cruising	Air-cooling system overcooling	

Diagnostics: Cooling system noises

With the engine running, what cooling system faults may be indicated by the following noises?

Noise	Possible fault
Screeching noise when engine revved	_____
Buzz or whistle near radiator	_____
Rattling noise near radiator	_____
Ringing or grinding noise at front of engine	_____
Gurgling from radiator	_____

ROUTINE MAINTENANCE

List the general rules and precautions to be observed when carrying out the following routine cooling system maintenance and running adjustments:

1 Dealing with hot systems.

_____

_____

2 Checking when the engine is running.

_____

_____

3 Antifreeze is spilt on paintwork.

_____

_____

 Antifreeze can be toxic.

 TIP When draining coolant from the system remove the radiator/header tank cap. This will allow the coolant to drain easily.

State THREE reasons for carrying out routine maintenance on the cooling/heater system:

1 _____

2 _____

3 _____

Multiple choice questions

Choose the correct answer from a), b) or c) and place a tick [✓] after your answer.

1 **The component which ensures a rapid warm-up when starting an engine from cold is the:**

   a) radiator [ ]

   b) thermostat [ ]

   c) pressure cap. [ ]

2 **Antifreeze is also:**

   a) a corrosion inhibitor [ ]

   b) a corrosion producer [ ]

   c) an active electrolyte. [ ]

3 **A piece of equipment used to check the coolant percentages is the:**

   a) pressure test [ ]

   b) chemical tester [ ]

   c) hydrometer. [ ]

4 **Which one of the following are the three methods of heat transfer?**

   a) radiation, warmth, convection [ ]

   b) convection, radiation, conduction [ ]

   c) conduction, light, radiation. [ ]

5 **A pressurised cooling system:**

   a) lowers the boiling point [ ]

   b) raises the boiling point [ ]

   c) maintains the boiling point. [ ]

SECTION 3

Engine lubrication systems

USE THIS SPACE FOR LEARNER NOTES

_____

_____

_____

_____

_____

_____

Learning objectives

After studying this section you should be able to:

● Identify and describe the main components, their construction and operation in light vehicle engine lubrication systems.

● Identify the key engineering principles and common terms used in light vehicle lubrication systems.

● Understand how to check, replace and test light vehicle engine lubrication system units and components.

Key terms

Wet sump   Oil is returned from the engine by gravity and carried in a sump below the engine.

Dry sump   Oil is supplied and returned to a separate oil tank away from the engine.

Full-fluid film   The film of lubricant is thick, where no metal-to-metal contact takes place.

Hydrodynamic lubrication   Using the natural movement of the oil 'wedge' to separate the surfaces of highly loaded bearings when shafts rotate.

Boundary   A term applied where the film of lubricant is applied by splash and mist.

Viscosity   The resistance to flow or 'thickness' of a liquid.

Multigrade   Oil which meets the viscosity requirements of several different single-grade oils.

Viscosity index   A number which indicates how the viscosity of a liquid changes with temperature.

www   http://auto.howstuffworks.com/engine6.htm

Remember: when working with engine lubrication systems you may be dealing with:

● very hot surfaces
● hot oil under pressure
● chemicals which are carcinogenic or toxic
● rotating components
● waste disposal.

ENGINE LUBRICATION SYSTEMS

In modern four-stroke engines, lubrication is carried out by pumping oil around the engine and also by oil splash and mist falling on parts of the engine. The principal bearing surfaces of the crankshaft and camshaft are supplied with oil under pressure. Other parts such as the cam lobes and cylinder walls are generally lubricated by splash or mist.

Label the diagram with the parts which make up the engine lubrication system from the list provided:

Main oil gallery
Oil pressure switch
Oil pump
Primary oil strainer
Main oil filter
Oil pressure transducer
Pressure relief valve
Oil feed restrictor

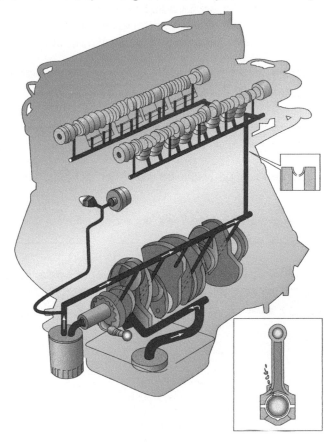

Tata (UK), Rover

Name the type of lubrication system shown:

With the aid of the diagram complete the description of how the oil is distributed by pressure, splash and mist by filling in the correct missing words from the list below.

tappets	strainer	filter	lobes	crankshaft
pump	squirt	camshaft	mist	revolving
bore	relief	gallery	pistons	splashes
pressure	thrown	crankcase	bearings	

The oil is drawn up to the _____ through the pick-up _____, it is then forced by the pump via the _____ valve, through the main _____ to the oil _____, where it is distributed under _____ to the _____ main and big end _____ and to the _____ bearings. The oil _____ from the sides of these bearings and is _____ around by the _____ shafts to lubricate the _____, cylinder walls and cam _____. This also creates a _____ which fills the _____. A _____ of oil may be directed at the cylinder _____ every crankshaft revolution. The oil flowing to the camshaft also pressurizes the hydraulic _____.

As well as a wet sump system, there are two other types of lubrication system. These are:

1 _____

2 _____

PURPOSE OF A LUBRICANT

All moving components usually have some form of lubricant placed between the surfaces to avoid metal-to-metal contact. The purpose of a lubricant is to reduce or remove friction from between the moving parts. This reduces heat and wear.

The types of lubrication in a motor vehicle may be classed as:

1 _____

2 _____

Lubrication should be based on maintaining a fluid film whenever possible and it is the behaviour of an oil under extreme load and temperature conditions which determines the usefulness of an oil.

The two drawings below each show a much-magnified representation of part of a bearing face. Add a similar sketch above each one to show what is meant by 'full-fluid film' and 'boundary' lubrication.

Full-fluid film lubrication   **Boundary lubrication**

_____   _____

When only boundary lubrication is present sometimes the layer of oil is only one molecule thick and could easily break down.

What would occur if the film of lubricant broke down because of the high temperatures and pressures that occur in the engine?

_____

PROPERTIES OF OIL

Oil possesses two main properties: 'body' and 'flow'. Body is the ability of an oil to maintain an oil film between two surfaces. Flow is the ability of an oil to spread easily over surfaces and to flow through pipelines and oilways. Briefly explain why these properties are important:

1 Body

_____

_____

TIP Worn bearings will reduce oil pressure. The increased clearances will reduce the resistance to oil flow and increase the volume of oil flowing around the engine – this in turn lowers the pressure!

2 Flow

_____

_____

How does the change in temperature affect these two properties?

_____

_____

Modern oils use many additives to improve their properties. List the common types of additives found in a modern engine oil:

1 _____

2 _____

3 _____

4 _____

5 _____

6 _____

Viscosity

The viscosity of an oil is a measure of its thickness (or body), or, more correctly, its a measure of the oil's ability to flow. This property varies with temperature and for a specific oil, for example a '30' grade.

Complete the following statements:

When the temperature is low the oil will be _____ and when the temperature is high the oil will be _____.

An oil which is said to be 'thin' is more properly described as _____ and an oil described as 'thick' should be called _____.

The viscosity of an oil is measured using a viscometer. What does this measure?

_____

_____

Viscosity index

It is important that oil viscosity (thickness) remains as stable as possible during changes in temperature.

The viscosity index of an oil is an indication of the oil's stability. Define viscosity index.

_____

_____

_____

Engine oil SAE viscosity classification

The viscosity of an oil is also expressed as a number with the letters SAE in front of it, for example: SAE 30.

SAE stands for the _____, which is the American organization that devised these viscosity standards.

State TWO typical engine oil viscosity numbers:

1 _____   2 _____

Which of the two typical engine oil numbers has the lower viscosity? _____.

Modern oils often have two viscosity numbers, and are called _____.

The reason for two numbers is that two measuring standards are used, one at engine working temperature and the other at a very low temperature.

The working viscosity is calculated with the oil at 99°C (210°F) and the grade of oil is expressed as:

SAE 5, 10, 20, 30, 40, 50, etc.

The most viscous of the above numbers is _____

The second viscosity range is calculated at −18°C (0°F).

These are very low viscosity oils and have the letter W (Winter) after them to indicate the measuring standard, for example:

SAE 5W, 10W, 20W.

The lowest viscosity in the list above is _____

A modern multigrade oil is an oil whose viscosity meets the flow standards measured at both temperatures. State why this is an advantage.

_____

_____

TIP The SAE number only signifies the viscosity of the oil at a specified temperature. It does not indicate the quality of the oil.

MULTIGRADE OILS

By improving the viscosity index of an oil (using suitable additives) it is possible for the oil to fall into two viscosity ranges when tested.

State TWO typical multigrade viscosities:

1 _____   2 _____

Why would a 5W30 oil be used in preference to a 15W40 oil?

_____

TIP Always refer to the manufacturer's data when deciding which grade of oil to use in an engine. This can be found in the owner's manual, oil supplier's information or a technical data sheet.

Lubricating oil (type and grade)

When selecting oil for an engine, it is important that as well as being of the correct grade it must be the correct type to make it suitable for a particular application.

In groups, carry out research on API, CCMC and ACEA ratings. Make notes and report back to others in your group with your findings. To help with this go to:

www.unitedoil.com.au/apiclass.htm and

http://www.carbibles.com/engineoil_bible.html

How may a diesel engine lubricating oil differ from a petrol engine oil?

_____

_____

What type of oil does a turbocharged engine require?

_____

_____

Synthetic oil

A conventional multigrade oil will lose its hot grade viscosity and therefore its lubrication protection as the engine mileage increases – it 'breaks down'.

TIP A 20W50 grade oil may reduce to a 20W30 grade oil after 6000 miles' engine use. This is because the long-chain viscosity improver additive, which maintains the oil's thickness when hot, gets chopped up by the shearing action of the moving parts.

A synthetic oil is created from crude oil by being refined in a more complex manner. The crude oil is more shear stable at high temperatures.

What are the advantages of using synthetic oil?

1 _____

2 _____

3 _____

OIL CONTAMINATION

Engine oil is subjected to extremely high pressures and temperatures. This affects both its viscosity and 'oiliness'. The engine oil also has to contend with other problems, the main one being contamination.

List below the principal contaminants and their source or causes.

Contaminant	Source or cause

Why is it necessary to change engine oil at regular intervals whereas other components are filled for life?

_____

_____

WET SUMP SYSTEM

The thick black lines on the diagram below indicate the oil passageways and bearing surfaces. Add names to the main components of the system. Draw in arrows to show the direction of oil flow within the system. Use the diagram on page 101 to help you.

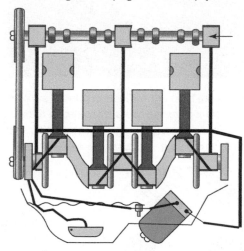

FILTRATION SYSTEMS

There are two types of filtration system. These are:

1 _____ and 2 _____

The system shown above is a _____ type.

Describe what is meant by a full-flow system.

_____

Briefly describe the operation of the system diagram opposite.

_____

_____

OIL PUMPS

Modern oil pumps tend to be driven from the crankshaft, either directly or by gears or chains. They are positive displacement pumps, whereby the same volume of oil going into the pump comes out. The volume is proportional to pump speed. The most commonly used pumps are rotor or gear types.

Rotor type

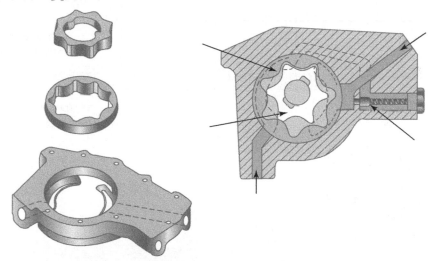

The sectional drawings show the direction of oil flow and pump rotation. Briefly describe the action of the pumps and add the following labels to the diagram above: **inner rotor, outer rotor, pressure relief valve, show inlet and outlet flow**

_____

_____

_____

Gear type

Label the diagram below to show the drive gear, oil inlet and outlet and indicate the direction of oil flow around the pump.

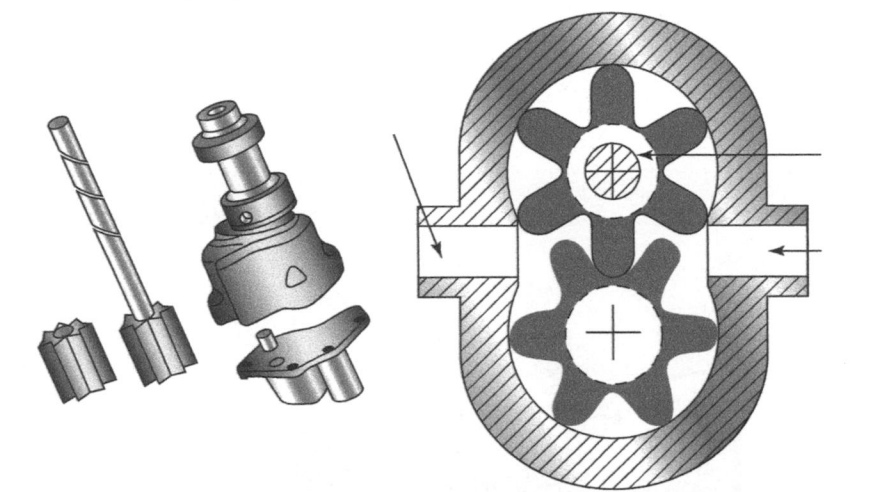

In one type (the diagram above) a gear is driven from the engine, the other gear meshes with this. Oil is drawn into the spaces between the teeth and is carried around the casing wall. In another (the photo below) the gears are eccentric (sometimes called a crescent type), the small gear is driven and the larger gear walks around it, carrying the oil in the space between.

In order to supply oil at high pressure the pump must maintain close working tolerances. This is even more critical when the pump is not submerged in the oil.

The diagrams below show a rotor-type pump and the measurements which need to be taken to determine the pump's serviceability.

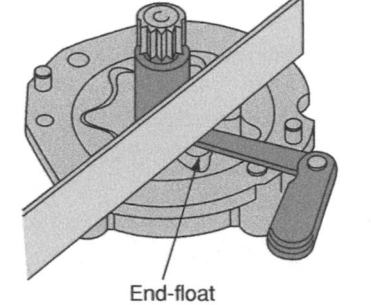

Lobe clearances

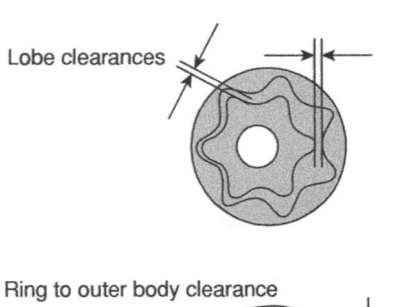

Ring to outer body clearance

End-float

Using vehicle manufacturer's data and a rotor-type oil pump, measure the clearances shown and record your findings.

Clearance (mm)	End-outer ring	Float rotor	Rotor lobe	Ring to outer body
Manufacturer's specification				
Measurement taken				

Is the pump serviceable/unserviceable? _____

OIL PRESSURE RELIEF VALVES

These are usually fitted in the system between the oil pump and the oil filter. Why is it necessary to fit a relief valve?

_____

_____

Describe their operation:

_____

_____

Complete the first sectioned diagram to show a **plunger type** oil pressure relief valve.

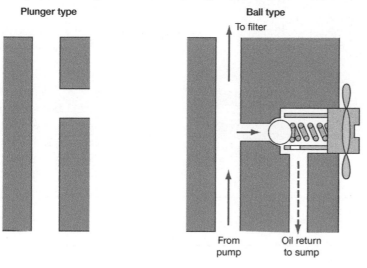

The sectioned view below shows the internal components of an oil pressure switch designed to operate a low pressure oil warning light. Name the main parts and describe how it operates.

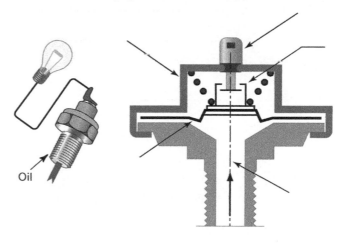

OIL PRESSURE WARNING LIGHT SWITCH AND GAUGE

The oil pressure switch (sometimes called a sender unit) monitors how much pressure your engine oil system is running at and sends out a signal. A single vehicle can have more than one oil pressure sender unit.

In many vehicles the oil pressure switch sends a signal to a gauge or light. If your car is equipped with an oil gauge you can see the status of the oil pressure. If you have a warning light it should light up when you turn on the ignition and go out when oil pressure builds (as engine and oil pump turn), then only light again if a low oil pressure condition is detected.

 On newer cars the oil pressure sending units can send a signal to a gauge or light and/or also send a signal to the vehicle's computer. The signal to the computer may lead to the computer shutting down the engine if low oil pressure is detected.

_____

_____

_____

State two positions where the switch may be located on the engine.

1 _____

2 _____

Connect a mechanical pressure gauge to a test engine and note the pressure when the oil light goes out. Compare with the pressure stamped on the sender unit.

Sender unit operating pressure	Actual measured pressure

OIL FILTERS

Most engines are fitted with two filters. The first (primary) filter is fitted around the pump intake in the sump and is of a coarse wire mesh type. The second (secondary) filter is of the replaceable element type.

Describe the basic construction of the filter element.

_____

_____

_____

Which way does the oil flow through the element?

_____

Describe how the filter should be removed and refitted.

_____

_____

_____

The diagram opposite shows a sectioned view of a full-flow oil-filter element and the bypass valve. Name the main parts and show the direction of oil flow.

What is the purpose of the full-flow filter bypass valve?

_____

_____

_____

_____

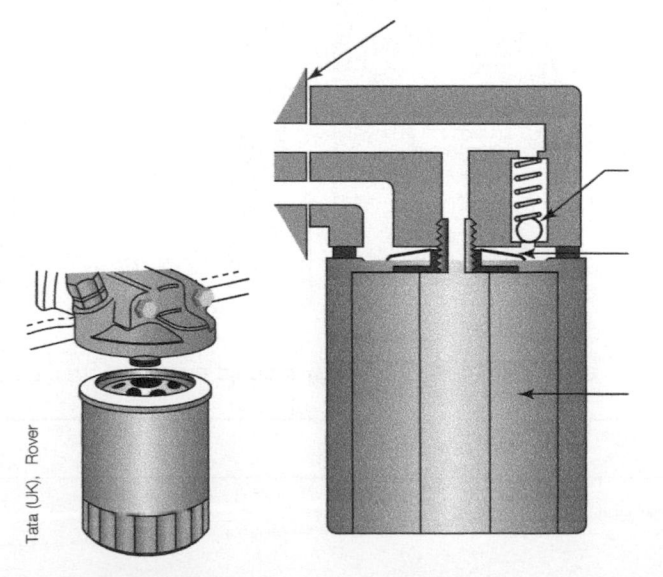

Tata (UK), Rover

OIL COOLER, PIPES AND HOSES

Many high-performance engines, because of very high operating temperatures, require extra oil cooling. This is achieved by use of a small external radiator similar to the one shown here.

Describe how the oil flow is controlled through the radiator.

_____

_____

_____

Describe the type of oil pressure hoses used.

_____

From engine

OIL LEVEL INDICATOR

The most popular oil level indicator is a 'hot wire' dipstick as shown below.

The dipstick is a hollow plastic moulding. A resistance wire is placed inside at a position between the oil level marks.

Describe how this is able to test the oil level.

_____

_____

_____

_____

Resistance (hot wire)

TYPES OF OIL-SEALING ARRANGEMENTS FOR ROTATING SHAFTS

Identify the types of engine oil-sealing arrangements shown and label the parts indicated:

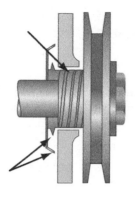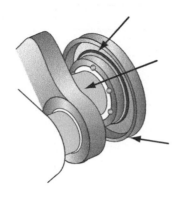

(1) _____

(2) _____

Explain how type (1) prevents oil leakage.

_____

_____

_____

Explain the positioning of type (2) in relation to the oil it retains.

_____

_____

 TIP Lip type seals may get hard and brittle with age and heat. This causes them to look serviceable but not flex which may cause leaks.

ENGINE LUBRICATION SYSTEM – PROTECTION DURING USE AND REPAIR

Describe how the system should be protected from the following hazards:

1 Ingress of dirt: _____

_____

2 Crankcase dilution: _____

3 Crankcase contamination: _____

Describe how the following should be maintained/checked to protect the lubrication system and its components:

1 Oil filter gauze in sump: _____

2 Correct engine operating procedures: _____

_____

3 Crankcase ventilation: _____

4 Sealing: _____

5 Use of oil: _____

6 Pressure relief valve: _____

MAINTENANCE

List three main precautions to be observed when carrying out routine maintenance:

1 _____

2 _____

3 _____

TIP Always use a good fitting socket when removing the sump plug. Replace the sump plug washer after carrying out an oil drain. Many technicians use a special vacuum unit to remove the engine oil from the dipstick hole whilst it is still hot.

DIAGNOSTICS: ENGINE LUBRICATION – SYMPTOMS, FAULTS AND CAUSES

State a likely cause for each symptom/system fault listed below.

Symptom	Probable fault	Likely cause
The oil warning light comes on when engine is warm	Low oil pressure	
Oil seems to be leaking from the engine seals and particularly from the crankshaft	High oil pressure	
Blue smoke is coming from the engine's exhaust after starting when cold	High oil consumption	
There are oil drops on the floor every time the vehicle is parked	Oil leakage	
The oil is very black when checked	Oil contamination	
The oil level on the dipstick is higher than normal	Overfilling Water in oil	

Multiple choice questions

Choose the correct answer from a), b) or c) and place a tick [✓] after your answer.

1 **The term multigrade when related to an engine oil means that:**
   a) the oil can be used in any engine type, all year round [ ]
   b) the oil viscosity gets thicker as the temperature goes up [ ]
   c) the oil does not change its viscosity noticeably as temperatures change. [ ]

2 **A common type of engine oil pump is called:**
   a) a piston type [ ]
   b) a gear type [ ]
   c) a ram type. [ ]

3 **What term is used to describe the type of lubrication given to an engine piston?**
   a) boundary [ ]
   b) full-flow [ ]
   c) pressurized. [ ]

4 **A relief valve is fitted to the main gallery of an engine. The purpose of this valve is to:**
   a) stop the oil flow to the bearings when the oil pressure is low [ ]
   b) limit the maximum oil pressure in the engine gallery [ ]
   c) maintain the supply of oil if the gallery becomes blocked. [ ]

5 **Which of the following is the correct sequence of flow of oil in an engine?**
   a) sump, oil pump, gallery, relief valve, crankshaft, disposable filter [ ]
   b) sump, relief valve, oil pump, disposable filter, gallery, crankshaft [ ]
   c) sump, oil pump, relief valve, disposable filter, gallery, crankshaft. [ ]

6 **When a single lip type oil seal is fitted it must be positioned with the:**
   a) coil spring and part number of the seal to face the oil [ ]
   b) coil spring and open part of the seal to face the oil [ ]
   c) coil spring and flat part of the seal to face the oil. [ ]

SECTION 4

Light vehicle fuel systems

USE THIS SPACE FOR LEARNER NOTES

_____

_____

_____

_____

_____

_____

_____

Learning objectives

After studying this section you should be able to:

- Describe the construction and operation of fuel systems.
- Identify the components and key engineering principles that are related to fuel systems.
- State common terms used in fuel system design.

Key terms

Carburettor A mechanical device for mixing fuel with air.
Stoichiometric Chemically correct ratio of fuel and air for complete combustion.
Single point A single injector system which sprays fuel for all cylinders into the air at one place, usually by the throttle body in the inlet manifold.
Multi-point An injection system in which each cylinder has its own injector. Only air enters the inlet manifold and injectors are situated in the inlet manifold close to the valve ports.
Absolute pressure True pressure (gauge pressure + atmospheric pressure).
Potentiometer A variable resistor.
ECU (ECM) Electronic control unit (module).
HC Hydrocarbon.
CO Carbon monoxide.
NO_x Oxides of nitrogen.
Lambda 1 The point at which the lowest values for CO, NO_x and HC are achieved.

 WWW http://www.aa1car.com/index_alphabetical.htm
http://www.motor.org.uk/

 Remember: when working with fuel systems you may be dealing with:

- very hot surfaces
- volatile and flammable fuel under pressure
- chemicals which are carcinogenic or toxic
- rotating components
- waste disposal.

LIGHT VEHICLE FUEL SYSTEMS

The fuel used for spark ignition engines is _____ and for compression ignition engines it is _____.

 **TIP** Remember: When working with fuel systems and inspection lamps make sure that they are designed to be used around fuels. Have the correct fire extinguisher nearby. Be aware of the dangers of static electricity.

PETROL INJECTION

In the past, the mixing of petrol and air was done by using a carburettor. The nearest thing to this, which may still be seen on some smaller engines today, is **single point** injection. Most modern petrol engines use **multi-point** systems, which have one injector situated in the inlet manifold just behind the back of the inlet valves for each cylinder. A small number of engines now use direct injection straight into the combustion chamber. In both single and multi-point systems, a

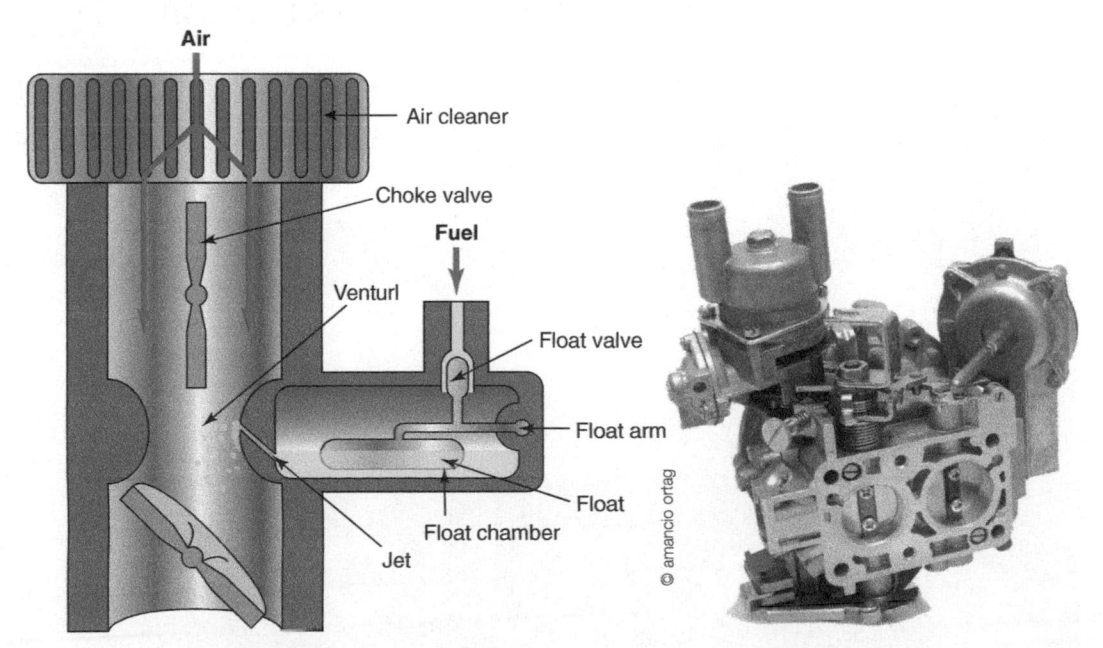

Basic carborator (cross section)

A basic carburettor

computer or electronic control unit (ECU) monitors the operating conditions of the engine and attempts to provide an ideal air fuel ratio (stoichiometric). The computer (ECU) controls the opening duration of the injectors, sometimes called the 'pulse width' which in turn controls the quantity of fuel supplied to the cylinders. Multi-point injection systems have injectors which pulse (inject fuel) all together, or in the firing order of the engine (sequentially).

The diagram below shows a schematic view of the Bosch Mono-Jetronic Layout Throttle Body Injection. In this system, fuel is injected intermittently by a solenoid operated single injector positioned above the throttle valve. The fuel distribution system via the inlet manifold is similar to that of the single carburettor layout. The injector is shown highlighted. Name the main components in the system.

State the main reason for using fuel injection.

_____

What does the 'stoichiometric mixture ratio' mean?

_____

Give three engine conditions where air fuel ratios change:

1 _____ 3 _____

2 _____

What is the stoichiometric ratio for petrol? _____

FUEL SUPPLY SYSTEM

The fuel supply system in today's vehicles is designed to provide clean fuel at the correct pressure to the injectors and also prevent harmful vapours from entering the atmosphere. There are two types of systems, name them and describe how they compare.

1 _____
_____
_____
_____

2 _____
_____
_____

The figures below show the fuel supply systems, name the components arrowed on each.

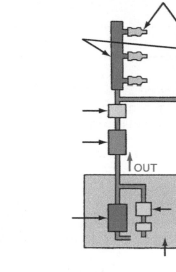

Fuel supply system type (2)

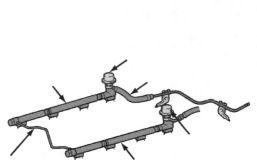

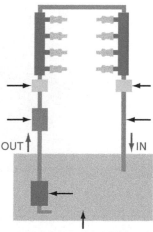

Fuel supply system type (1)

TIP When removing and replacing fuel pipes, make sure that any threaded fittings are carefully aligned before tightening. It is easy to cross-thread the fittings and leaks will inevitably result. Use the correct tools on quick connect and crimp fittings.

Most fuel pumps are now installed inside the tank. Fuel tanks also include devices which prevent vapours from leaving the tank, such as overfill limiting valves and liquid vapour separators.

The fuel tank will normally have vent valves connected via hoses to a charcoal canister. What does this do?

_____
_____

TIP Always drain petrol from tanks by using a fuel retriever. For information on safety go to **http://www.hse.gov.uk/mvr/priorities/fire.htm**

A modern tank will also have a filler cap that is non-venting. It is fitted with a pressure/vacuum relief valve which opens under extreme conditions but will not normally allow vapours to flow into the atmosphere.

A cutaway of a pressure-vacuum gasoline filler cap

 When replacing tank filler caps, use original equipment. This ensures correct filling and venting system operation.

Fuel pumps

Electric fuel pumps can be located inside or outside the tank. State two advantages of having a pump fitted inside the tank:

1 _____

2 _____

 Whilst it is dangerous to have a spark near petrol, the in-tank fuel pump is safe, because there is not sufficient oxygen to support combustion in the tank.

Older types of electric pump used rollers which were mounted on an offset disc and were driven by a motor which rotated inside a steel ring. Fuel was captured in the spaces between the rollers. This type of pump is called _____

This type of pump can generate very high pressure and the output tends to be constant. However, the output comes in pulses and a pulsation damper is needed in the fuel line.

Many newer systems use a 'turbine' style fuel pump as shown in the figure below.

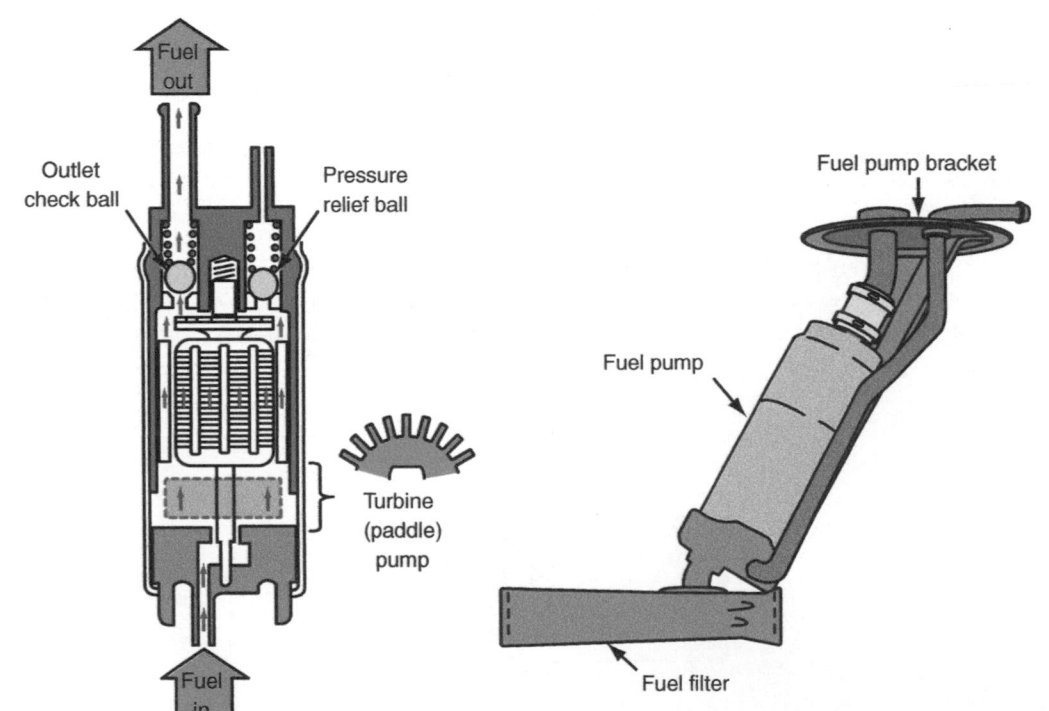

A turbine pump has an impeller ring attached to the motor. The blades in the impeller push the fuel through the pump as the impeller spins. This is not a positive displacement pump so has a number of advantages. List three of these advantages:

1 _____

2 _____

3 _____

Complete the description of how the fuel supply system works by adding the correct missing words from the list below.

rail	mesh	relay	voltage	continuously
valve	regulator	varied	motor	timer
injectors	starts	speed	constant	pressure

The ignition energizes a _____ and _____ is supplied to the pump. The _____ starts to spin until system _____ is built up. An ECU controlled _____ limits run time, until the engine _____. Fuel is drawn into the pump through a _____ filter sock and exits through a one-way _____. It then flows to the fuel _____ via another filter and is routed to the individual _____. There may be a pressure _____ on the fuel rail, or it may be in the tank. The fuel pump runs _____ whilst the engine is running and the ignition is on.

What is the function of the one-way valve in the fuel pump?

_____

What happens to the pump if the engine stalls?

_____

What would an 'inertia switch' be used for in an EFI system?

_____

Fuel injectors and fuel rail

The fuel injectors are electro-mechanical devices that meter and atomize the fuel. They are supplied with fuel from the fuel rail. There may be a regulator attached to the fuel rail and a return of excess fuel to the tank, or a regulator in the tank itself. An external filter cleans the fuel prior to it entering the fuel rail. The figure opposite shows a fuel rail from a 'return fuel' system. Name the components arrowed.

Fuel injectors resemble a spark plug in size and shape. 'O' rings seal the injectors in the manifold.

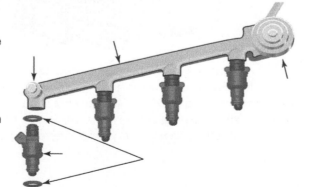

What is another important function of the 'O' rings?

_____

_____

Study the diagram of the fuel injector below and complete the description of its operation by filling in the missing words from the list below.

solenoid	collapses
back	winding
shutting off	field
needle	on
sprays	spring

Most injectors consist of a _____, a needle valve and nozzle. The ECU-controlled solenoid _____ creates a magnetic _____ when switched _____, which draws _____ the armature and pulls the _____ from its seat. Fuel then _____ from the nozzle. When the solenoid is de-energized, the magnetic field _____ and a helical _____ forces the needle valve back on its seat, _____ the fuel supply.

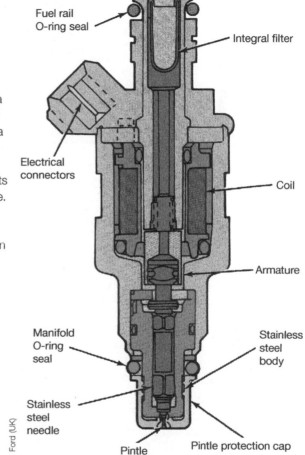

The earth circuit is controlled by the ECU to facilitate injector switching. Is this true or false? _____

Mixture control

The ECU or microprocessor switches the injector(s) on and off according to the conditions that the engine is working under. In order for it to do this, it has to have the relevant information to reference to a fuel mixture map, which it holds in its memory. The figure below shows a three-dimensional fuel mixture map held in the memory of the ECU which gives one million pre-determined reference points that the ECU can detect and reset to, in milliseconds.

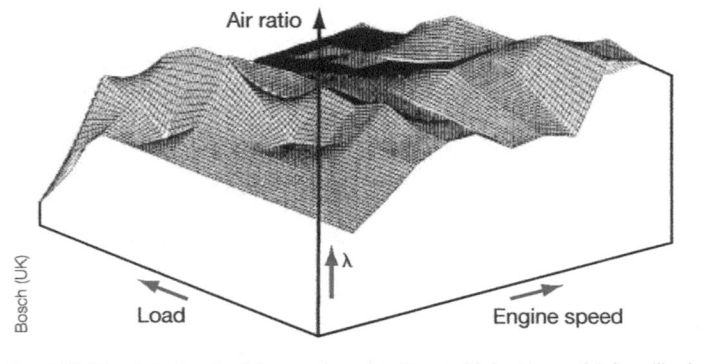

From this map the ECU is able to decide on the duration of injection which will give a stoichiometric mixture ratio. A number of sensors feed information to the ECU. The figure below shows many of the sensors that may be used on an electronic fuel injection system.

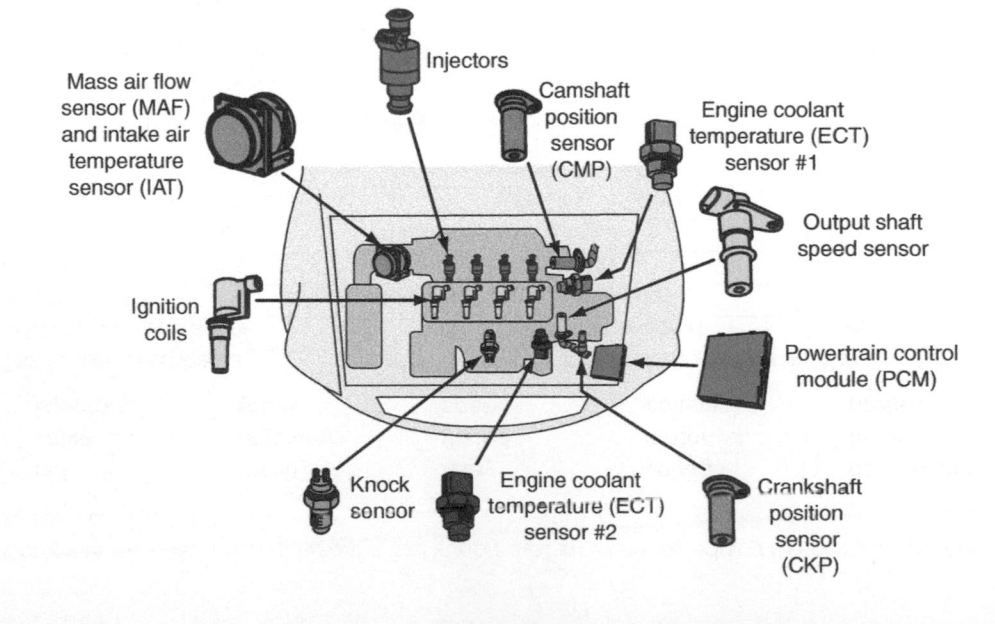

The incomplete block diagram (see figure below) shows the layout of a basic electronic petrol fuel injection system, with the information being fed to the ECU. Place the component names (see the list below) in the boxes, in the correct order which make up the fuel supply and final air/fuel mixture system.

Note: The **pressure regulator** should be shown in both of its possible positions.

inlet manifold	fuel tank	fuel pump
fuel filter	ECU	pressure regulator
fuel rail and injectors	cylinders	

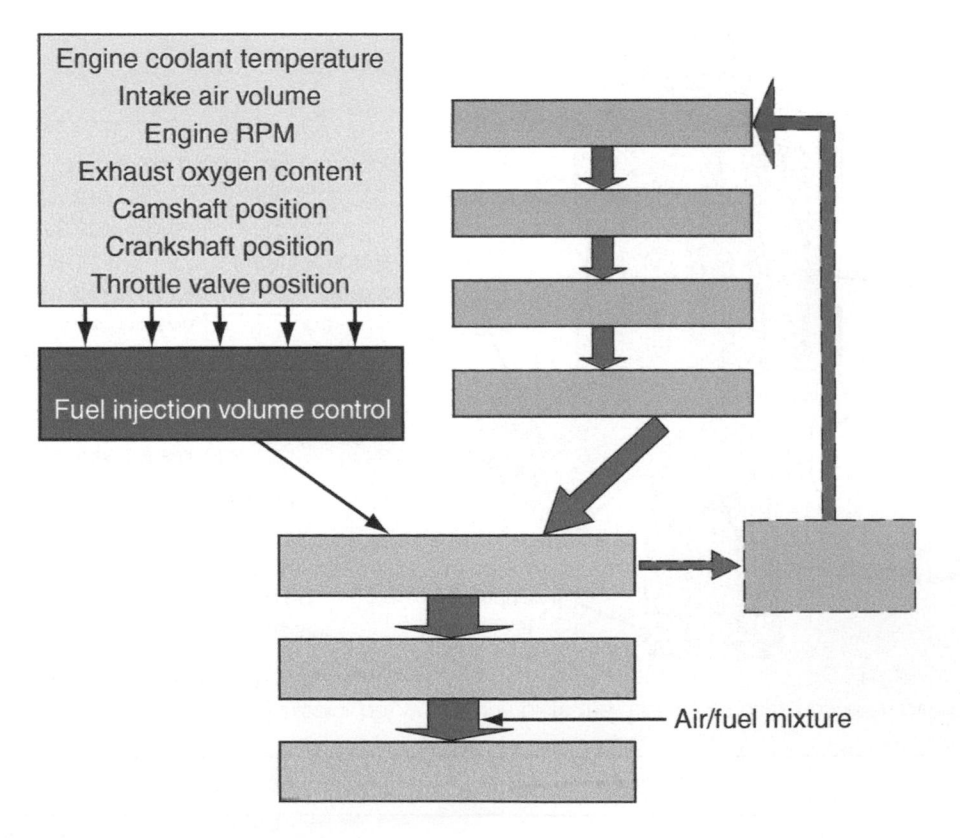

SENSORS AND THEIR OPERATION

Engine coolant temperature

This sensor is a relatively simple unit but is very important within the system. It measures and provides accurate information on the engine temperature. If it feeds incorrect information to the ECU many of the fuel, ignition, emission and drive-train functions handled by the ECU can be affected. Most are NTC thermistors where the resistance drops as the temperature goes up. The ECU supplies a known voltage to the sensor when the engine is cold, the return voltage is reduced as the engine warms. The engine temperature is then calculated and related to the fuelling map.

How many electrical connections are there on an Negative Temperature Coefficient (NTC) sensor?

_____

What is a typical supply voltage to the sensor? _____

Examine and sketch an Engine Coolant Temperature sensor in the space below

A mass air flow (MAF) sensor   Engine Coolant Temperature sensor

Air intake measurement

A mass air flow sensor (MAF) is used on most modern systems to measure the 'amount' of air entering the engine. Its function is to measure the volume and density of the air entering the engine, so that the ECU can calculate the quantity of fuel needed to maintain the correct mixture (in conjunction with other information).

They have no moving parts unlike the older type of vane air flow meter. The MAF sensor uses electrical current to measure air flow. The sensing element is either a 'hot wire' of platinum or a 'hot film' of nickel foil which is heated electrically to maintain a certain temperature above the incoming air.

What temperature above ambient are the sensing elements heated to?

Hot wire _____

Hot film _____

The element is cooled as air flows past it, the density and humidity will effect how much cooling takes place. The amount of current needed to keep the element at its working temperature will change and is directly proportional to the air 'mass' entering the engine.

The hot wire sensor has a 'self-cleaning' cycle.

Oxygen (O$_2$) sensor

© Juanmonino

Known also as a lambda sensor, the oxygen sensor is the key sensor in the feedback mixture control loop. It monitors and informs the ECU of air/fuel mixture.

What happens if the mixture is incorrect?

_____

_____

Some systems have two oxygen sensors, why is this?

_____

_____

Engine RPM (speed sensor)

Often known as the crankshaft position sensor (CKP), it monitors engine speed and advises the ECU to adjust injector pulse width accordingly. There are two types; magnetic or 'hall effect'.

What is the difference?

Bosch (UK)

_____

_____

The signal is also used to synchronize the injectors with events in the cylinder when used in conjunction with the camshaft position sensor (CMP).

Camshaft position (phase sensor)

This sensor is used to synchronize the firing of the injectors with the individual cylinders in the engine. It is used for fuel injection timing.

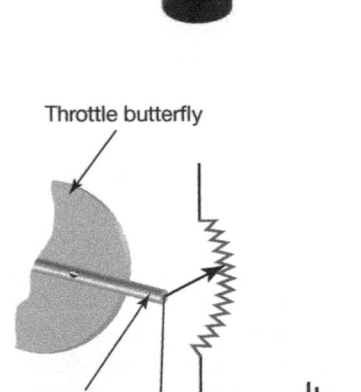

Bosch (UK)

Throttle position sensor (TPS)

The figure opposite shows the basic principle of the TPS. Indicate on the diagram the reference voltage, signal voltage and ground (earth) connections.

The throttle position sensor (TPS) allows the ECU to monitor the throttle position. The signals clarify the engine load and operating conditions. It is a 'potentiometer' which reacts to the position of the throttle plate. The ECU can momentarily enrich the mixture if there is a sudden opening of the throttle.

What is another name for a potentiometer?

Throttle butterfly

Throttle spindle

_____

What other sensor could be used to measure engine load? What else can it do?

_____

_____

Idle speed control

Idle speed is controlled by the ECU based on operating conditions and various inputs from sensors.

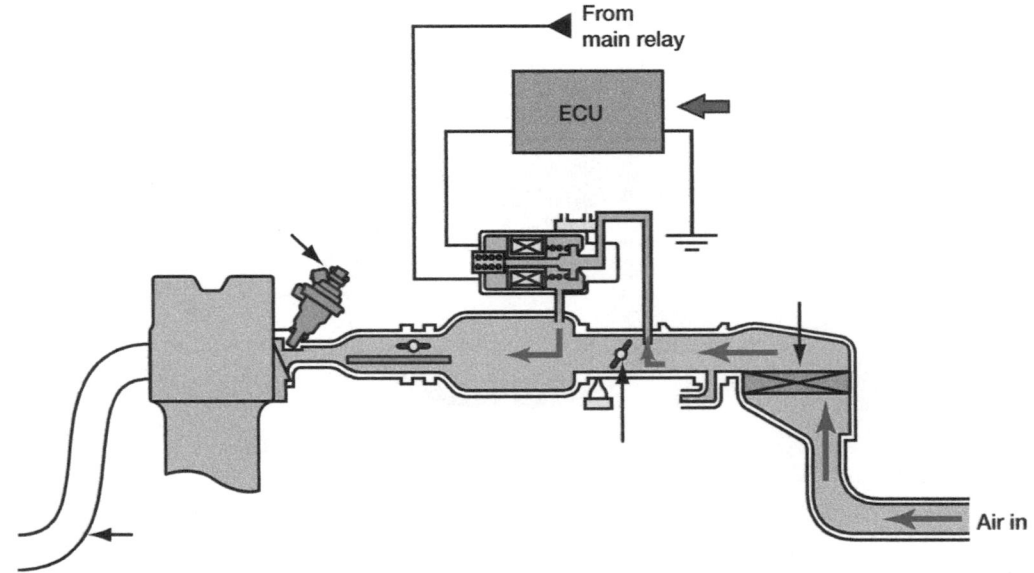

From main relay

ECU

Air in

Idle air control system

The idle air control (IAC) valve system is a stepper motor or actuator that positions the IAC valve in the air bypass channel around the throttle valve.

The ECU calculates the amount of air needed for smooth idling and controls the IAC accordingly.

Name the components indicated on the diagram and list three sensor inputs. What happens to the valve if the idle speed is too low?

_____

_____

_____

The figure below shows the basic principle of the IAC. Sketch the valve in the fully open position to show maximum air flow being diverted.

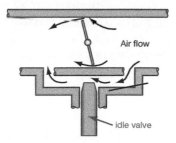

Air flow

idle valve

DIRECT INJECTION (PETROL)

Direct injection has been with us for a number of years on diesel engines, and more recently this type of system is being used on some petrol engine vehicles. There are a number of advantages of this system for petrol engines when compared to indirect injection.

List three advantages below:

1 _____

2 _____

3 _____

The figure below shows a direct injection system. Draw lines from the names of the main components in the boxes to their respective parts.

Fuel rail pressure regulator
Pistons
Injectors
Fuel rail

Bosch (UK)

High pressure pump
ECU

Bosch direct injection petrol system

In direct gasoline-injection engines, the air/fuel mixture is formed directly in the combustion chamber. During the intake stroke, if the system has a lean burn mode (normally during idle and low engine loads), only combustion air is drawn in through the open intake valve, the fuel being injected at high pressure into the combustion chamber by the special high pressure injectors at the end of the compression stroke.

What would a typical fuel pressure be for this type of system? _____

The fuel is aimed at the spark plug. It cannot penetrate the highly compressed air easily in the outer parts of the cylinder, hence a mixture ratio of around 14:1 is formed around the plug. The layer of surrounding gas insulates this mixture from the cold cylinder walls, increasing the thermal efficiency of the engine.

What is the layer of air/fuel mixture and the layer of surrounding gas called?

_____

Under stoichiometric (medium load) and full power mode (heavy loads and hard acceleration) the fuel is injected during the inlet stroke. The precise metering, timing, preparation and distribution of the intake air and the injected fuel for every combustion stroke and engine requirement at each moment in time, leads to low fuel consumption figures and low emission levels. These systems normally use a 'drive-by-wire' or electronic throttle controlled by the ECU. Some systems have been designed where individual cylinders can be cut to decrease fuel consumption even more, under certain conditions.

TIP For more on direct petrol injection systems go to **http://www.autospeed.com/ cms/title_Direct-Petrol-Injection/A_107830/article.html**

State the air/fuel ratio for the three modes listed below:

Lean burn. _____

Stoichiometric. _____

Full power. _____

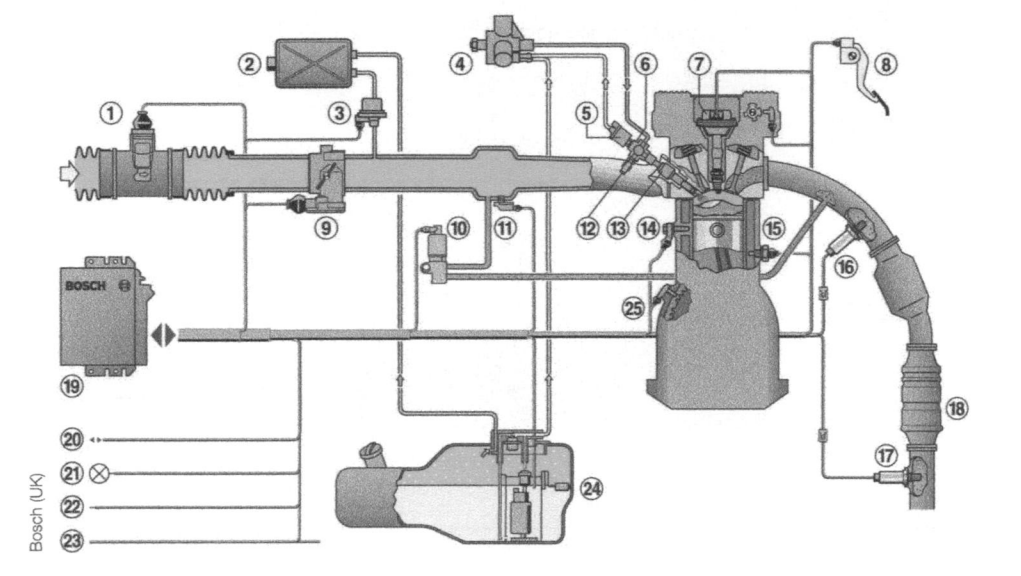

An overview of a complete direct petrol injection system

Bosch (UK)

DIESEL FUEL INJECTION SYSTEMS

One of the biggest differences between the petrol and diesel engine is how the fuel is ignited. Describe how this takes place in the diesel (CI) engine.

_____

_____

What are typical compression ratios, cylinder pressures and temperatures before combustion for CI engines?

_____

_____

Traditionally, injectors have either been positioned in a pre-combustion chamber, which is a small upper chamber in the cylinder head, connected to the main combustion chamber by a small passage or positioned so that they inject directly into a combustion chamber which is normally part of the piston crown.

What are these two types of injection called? _____

High pressure fuel pumps and injectors are required for this task and on older diesel systems these were completely mechanical. The systems then progressed to using electronically controlled mechanical injection pumps or unit injectors. Today, virtually all late model light-to-medium-duty diesel systems, are 'common rail' direct injection.

Common rail diesel control systems use many of the same sensor inputs as for petrol injection and are quite similar in many ways to the direct injection petrol systems.

The figure below shows a schematic block and line diagram of a typical common rail diesel system.

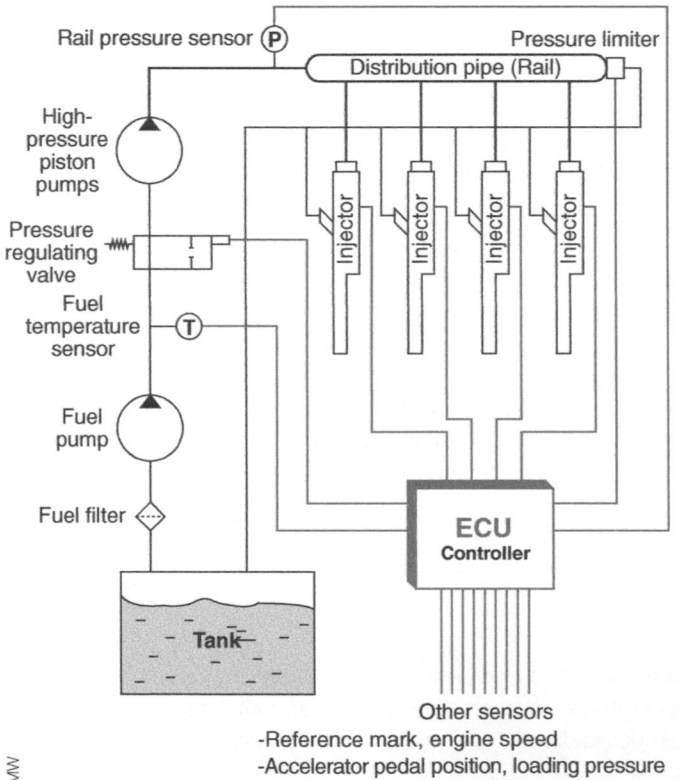

BMW

Typical common rail diesel system layout

Study the diagram with a colleague, compare it with the direct injection petrol system shown previously and discuss the similarities.

The reason for using the common rail system, rather than the more traditional mechanical and electrically controlled pumps and injectors is that it:

1 Enables stringent emission regulations to be met.
2 Gives better fuel economy than other systems.

Other advantages have also made it the choice of many manufacturers today.

List three advantages below:

1 _____

2 _____

3 _____

Common rail system and its operation

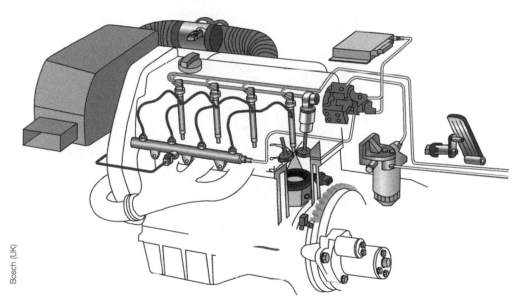

Bosch (UK)

Identify and label the main components on the diagram:

Show the fuel flow from the tank and its return, by drawing arrows on the fuel lines.

Operation

Complete the following descriptions using the correct words selected from the word list below.

pre-filter	excess	loaded	proportional	solenoid
data	lubricated	limiter	radial	volume
turbine	ECU	triple	engine	pulses
accumulator	external	high	tank	camshaft

Low pressure delivery

Fuel is drawn through a _____ by a low pressure roller cell or _____ pump mounted in the fuel _____. It is then delivered via an _____ filter to the _____ pressure pump.

High pressure delivery

A _____ piston, high volume, _____ pump which is driven at half _____ speed delivers the fuel to the fuel rail at pressures of between 130–1750 bar. It is _____ by diesel fuel and may be fitted on the end of the _____. Pump delivery rate is _____ to the speed of the engine. A spring _____ ball valve returns _____ fuel to the fuel tank. Initial spring pressure controls fuel pressure at approximately 100 bar. An ECU controlled _____ varies the force on the spring depending on operating conditions.

The fuel rail is common to all cylinders and is an _____ which holds large volumes of fuel at high pressure. The large _____ dampens the pressure _____ from the pump. There is a pressure sensor and a pressure _____ valve fitted, to return fuel to the tank if there is excess pressure. Some systems have a flow limiter valve fitted between the rail and the injector, this ensures that fuel will not continually flow from the injector if it should remain open. The _____ controls the system by taking information from sensors, comparing it to pre-programmed _____ and acting accordingly. Other systems can feed data to the ECU.

Triple piston high pressure radial pump

Name the components indicated.

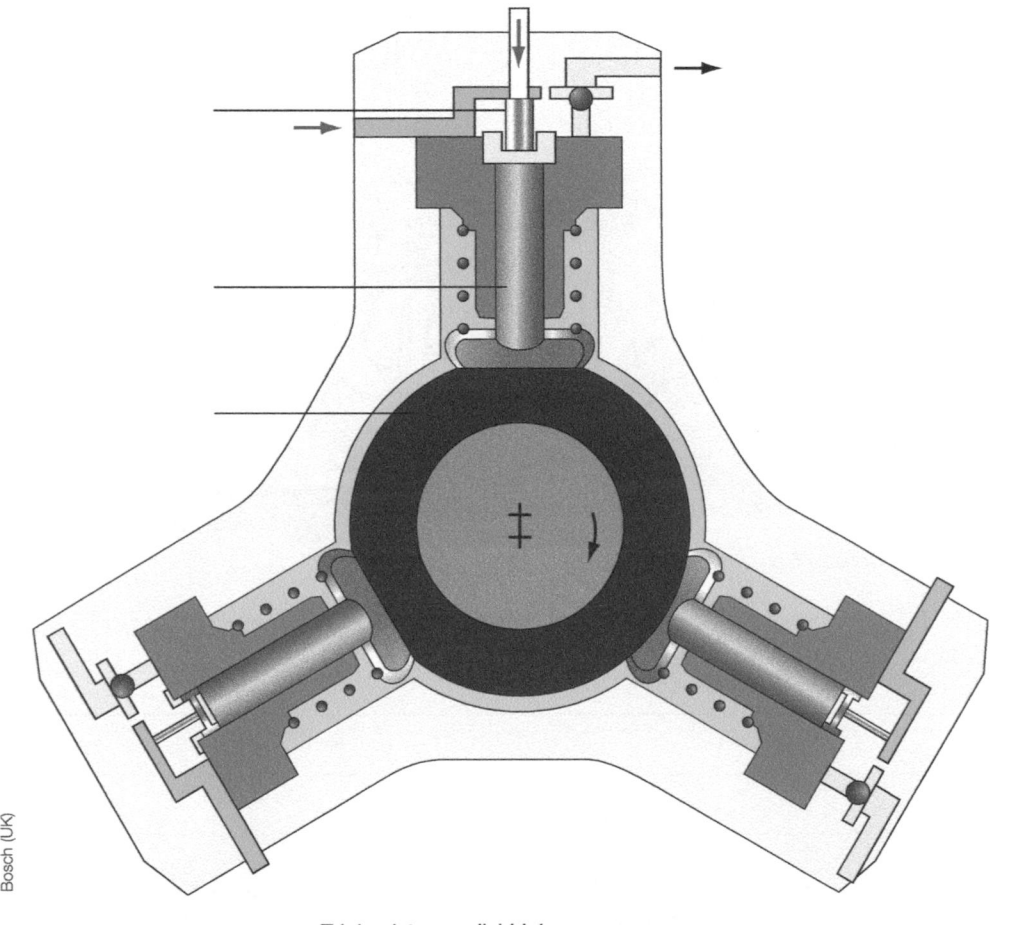

Bosch (UK)

Triple piston, radial high pressure pump

So that pump flow can be varied with engine load, one of the pistons of the pump can be shut down. This is achieved by using a solenoid to hold the intake valve of that piston open. However, when a piston is deactivated, the fuel delivery pressure fluctuates to a greater extent than when all three pistons are in operation.

Pressure control valve

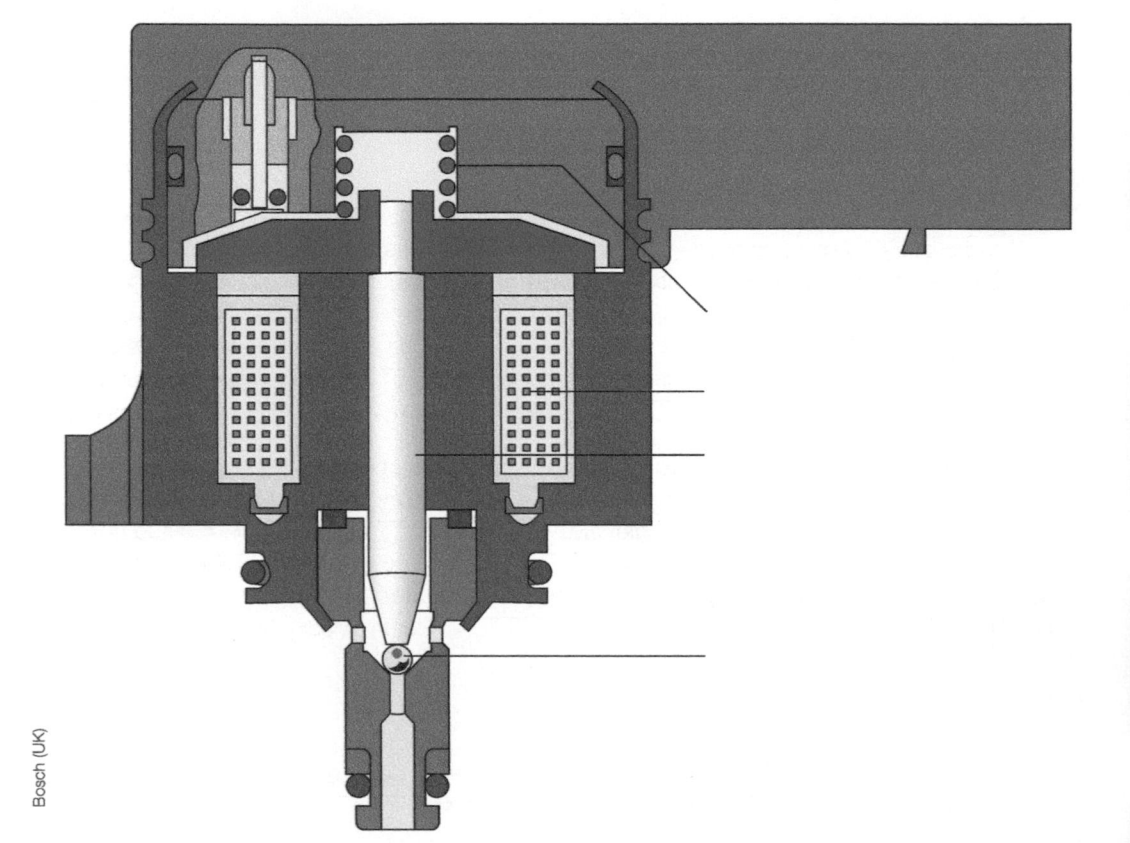

Bosch (UK)

Label the components marked on the diagram.

The fuel pressure control valve comprises a fuel-cooled solenoid valve. When the pressure control valve is not activated, its internal spring maintains a fuel pressure of about 100 bar. When the valve is activated, the force of the electromagnet aids the spring, reducing the opening of the valve and so increasing fuel pressure. The fuel pressure control valve also acts as a mechanical pressure damper, smoothing the high frequency pressure pulses coming from the radial piston pump when only two pistons are activated.

Injector design and operation

The diagram shows a common rail diesel injector. Because of the very high fuel rail pressure, the injector uses a solenoid operated, hydraulic servo system to open and close the nozzle. In this design, the solenoid controls the movement of a ball valve which regulates the flow of fuel from a control chamber within the injector.

Describe the operation of the injector when the current to the solenoid is switched off.

_____

_____

_____

_____

_____

_____

_____

_____

_____

_____

Instead of the solenoid shown, some of the latest injectors use piezoelectric crystal wafers. A voltage is sent to the wafer from the ECU causing the wafer to flex, which then controls the ball valve as above. The piezo injector allows the implementation of very short and rapid fuel delivery characteristics. It allows consistent and accurate delivery of diesel with up to seven injection cycles per engine combustion cycle.

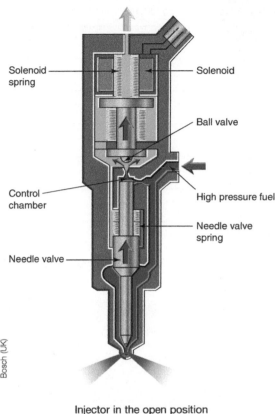

Solenoid spring — Solenoid

Ball valve

Control chamber — High pressure fuel

Needle valve spring

Needle valve

Bosch (UK)

Injector in the open position

TIP Some common rail injection systems operate at voltages of well over 100 volts and fuel pressures of up to 30, 000 psi.

Do not 'feel' for leaks with your hands around the engine or loosen high pressure injection pipes.

Multiple choice questions

Choose the correct answer from a), b) or c) and place a tick [✓] after your answer.

1 **A 'hot film' device is used as part of the:**

a) mass airflow sensor [ ]

b) coolant temperature sensor [ ]

c) exhaust gas recirculation system. [ ]

2 **One advantage of a common rail diesel fuel system is:**

a) injection is sequential [ ]

b) much lower compression [ ]

c) lower NO_x emissions. [ ]

3 **The term MAP when applied to a sensor stands for:**

a) Manifold Air Pressure [ ]

b) Manifold Atmospheric Pressure [ ]

c) Manifold Absolute Pressure. [ ]

4 **The chemically correct air/fuel ratio of a petrol engine for complete combustion is:**

a) 15.7:1 [ ]

b) 14.7:1 [ ]

c) 17.7:1. [ ]

5 **The usual type of diesel fuel injection system on modern vehicles is:**

a) rotary pump [ ]

b) common rail [ ]

c) unit injection. [ ]

6 **The term NTC when used relating to an engine sensor stands for:**

a) negative temperature coefficient [ ]

b) negative thermal compensation [ ]

c) negative temperature compensation. [ ]

7 **The 'lambda sensor' is usually located:**

a) in the exhaust front pipe [ ]

b) in the exhaust silencer [ ]

c) in the exhaust port. [ ]

SECTION 5

Light vehicle ignition systems

USE THIS SPACE FOR LEARNER NOTES

_____

_____

_____

_____

_____

_____

Learning objectives

After studying this section you should be able to:

- Identify ignition system components.
- Describe the construction and operation of distributor and distributor-less ignition systems.
- Compare components and assemblies against alternatives to identify differences in construction and operation.
- Identify the key engineering principles and common terms that are related to ignition systems.

Key terms

Static and dynamic timing Ignition timing set when the engine is stationary and when running.

Closed loop An electronic control system using feedback to maintain correct ignition timing/mixture strength.

Amplifier A device used to increase the electrical signal in an electronic ignition system.

Transistor A semi-conductor which can be used to switch electronic circuits and also amplify voltage.

DIS Distributor-less ignition system.

Distributor A mechanical component, which distributes high voltage to the spark plugs.

Contact breaker A mechanical switching device used in older ignition systems.

Air gap The gap between the electrodes of a spark plug or reluctor and pick-up.

WWW http://www.magnet.fsu.edu/

http://auto.howstuffworks.com

Remember – when working with ignition systems you may be dealing with:

- very hot surfaces
- high voltages
- components which are carcinogenic or toxic
- rotating components
- waste disposal.

SPARK IGNITION SYSTEMS

Ignition systems have changed quite considerably in recent years. Prior to the 1980s, a number of mechanical components connected to the engine were used to support the generation, distribution and timing of the high voltage spark which ignited the mixture in the combustion chamber. Today, much of this is done by the use of electronics, microprocessors, sensors and engine management systems. One thing that has changed very little is the construction and operation of the ignition coil, which is used to generate the spark.

All ignition systems consist of two interconnected electrical circuits, low voltage and high voltage.

Name the two circuits in the boxes below:

Ignition system components

Ignition switch
Spark plugs
High voltage cables (not all systems)
Ignition coil primary winding
Triggering device and control module
Battery
Distributor cap and rotor (older systems)
Ignition coil secondary winding

Draw arrows from the components listed in the centre box to show which circuit they belong to.

Ignition coil operating principles

An ignition coil is a 'step up' pulse transformer, which has no moving parts. It is able to transform battery voltage to the very high voltages required for a spark to cross the electrode gap of a spark plug. The increase in voltage is carried out by a process called 'electromagnetic induction'. Very simply, if an electric current is passed through a conductor (wire), a magnetic field is produced. The lines of force of the magnetic field radiate out from the conductor. If a second conductor is placed in this field, as the lines of force pass through it, a voltage is produced in the second conductor. If the number of conductors and the speed of build-up of the magnetic field are increased so too will be the induced voltage.

What is the system voltage? _____

What would a typical spark plug (high tension) voltage be?

_____

Study the sectioned diagram below, which shows a basic coil wiring arrangement.

Name the parts indicated.

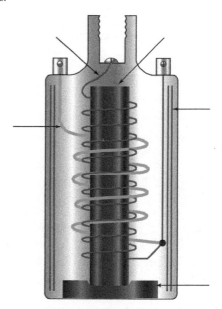

Coil output

When the primary circuit is closed, current flows around the primary windings in the coil, creating a magnetic field.

How can this be strengthened?

_____

When the primary circuit is switched off, the magnetic field collapses quickly and induces an electromotive force (EMF) into the secondary winding.

The coil's secondary voltage builds up until it creates a spark at the plugs. With modern systems and lean burn technology, the coil is easily capable of producing 30–60 kV. This does not mean that this will be needed, as the amount of voltage required for the spark to bridge the gap of the plug varies with operating conditions.

State two factors which influence the ignition coil's output:

1 _____

2 _____

Primary circuit connections

Most single ended coils are wound to produce a negative spark at the spark plugs.

The effects of wiring the coil to give incorrect polarity would be:

_____

_____

_____

Show the coil top markings when the primary circuit is wired correctly.

MUTUAL INDUCTION

 Set up a simple rig as shown below. It demonstrates how an EMF can be induced into a circuit by varying the current flow in a neighbouring but separate circuit. Describe how to induce and vary an EMF.

_____

_____

_____

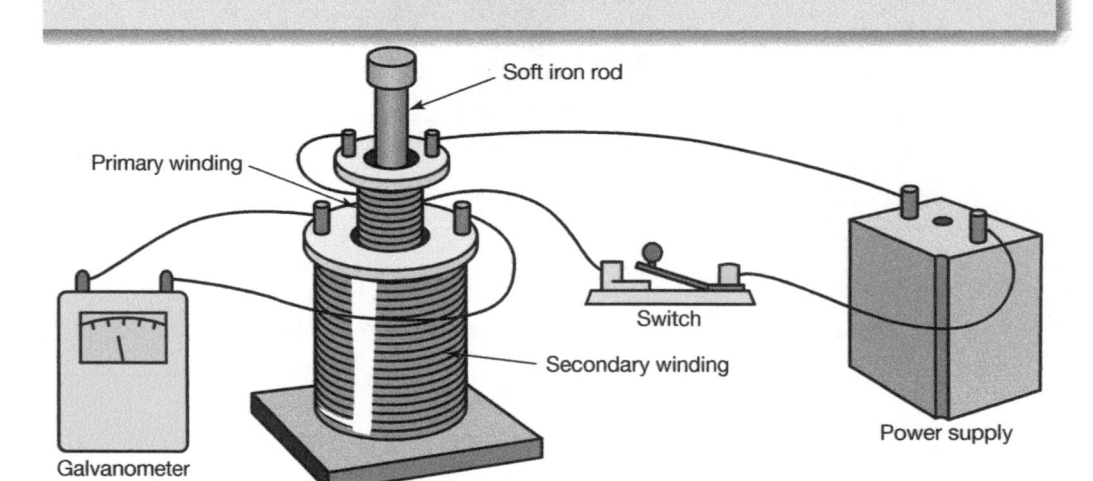

Describe how an EMF is induced by mutual induction:

_____

_____

_____

In what ways does the construction of an ignition coil differ from the apparatus above?

1 _____

2 _____

3 _____

_____

_____

DISTRIBUTOR UNITS (EARLY ELECTRONIC SYSTEMS)

The purpose of a distributor can be divided into three distinct roles. State these below:

1 _____

2 _____

3 _____

The method of interrupting the flow of primary current creates considerable variations in design, as can be seen on the distributors shown in the figure below. However, the methods of high-voltage distribution and the means of advancing the point of ignition remain basically the same on each distributor.

Name the main parts of the distributor units shown and identify the type of ignition system for which they would be suitable.

Ford (UK)

_____   _____

DWELL PERIOD

When maximum current flow is present in a coil winding it is said to be 'saturated' and the strength of the magnetic field will be at its maximum.

What is meant by 'dwell period'?

_____

_____

State what happens if the coil:

1 Does not reach saturation point.

_____

_____

2 Has current applied longer than needed.

_____

MECHANICAL ADVANCE AND RETARD MECHANISMS

Identify and label the main components in the diagram opposite.

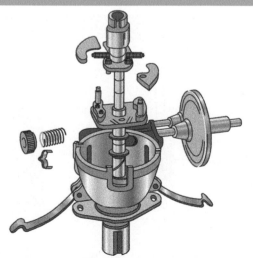

TRW, lucas

When the engine speed or load is varied it is usual to adjust the ignition timing automatically. State the two ways in which this may be done on older units:

1 _____

2 _____

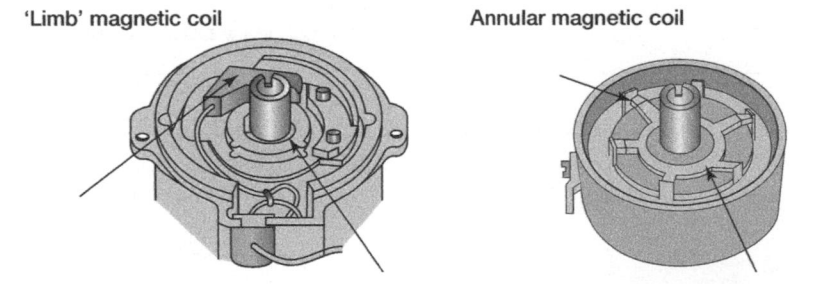

TIP For more on ignition timing go to **http://auto.howstuffworks.com/ignition-system.htm**

Complete the description of spark timing below using the correct words from the word list.

Power	pistons	increases	retarding	cylinder
constant	earlier	advance	nitrogen	temperatures
spark	time	work	emissions	oxides
burn	faster	more	speed	knocking

The timing of the _____ is important. The _____ that the air/fuel mixture takes to _____ is roughly _____. However, the _____ of the _____ increases as the engine speed _____. This means that the faster the engine goes, the _____ the spark has to occur in order to get the optimum amount of _____ from the engine. This is called spark _____: the _____ the engine speed, the _____ advance is required. When maximum _____ is not required other things such as _____ need to be considered. For instance, by _____ the spark timing (moving the spark closer to the top of the compression stroke), maximum _____ pressures and _____ can be reduced. Lowering temperatures helps reduce the formation of _____ of _____ (NO_x), which is a harmful pollutant. Retarding the timing may also eliminate _____; some cars that have knock sensors will do this automatically.

ELECTRONIC IGNITION SYSTEMS (DISTRIBUTOR TYPE)

Electronic ignition systems used in conjunction with distributors utilize power transistors as high-speed switches which carry the primary circuit voltage. The triggering system (trigger wheel, timing rotor, reluctor or photoelectric cell) signals the transistor in order for it to 'switch' the coil on and

off. The only moving parts in the distributor that may then wear are the advance-retard mechanisms.

Collector base emitter — From triggering device

TRIGGERING SYSTEMS

The diagrams below show two types of inductive coil or permanent magnet triggering designs.

Name the operating parts and describe their switching action.

'Limb' magnetic coil   Annular magnetic coil

'Limb' magnetic coil

_____
_____
_____

Annular magnetic coil

_____

_____

_____

'Hall' effect system

The 'Hall' effect generator has a trigger which incorporates a semiconductor material enclosed by a permanent magnet. When voltage passes through the semiconductor it generates an independent small current at right angles to the magnetic field. This is supplied to the amplifier.

What then occurs as the drive shaft turns?

_____

_____

Rotor ——

—— Vanes

Hall pick-up ——

Amplifiers

Amplifier ——

What is the purpose of an amplifier?

_____

_____

TIP Some amplifier units contain beryllium and should not be crushed or opened in any way. When fitting a new amplifier unit check that the earthing surface is clean and undistorted. Use silicone heat sink compounds if specified.

HIGH-TENSION CIRCUIT (DISTRIBUTOR SYSTEM)

The high-tension circuit consists of the coil, distributor cap, rotor, plugs and plug leads. Describe the flow of the high voltage current.

_____

_____

_____

Rotor

On certain older high-speed engines the maximum speed is controlled by a rotor cutout. How does this operate?

_____

_____

_____

SUPPRESSORS

Whenever an electrical spark occurs, waves of electrical energy are radiated out from the spark source. This energy causes external electrical interference which by law must be suppressed.

Resistors that have a value between 5000 and 25 000 ohms are required; these may be incorporated in at least two components. Where might these be found?

1 _____

2 _____

State the statutory requirements relating to installation and functioning of the ignition system in terms of radio interference suppression.

_____

_____

SPARK PLUGS

On the sectioned view of the spark plug shown, the leader lines indicate certain special features. Comment on these features.

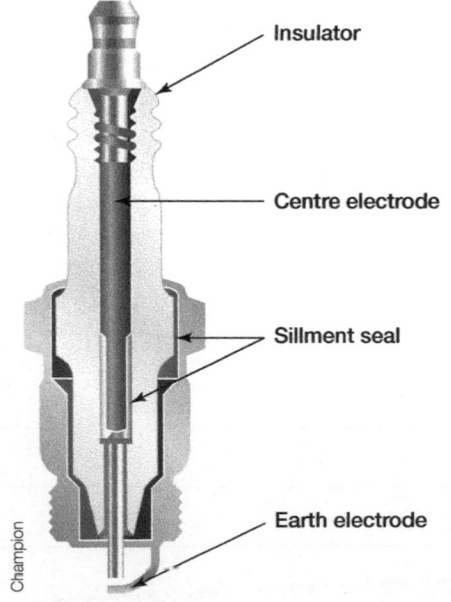

Insulator

Centre electrode

Sillment seal

Earth electrode

Champion

What are the effects of running an engine with the plug electrodes set:

1 Too wide? _____

2 Too close? _____

_____

The high-voltage energy produced by the coil is dissipated in the form of a spark across the spark plug electrodes. This ignites the petrol/air mixture in the combustion chamber.

The spark plug must operate efficiently under widely varying conditions of pressure and temperature and must be designed to suit the type of engine to which it is fitted.

State the importance of the following variations in plug design:

Plug reach (long or short) – _____

_____

Thread diameter – _____

_____

Methods of sealing plug seat – _____

_____

(1) Gasket type   (2) Taper seat type

TIP It is important to tighton plugs properly. Always use a spark plug socket and a correctly set torque wrench. Be careful not to snap the ceramic insulator. NEVER fit taper seat plugs to compression washer type cylinders (and vice versa). Always use the correct reach of plug and NEVER over-tighten them.

Variation in plug design

Heat range

For any particular application, spark plugs must be selected that do not foul at slow speeds nor get red hot at high speeds. Complete the description of how heat range is designed into a spark plug by filling in the correct missing words from the list below.

hard   range   inside   low   tip   quickly   soft   central   length   hot   tip

Heat _____ is determined by the _____ of the insulator ____ (nose) extending below the insulator seating gasket _____ the plug. A cold _____ plug has a _____ insulator seat. The short nose dissipates the heat _____ from the centre insulator and the plug is said to run cold. It is suitable for _____ running engines. A hot or _____ plug has a high insulator seat. The long _____ nose makes it difficult to conduct the heat quickly away from the ____ and so the plug runs hot.

State which plug in the figure below is the hot and cold plug, and indicate with arrows the heat flow path on both plugs.

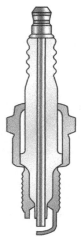

TIP For more on spark plugs go to:
http://www.sparkplugs.co.uk/pages/technical/glossary.htm
www.ngk.co.uk/

Research, name and sketch two different types of spark plug tip in the space below.

ELECTRONIC IGNITION SYSTEMS (DISTRIBUTOR-LESS)

Virtually all modern spark ignition engines are equipped with distributor–less (DIS) electronic ignition. Name the two types of system.

1 _____

2 _____

In both cases an ignition module controlled by the ECU dictates firing order and ignition timing. A crank sensor is used to trigger the ignition system. Other sensors feed information to the ECU which then calculates the optimum settings from a three-dimensional map in its memory.

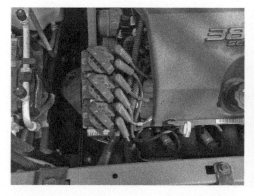

A coil pack for a double-ended or waste spark ignition system

A coil per cylinder ignition system

List five advantages of DIS systems over those with a distributor:

1 _____

2 _____

3 _____

4 _____

5 _____

A DISTRIBUTOR-LESS IGNITION SYSTEM (DIS) USED BY FORD

The flywheel has (36–1) 35 sensor teeth. The missing tooth space passes the variable reluctance sensor (VRS) at a point 90° before top dead centre (btdc) of No. 1 cylinder. This positioning and the speed of the pulse generated as the flywheel turns, give information for spark plug firing and engine speed.

Flywheel teeth at 10 degree intervals

Missing tooth

VRS sensor

Ford (UK)

DIS coil pack

Describe the action of the ignition coil and name the parts on the diagram below:

_____

_____

_____

_____

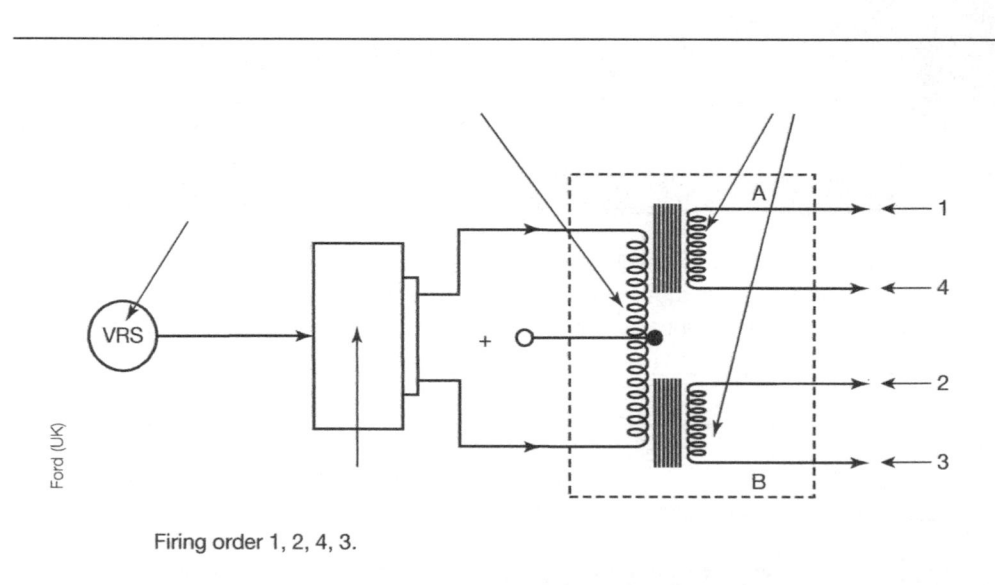

Ford (UK)

Firing order 1, 2, 4, 3.

COIL PER CYLINDER IGNITION

With this system a separate coil is used for each cylinder. It may be fitted directly to the plug, eliminating the need for a high voltage lead or it may be close to the plug.

State three advantages that this gives:

1 _____

2 _____

3 _____

A coil-near-plug system

In a COP system, the coil is mounted directly above the spark plug

On these modern systems, the ECU totally controls ignition timing and it is not adjustable. When the engine is cranked for starting, the ECU sets the timing at a fixed value. Once a predetermined speed is reached the ECU looks at values sent in from sensors (often common with the fuel injection system) and makes adjustments accordingly. All ECUs have limits as to how far timing can be advanced or retarded.

List four critical sensor inputs.

1 _____

2 _____

3 _____

4 _____

If the ignition timing is incorrect, detonation may cause ignition 'knock'. A piezoelectric knock sensor will detect this if fitted to the system. What happens if an ignition 'knock' is detected by the sensor?

_____

_____

For an ignition system quiz go to
http://auto.howstuffworks.com/ignition-system-quiz1.htm

Multiple choice questions

Choose the correct answer from a), b) or c) and place a tick [✓] after your answer.

1 **A triggering device is most commonly found in:**

   a) an electronic ignition system [ ]

   b) an ECU ignition module [ ]

   c) a fuel injection system. [ ]

2 **The ignition 'dwell period' is the period during which:**

   a) no current flows through the coil primary winding [ ]

   b) the secondary circuit in the coil is closed [ ]

   c) a magnetic field builds up in an ignition coil. [ ]

3 **An engine which runs hot requires a:**

   a) cold soft plug [ ]

   b) cold hard plug [ ]

   c) hot hard plug. [ ]

4 **When an engine is idling, the high tension voltage at the spark plug is approximately:**

   a) 40–50kV [ ]

   b) 12–14V [ ]

   c) 10–12kV. [ ]

5 **A higher voltage is required to cause a spark at the spark plug gap when the:**

   a) gap is reduced [ ]

   b) gap is increased [ ]

   c) cylinder pressure is low. [ ]

6 **An engine misfires at all engine speeds. One probable fault is a:**

   a) defective coil [ ]

   b) high resistance plug lead [ ]

   c) defective knock sensor. [ ]

7 **The sequence in which the high tension spark flows is:**

   a) distributor cap, rotor, cap segment, plug lead [ ]

   b) distributor cap, cap segment, rotor, plug lead [ ]

   c) cap segment, rotor, distributor cap, plug lead. [ ]

8 **In an ignition coil:**

   a) there are fewer turns in the secondary than primary winding [ ]

   b) there are fewer turns in the primary than secondary winding [ ]

   c) the primary and secondary windings have an equal number of turns. [ ]

Exhausts, air filtration and emission control systems

USE THIS SPACE FOR LEARNER NOTES

_____

_____

_____

_____

Learning objectives

After studying this section you should be able to:

- Describe the construction and purpose of air filtration systems.
- Explain the operating principles of air filtration systems.
- Describe the construction and purpose of the exhaust systems.
- Explain the operating principles of the systems.
- Describe the function of catalytic converters.
- Describe the construction and function of inlet and exhaust manifolds.
- Describe the procedures used when inspecting induction, air filtration and exhaust systems.
- Identify symptoms and faults associated with air and exhaust systems.
- Explain the methods employed to reduce emissions; PCV, EGR, particulate filters, additives.
- Explain the composition and effects of exhaust emissions.

Key terms

Catalytic converter A honeycomb construction which converts harmful gases into harmless gases.

Exhaust gas recirculation (EGR) A system used to reduce NO$_x$ and the levels of unburnt hydrocarbon.

Exhaust manifold Provides a means for the burnt gases to be directed to the exhaust system.

Inlet manifold Allows the air/fuel mixture to be distributed from a single duct to the many required branches.

Positive crankcase ventilation (PCV) A system which reduces hydrocarbon pollution by drawing crankcase vapours into the engine to be burnt and form water and carbon dioxide.

WWW

http://www.eurocats.co.uk/images/eec-information.pdf

http://www.mtsspa.net/it/prodotti/silenziatori.aspx

http://www.thegreencarwebsite.co.uk

AIR CLEANER-SILENCER BASIC TYPES

Most modern air cleaner-silencers are of the paper element type. They are usually fixed to the side of the engine compartment. Some older carburettor models and single point injection have them mounted on top of the engine intake.

Name four types of air cleaner:

1 _____   3 _____

2 _____   4 _____

Three requirements of the assembly are:

1 _____

2 _____

3 _____

What would be the effects of a partially blocked air cleaner?

_____

INLET MANIFOLDS

Fill in the missing gaps by using the words from the list below (there are four extra distracter words):

volumetric   iron   length   performance   mixture   diameter   plastic   weigh

Inlet manifolds allow the air/fuel _____ to be distributed from a single duct to the many required branches necessary for multi-cylinder engines. The engine _____ can be affected by the passage that the air takes to the inlet port. Each individual manifold tract ideally should be of the same _____ and of equal _____ to the inlet port.

Inlet manifolds also serve as a mounting point for fuel injectors, fuel rail, throttle body and EGR valve.

What is a plenum chamber?

_____

_____

_____

Why may some form of heating of the inlet manifold be considered necessary?

_____

EXHAUST SYSTEMS

The basic purpose of the exhaust system is to silence the noise created by the high velocity (speed in a given direction) of the exhaust gas (around 100 m/s) as it leaves the engine and directs the waste gases away from the vehicle. The noise is in the form of varying frequencies, and different types of boxes are designed to cope with these frequencies.

The complete system must be flexibly mounted on the chassis and be gas tight throughout its length, especially at the various joints.

 Exhaust system components get very hot when the engine is running. Contact with them can cause severe burns.

A complete exhaust system is shown below.

Name the parts indicated.

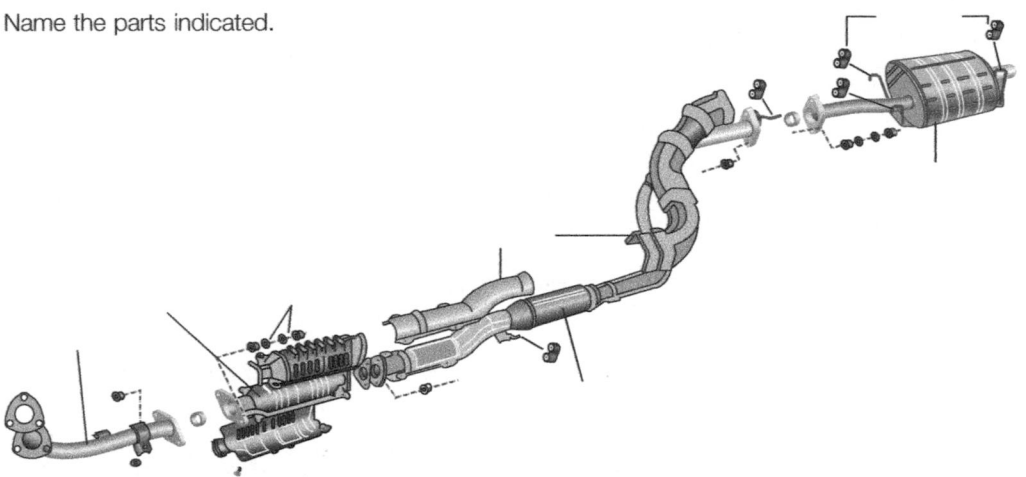

Why is a flexible pipe placed in the position shown?

_____

The flexible pipe is constructed of woven metal interlock which covers a heat resistant packing.

Why are heat shields fitted?

_____

_____

Which noise frequencies cause the most concern?

_____

The complete system must be flexibly mounted on the chassis using rubber mounts to absorb vibrations allowing the system to have some form of movement. At the attachment to the exhaust manifold there must be a gas-tight seal. Various types of clamps are used to join the different sections of the exhaust system together.

An exhaust system corrodes on the outside. Why does it also corrode from the inside outwards?

_____

_____

Exhaust system – protection during use

Describe how the exhaust system is protected from the following hazards during use or repair:

1 Preventing corrosion.

_____

_____

2 Preventing external mechanical damage.

_____

_____

3 Avoiding overheating of other parts.

_____

 **TIP** Before attempting to remove the exhaust system, use a wire brush on nuts and bolts, then spray them with penetrating oil.

Routine maintenance

State the reasons for carrying out routine maintenance on exhaust systems:

1 _____

2 _____

3 _____

List some general rules/precautions to be observed while carrying out routine air-fuel/exhaust system maintenance when:

1 Running an engine in a confined space.

_____

_____

2 Handling hot components.

3 Working over unguarded air intakes.

Statutory requirements

List any statutory requirements of exhaust systems relating to installation and functioning of the system when considering the following:

Condition, security and serviceability – _____

_____

Noise emission – _____

Modifications – _____

_____

Exhaust manifold

The exhaust manifold provides a means for the burnt gases to be directed to the exhaust system. They need to be able to withstand temperatures in excess of 800°C. Vehicle exhaust manifolds are commonly made of cast or nodular iron, although many modern vehicles use stamped, heavy gauge sheet metal or stainless steel manifolds.

Exhaust manifolds are tuned to specific engine and chassis combinations. Exhaust system length, pipe size and silencer type are used to tune the exhaust gases in the exhaust system. Correctly designed systems will improve the engine's volumetric efficiency.

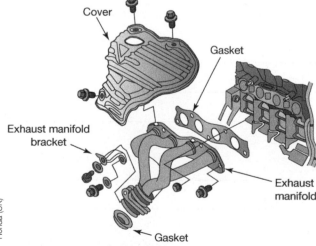

Cover

Gasket

Exhaust manifold bracket

Exhaust manifold

Gasket

Honda (UK)

POLLUTANTS CREATED BY THE MOTOR VEHICLE

The creation of some pollution is inevitable when power is obtained by internal combustion using a hydrocarbon fuel. When an engine is not running in tune, an excess of harmful emissions is passed into the atmosphere.

In general, a well-tuned SI engine is likely to emit more pollutants than a well-tuned CI engine. The modern motor vehicle engine is carefully monitored and controlled by electronics, which help to significantly reduce harmful pollutants and emissions.

Name the five pollutants produced by an engine and state why they are there:

Pollutant	Pollutant symbol	Reason for pollutant
_____	_____	_____
_____	_____	_____
_____	_____	_____
_____	_____	_____
_____	_____	_____

Pollution effects on the environment

Fuel which is burnt in the internal combustion engine is carbon-based. By-products of the combustion engine can be harmful to humans and the atmosphere. These gases contribute to what is known as the greenhouse effect, which is said to be increasing global warming. This needs to be reduced to help preserve our planet.

Motor vehicles produce approximately 13 per cent of the UK's total greenhouse gas emissions. One of these gases, CO_2, has increased the most since the industrial age (eighteenth–nineteenth century), which is largely due to the burning of fossil fuels, e.g. petrol and diesel.

Passenger cars currently have a recommended maximum target level for CO_2 emissions of 140 g/km. This equates to fuel consumption for petrol vehicles of 5.8/100 km or 5.3/100 km for diesel engines.

EXHAUST EMISSIONS

Exhaust gas is the product of the combustion of air and fuel mixture.

Air consists mainly of _____

Fuel consists mainly of _____

State how these products combine to form the exhaust gas.

_____

_____

_____

State the harmless exhaust gas products:

1 _____

2 _____

3 _____

Complete the chart to state the effects of the main pollutants produced by motor vehicles.

Pollutant	Symbol	Effects on people	Effects on the environment
Carbon monoxide	_____		
Carbon	_____		
Hydrocarbon	_____		

Pollutant	Symbol	Effects on people	Effects on the environment
Lead (this is no longer added to petrol due to the following effects)	_____		
Oxides of nitrogen	_____		
Carbon dioxide	_____		

CATALYTIC CONVERTER

A catalytic converter provides a means of considerably reducing harmful exhaust emissions to below the legally required limit. When operating, the catalytic converter acts like a red hot furnace which causes a high proportion of the harmful gases to re-burn and react with one another, rendering themselves harmless.

All modern vehicles are required to fit catalytic converters.

Why must the converter be fitted near the front of the system?

_____

_____

Describe the contents of the converter and the chemical reaction that occurs.

_____

_____

_____

_____	_____
_____	_____
_____	_____

TIP A two-way convertor does not convert NO_x.

How is the catalytic converter affected by temperature?

_____

_____

_____

TIP A spark plug misfiring can allow unburnt fuel to enter the converter causing catastrophic overheating damage.

What are the disadvantages of catalytic converters?

_____

_____

Research and find out what selective catalytic reduction (SCR) is and how it helps to reduce harmful exhaust emissions. Describe the use of urea water solutions and ammonia in SRC.

What precautions should be observed to ensure long catalytic life?

1 _____

2 _____

3 _____

4 _____

⚡ When working underneath a vehicle always wear safety glasses or goggles.

LAMBDA (OXYGEN) SENSOR

To ensure that the exhaust gas contents of a spark ignition engine are kept within reasonable limits (do not become rich) a lambda sensor is fitted to the manifold where it joins the exhaust system. By checking the oxygen content of the exhaust gases the lambda sensor feeds back to the ECU, which adjusts the fuel delivery and ignition timing if required.

Why is a post-catalytic lambda sensor commonly fitted?

_____

What would pipe 'A' be connected to when the blanking plug was removed?

_____

Fill in the missing gaps below with these words (there are two extra distracter words):

stoichiometric   variable   oxygen   voltage   chemically   ratio   fluid

If the air fuel _____ entering the combustion chamber is too rich the exhaust gasses will be low in oxygen and the sensor will send a low _____ signal to the ECU, as the mixture weakens more

_____ becomes present and the voltage signal rises. The aim of the system is to maintain a _____ correct air/fuel ratio of 14.7:1 by weight (the _____ ratio) at all times.

What is lambda 1? _____

A system which can continually monitor and adjust itself is known as a _____

What is a lambda sensor also known as? _____

EVAPORATIVE EMISSION CONTROL

Having recognized the need to control pollution, it is relatively easy to control crankcase emissions and fuel loss through evaporation.

How are these emissions controlled?

_____

_____

A charcoal (carbon) canister, shown below is normally located in the engine compartment. Briefly describe how it is used to control evaporative emissions:

_____

_____

_____

_____

_____

_____

A charcoal canister

Vacuum-pressure
connection

Carbon canister

Electrical
connection

Positive crankcase ventilation (PCV)

A source of hydrocarbon pollution is the engine oil, which can be drawn into the combustion chamber, either by passing by the piston or from the inlet valve stem. Oil vapours can form in the crankcase and escape to the atmosphere. A solution to this is for the vapours to be drawn into the engine and burnt to form water and carbon dioxide. This system is known as positive crankcase ventilation (PCV).

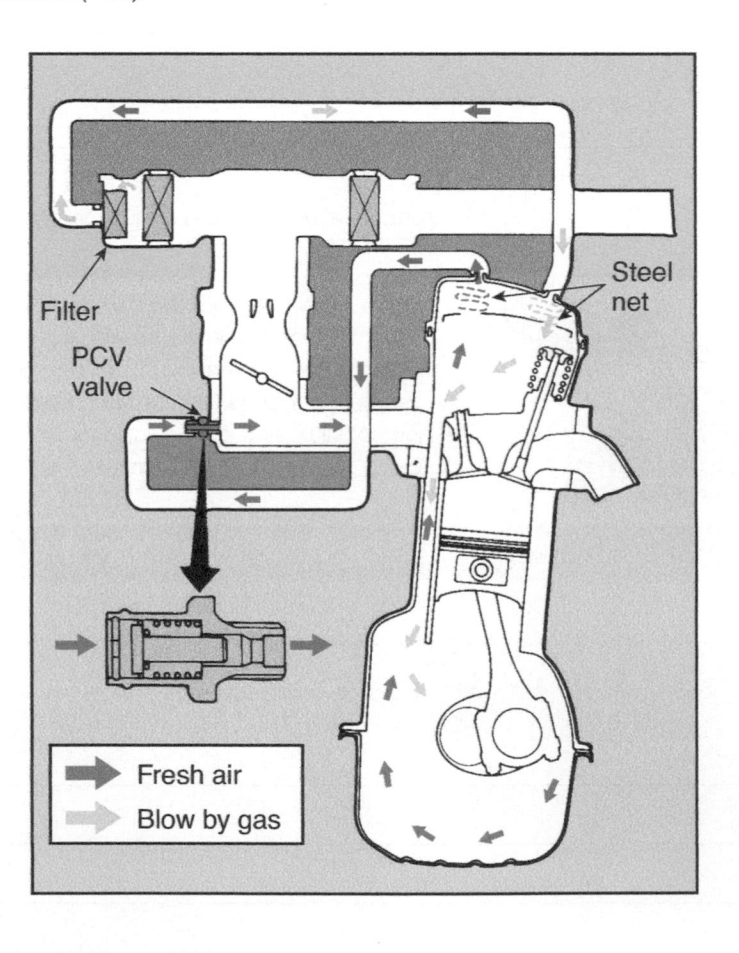

Filter

PCV
valve

Steel
net

Fresh air

Blow by gas

Give two reasons why a flow of air through the crankcase is necessary:

1 _____

_____

2 _____

_____

Exhaust gas recirculation (EGR)

This is an emission control commonly used to allow up to 15 per cent of the exhaust gas to recirculate back into the inlet manifold. Why is this considered necessary?

_____

_____

Computer controlled EGR layout

The exhaust gas recirculation (EGR) system consists of an EGR diaphragm valve, connected by pipes between the exhaust and inlet manifold, and an EGR solenoid (modulator) valve and connecting vacuum hoses. The solenoid valve is controlled by the engine ECU which will operate the EGR valve and can vary the flow relative to engine speed, load and temperature requirements. How is this flow control variation achieved?

_____

_____

Secondary air injection system

To decrease exhaust emissions when the engine is cold and to ensure a quick warm-up of the catalytic converter, air can be blown into the exhaust manifold at points near to each exhaust valve. This secondary air injection system consists of a pump, air cut-off valve and solenoid valve and exhaust non-return valve. The system is controlled by the engine's ECU. After warm-up the pump switches off. How does this air cause a rapid warm-up during the engine's warm-up cycle?

_____

_____

Particulate filters

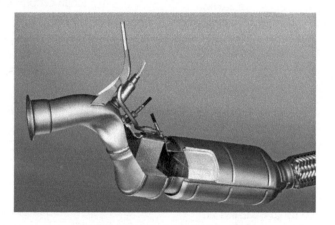

A particulate filter for a diesel engine

On diesel engine vehicles a particulate filter stops soot and smoke particles being ejected from the exhaust. It is positioned after the oxidation catalyst converter box. The filter is made of silicon carbide. The exhaust gas is forced through the porous structure which collects the contaminates before the gas leaves the system.

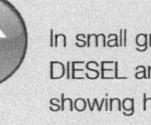

In small groups research and find out about the European emission standards for DIESEL and PETROL passenger cars. Draw a chart for both, starting at Euro 1, showing how the standards have changed over several years. Compare early and current diesel particulate requirements and similarly for NOx emissions. Find out when the regulations are next due to change.

Pollution Control

The complete regeneration (filter cleaning) takes 2 to 3 minutes.

What two service procedures should be carried out every 50 000 miles?

_____

_____

Multiple choice questions

Choose the correct answer from a), b) or c) and place a tick [✓] after your answer.

1 Air consists mainly of oxygen and:
   a) hydrogen [ ]
   b) carbon [ ]
   c) nitrogen. [ ]

2 What percentage of the exhaust gases are re-circulated back into the inlet manifold, on an engine which uses EGR?
   a) 5% [ ]
   b) 25% [ ]
   c) 15%. [ ]

3 Which one of the following statements is correct?
   a) carbon monoxide is the main contributor to the greenhouse effect [ ]
   b) hydrocarbon is the main contributor to the greenhouse effect [ ]
   c) oxides of nitrogen is the main contributor to the greenhouse effect. [ ]

4 Expansion, resonate and absorption are types of:
   a) catalytic converters [ ]
   b) exhaust boxes [ ]
   c) lambda sensors. [ ]

5 The purpose of the charcoal canister is to temporarily store or absorb the fuel vapour that builds up in the fuel tank:
   a) true [ ]
   b) false. [ ]

6 The purpose of an air filter is to clean and:
   a) silence the air [ ]
   b) dry the air [ ]
   c) direct air flow. [ ]

7 Fuel consists mainly of:
   a) hydrogen and nitrogen [ ]
   b) hydrogen and carbon [ ]
   c) oxygen and nitrogen. [ ]

8 A two-way catalytic converter converts:
   a) CO to CO_2 and HC to H_2O [ ]
   b) HC to H_2O and NO_x to N_2 [ ]
   c) NO_x to N_2 and HC to H_2O. [ ]

PART 4
CHASSIS

USE THIS SPACE FOR LEARNER NOTES

_____

_____

_____

_____

_____

_____

_____

SECTION 4
Suspension systems 173

SECTION 5
Steering systems 183

SECTION 1

Vehicle construction

USE THIS SPACE FOR LEARNER NOTES

_____

_____

_____

_____

_____

_____

_____

Learning objectives

After studying this section you should be able to:

- Identify and describe light vehicle body construction features.
- Identify methods of vehicle construction and names of main panels.
- Identify and describe light vehicle safety features, including the function and operation of airbags and seat belt restraint systems.

Key terms

Passive restraint   A safety restraint system which operates automatically.

Active restraint   A safety restraint system which the vehicle's occupants must make a manual effort to use.

Composite   Vehicle construction which utilizes a separate chassis and body.

Integral   Vehicle construction with no separate frame, using panels joined together to give overall strength. Can be called monocoque.

Space-frame   A strong chassis construction used for sports cars.

Sub-frame   Detachable assembly that is mounted to the underbody of the car to support the engine, transmission, suspension, etc.

Compliance   Slight controlled movement within the suspension bushes on the vehicle.

Crumple zones   Zones of controlled deformation under impact which are built into the front and rear of modern vehicles.

Safety cage   The reinforced central section of the car body which acts as the passenger compartment.

Bulkhead   Panels at the rear of the engine compartment and also separating the passenger compartment from the boot, which span the full width of the body.

Airbag   A bag which inflates on impact and saves the driver and passengers from injury.

Pre-tensioner   Tensions seat belts under impact to prevent occupants from sliding under the seat belts on impact.

Vehicle construction

```
J X E P E I P B M A O C V E A
N I U R H T Y A I U R L M L C
L O I E K A I R S U X A M Q T
M G K T E B B S M S R B Q C I
Z U K E C A F P O F I K T L V
R W U N G V L A B P J V R A E
F V F S R E Y U S U M Q E R H
G O E I Z F S H N N S O F G R
R T T O B U L K H E A D C E A
Q E N N B T Y Z P N S W X T D
L E Y E C O O V X U W W Y N O
M R Y R S U U S F W W Y K I Z
U L D M J O P Y W Y D H D N M
E C N A I L P M O C N Z N Z O
N F J G Z I Q T U F E A V X A
```

ACTIVE
AIRBAG
BULKHEAD
COMPLIANCE
COMPOSITE
CRUMPLEZONE
INTEGRAL
PASSIVE
PRETENSIONER
SUBFRAME

www.howstuffworks.com

www.thatcham.org

www.roadtransport.com

www.euroncap.com

VEHICLE CONSTRUCTION

The way that a vehicle is constructed is really a compromise of a number of different requirements. To a degree it will depend on the function of the vehicle as to which requirements come to the fore. List five important requirements for a modern family car:

1 _____

2 _____

3 _____

4 _____

5 _____

METHODS OF CONSTRUCTION

Construction methods have been mostly divided into two main types. These are either a separate chassis with a body mounted to it, or a chassis and body made from individual panels, which are joined as one complete assembly. State the names of these two types of design:

1 _____

2 _____

Which type does the mass-produced modern vehicle use today?

_____

What are the main advantages of this design?

_____

The figure below shows an integral body as used on a light vehicle, which is part-assembled. Identify and label the vehicle's panels using the words listed below.

front wing	bulkhead	sill	door pillar
rear wlng	floor pan	roof rail	parcel shelf

Chassis member

Floor section

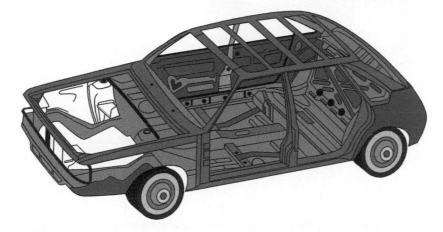

The thickness, composition and joining methods of the panels will depend on what they have to do and what stresses they are under. In some cases the vehicle systems may be mounted on sub-frames which are then attached to the body. It is also important that the vehicle can be raised without damage when being maintained – reinforced jacking or lifting points are part of the structure.

The cross-sectional shape of key structural body members is an important factor with regards to body strength and rigidity. Make sketches below, to show the cross-sectional shape of the typical body members listed.

Inner and outer sill

Door pillar

JACKING POINTS

The jacking points are the places on a vehicle body at which the jack is applied in order to lift a wheel (or wheels) clear of the ground. The body structure at the jacking points is made sufficiently strong to withstand the concentrated load imposed when the vehicle is jacked up.

How many jacking points are usually built into the vehicle body?

_____

These jacking points are designed for use when changing a wheel beside the road. Give two reasons why a vehicle's own jacking points are not normally used during servicing by a garage:

1 _____

_____

2 _____

List five important aspects to consider when instructing a customer on jacking procedures using a vehicle side jack on their car:

1 _____

2 _____

3 _____

4 _____

5 _____

What do the drawings in the figure below illustrate?

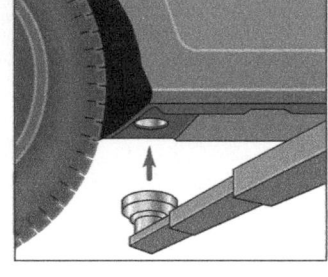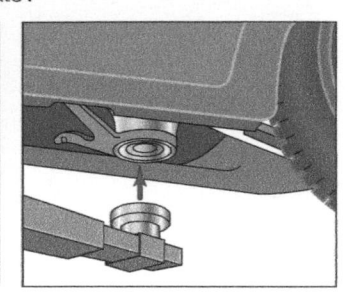

_____

_____

How is damage to the underbody avoided?

_____

_____

_____

SUB-FRAMES, SOUND INSULATION AND MOUNTINGS

Sub-frames

On a car the major components such as engine, transmission and suspension are sometimes not directly attached to the main integral body/chassis. The loads and stresses imposed on the car body by such components can be somewhat isolated from it by attaching the components to a rigid frame known as a _____.

The figure opposite shows a typical front sub-frame assembly. Name the various mounting points on the diagram:

a _____

b _____

c _____

d _____

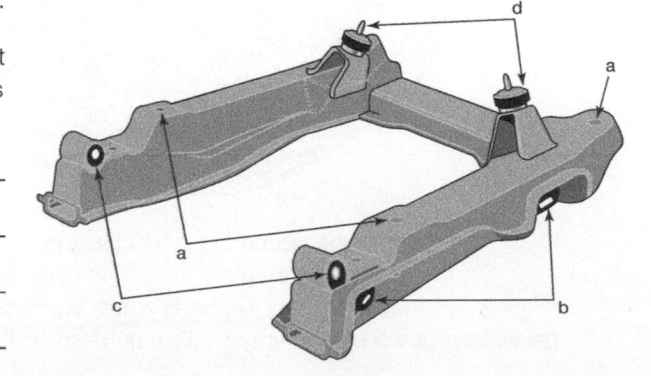

Sound insulation

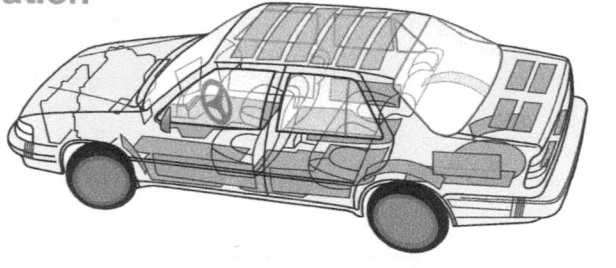

This is important in the modern vehicle. It reduces driver and passenger fatigue as well as being a good selling point for the vehicle. The shaded grey areas of the car diagram above illustrates one type of sound insulation and typical places where it is applied. State what material is likely to be used and its intended effect.

Material – _____

Effect – _____

TIP Plastic foam, thick felt or similar may be used additionally to dampen out resonances in body cavities and the like.

Flexible mountings

The major units on a vehicle are usually attached to the chassis or body structure through flexible rubber mountings. Name and label the mountings shown, which are for engine/transmission and suspension systems.

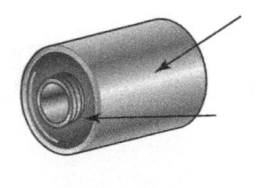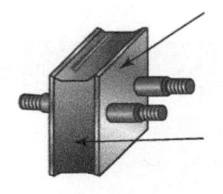

Give reasons for using these mountings:

_____

_____

BODY (OCCUPANT SAFETY)

Manufacturers of the modern car have vastly improved safety by enclosing the driver and passengers in a 'safety cage' or rigid cell. At the front and rear of the cell there are 'crumple zones' which are designed to absorb the impact of collision. Side impact protection is also considered and can be in the form of crumple zones or reinforced bars in the door frames. It has been proved that by reducing the deceleration on the occupants, many internal injuries can be avoided.

Front and rear crumple zones

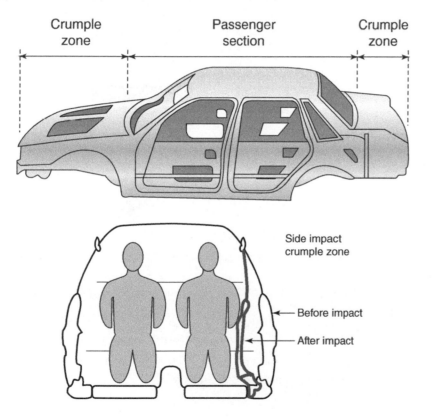

Crumple zone   Passenger section   Crumple zone

Side impact crumple zone

Before impact

After impact

Seat belts

The restraining force required to prevent car occupants from being thrown forward when a collision occurs is extremely high. Seat belt anchorage points on the vehicle body must therefore be sufficiently strong to withstand the loads involved.

The seat belt is normally wound on a spring-loaded reel whon it is not in use. An amount of belt is drawn off the reel according to the seat position and the size of the wearer when it is secured in place.

It allows the wearer freedom of movement when belted in and locks into position in the event of a sudden stop. It is known as the _____ system.

Add arrows to the figure below to indicate the seat belt anchorage mounting points.

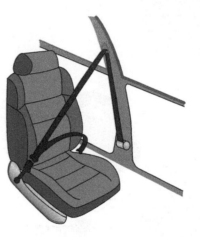

 TIP   When cleaning seat belt webbing, only use a mild soap solution and water.

149

Seat belt pre-tensioner and airbags

The figure below shows the action of a device which is part of the seat belt mechanism.

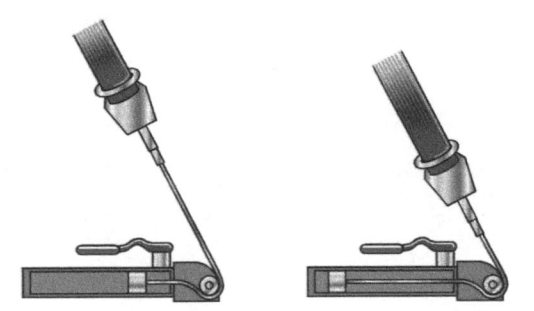

Name the device and state its purpose. _____

_____

_____

The system is shown below. Name parts 1, 2, 3 and 4:

Airbags – supplementary restraint systems (SRS)

In the event of a severe frontal impact above 15 to 18 mph, a driver's airbag will spontaneously inflate to reduce head and chest injuries which even a properly restrained driver can suffer. Airbags can also be provided for the front seat passengers and for side impact.

What happens when an accident occurs?

There are six statements below which describe what happens when an accident occurs and airbags are deployed. Place them in the correct order by using numbers 1–6 beside each statement.

- Seat belts lock or tension. ___
- At approximately 30 ms the airbag unfolds as the driver moves forward due to the crumple zone collapsing. ___
- At 120 ms the driver will move back into the seat – the airbag is almost deflated through the side vents. Driver visibility is returned. ___
- At approximately 15 ms after impact deceleration is sufficient to trigger the airbag. ___
- At 40 ms the airbag is fully inflated – the driver's momentum is absorbed by the airbag. ___
- The igniter sets light to the fuel pellets in the inflater. ___

The main parts of the vehicle airbag system are listed below. Describe their function.

Airbags: _____

_____

_____

Warning light: _____

Passenger seat switches: _____

_____

Igniters and inflator: _____

_____

Deceleration sensors: _____

_____

ECU: _____

_____

Multiple choice questions

Choose the correct answer from a), b) or c) and place a tick [✓] after your answer.

1 The term 'integral', when applied to vehicle construction means that:

   a) a strong chassis with a separate body is used to build the vehicle [ ]

   b) the inside of the vehicle is made of materials which are integrated together [ ]

   c) the vehicle is made up of a number of body panels with no separate chassis. [ ]

2 The engine and transmission in a vehicle of unitary construction is supported by a:

   a) ladder chassis [ ]

   b) sub-frame [ ]

   c) space frame. [ ]

3 A pre-tensioner is part of the:

   a) seat belt system [ ]

   b) airbag system [ ]

   c) braking system. [ ]

4 A capacitor used in the airbag circuit:

   a) operates the fault warning light circuit [ ]

   b) produces a spark which ignites fuel tablets [ ]

   c) senses the impact velocity in a crash. [ ]

5 What is a typical time in which a driver's airbag is fully inflated?

   a) 10 milliseconds [ ]

   b) 120 milliseconds [ ]

   c) 40 milliseconds. [ ]

6 Which gas is used to inflate an airbag?

   a) hydrogen [ ]

   b) nitrogen [ ]

   c) helium. [ ]

7 When any work on airbags is carried out:

   a) the back-up power supply must be depleted [ ]

   b) only the yellow and black wires can be disconnected [ ]

   c) the ignition must be switched on and off three times. [ ]

8 The design features of a modern vehicle which absorb the impact of collision are known as the:

   a) sub-frames [ ]

   b) crumple zones [ ]

   c) safety cages.

SECTION 2

Wheels and tyres

USE THIS SPACE FOR LEARNER NOTES

_____

_____

_____

_____

_____

_____

_____

Learning objectives

After studying this section you should be able to:

- Identify light vehicle wheel and tyre components.
- Describe the construction and use of different types of light vehicle wheels, tyres and tread patterns.
- Identify tyre markings and key engineering principles related to light vehicle wheel and tyre systems.
- Recognize defects associated with tyres and wheels.

Key terms

Radial A tyre in which the plies are placed at right angles to the rim.
Cross ply A tyre in which the plies are placed diagonally across each other at an angle of approximately 30 to 40 degrees.
Bias belted A design of tyre which is a hybrid, a cross between a ply and radial design.
Well base rim A rim with a centre channel which enables easy removal and re-fitting of the tyre.
Aspect ratio The ratio between the height and width of a tyre (expressed as a percentage).
Un-sprung weight The weight of those parts of the car which are not carried by the suspension.
Dynamic imbalance Lack of balance of a rotating part such as a wheel in motion which can cause vibration or judder.
Static balancing Checking a wheel's balance by seeing if it stops in the same position when rotated. If it does the wheel is imbalanced so a small weight is attached to the rim opposite the heavy spot to counter the imbalance.

Remember – when working with wheels and tyres you may be dealing with:

- Jacking and supporting vehicles.
- High air pressures and pressurized components.
- Rotating machinery.
- Chemicals and solvents.
- Waste disposal.

www.pirelli.co.uk

www.aa1car.com

www.dunlop-tires.com

The road wheel assembly (tyre and metal wheel) transmits the drive from the drive shafts to the road surface, and back the opposite way during braking. It also provides a degree of springing to accommodate minor road irregularities.

Types of road wheel

Steel wheels are strong and relatively cheap. Wheel trims are usually fitted to them to improve their appearance.

Give three reasons for using alloy wheels:

1 _____

2 _____

3 _____

Steel wheel

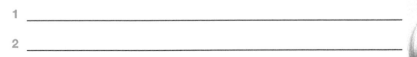

Aloy wheel

Some vehicles may also have a 'space-saver' spare wheel. These reduce the space taken up by the spare wheel and increase the luggage capacity if carried in the boot.

Space saver wheel

Courtesy of NRMA Motoring & Services

TIP Make sure customers know that space saver wheels are only a temporary tyre. Continued use can result in tyre failure, differential wear and affect anti-lock braking and traction control.

Wheel rim design

Three design features are particularly important, these are the parts of the rim which:

1 Allow fitting and removal of the tyre.

2 Tighten the tyre bead on to the rim during inflation to provide an airtight seal and to secure the tyre on to the rim.

3 Prevent the tyre bead becoming dislodged in the event of a sudden deflation.

Indicate, using arrows, where these three rim features are on the basic car rim profile shown below.

Car wheel rim

Most cars use tubeless tyres on 'one piece' wheel rims. A selection of rim profiles are shown in the figure below. It is important to remove and fit tyres from the correct side of the rim. The arrows on the rim profiles below show which side this is.

Why is this so important?

_____

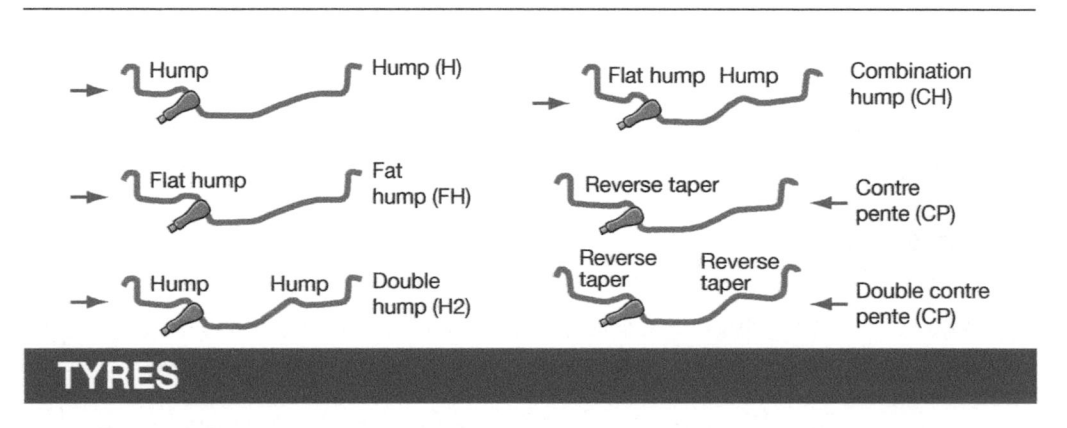

TYRES

Pneumatic tyres fulfill a number of functions. List three below.

1 _____

2 _____

3 _____

Tyre construction

The tyre is a flexible rubber casing which is reinforced or supported by other materials, for example, rayon, cotton, nylon and steel.

What makes a tyre suitable for its application?

_____

_____

Tubed and tubeless tyres

Originally an inner tube together with its valve provided an airtight seal. Virtually all modern vehicles use tubeless type tyres. The section below shows how the tyre, rim and valve combined to form an airtight seal.

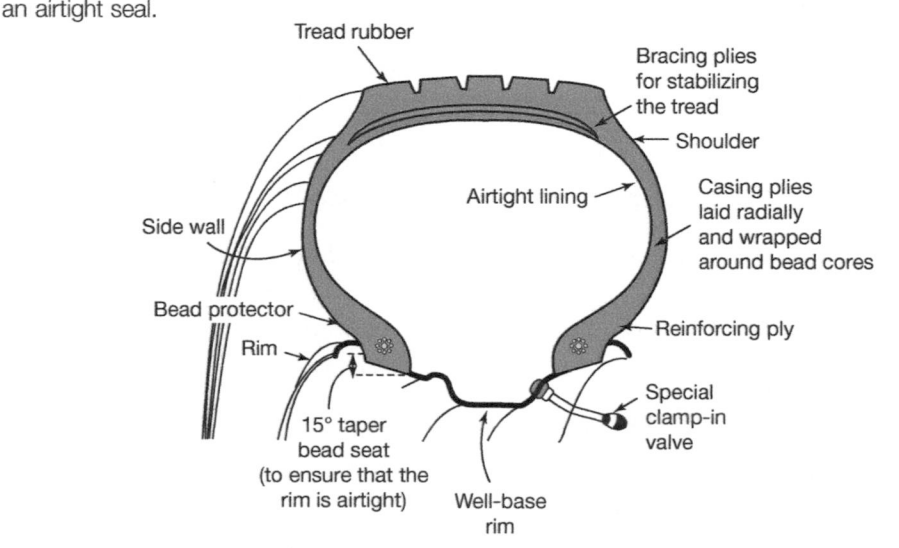

The figure below shows tubed and tubeless tyre valves. Label the parts of the sectioned valve and valve core.

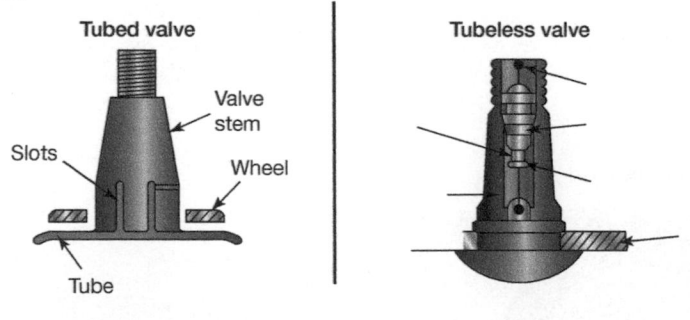

What is the purpose of the slots in the outer stem of the tubed valve?

_____

Name the type of valve core used in both valves.

_____

Types of tyre construction

The two principal types of tyre construction used on road vehicles are:

1 _____

2 _____

Today nearly all cars are fitted with radial tyres. A tyre casing consists of plies which are layers of material looped around the beads to form a case. The basic difference in structure between radial and cross ply tyres is in the arrangement of the plies.

With diagonal (or cross) ply construction the ply 'cords' form an angle of approximately _____ to the tyre bead.

Radial ply cords form an angle of _____ to the tyre bead.

Draw lines on the two outlines below to show the position of the cords in relation to the tyre bead.

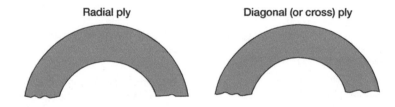

Radial ply   **Diagonal (or cross) ply**

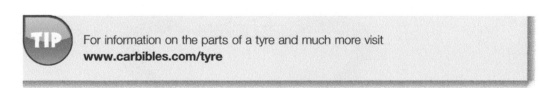

TIP   For information on the parts of a tyre and much more visit
www.carbibles.com/tyre

One feature of the radial ply tyre is that the walls are more flexible than the diagonal (or cross) ply tyre. Why is this?

_____

_____

State three advantages of radial ply tyres when compared to a cross ply:

1 _____

2 _____

3 _____

Run flat tyres

There are several types of run flat tyres. List the three main categories that they fall into:

1 _____

2 _____

3 _____

TIP   For more information on run flat tyres go to **http://www.michelinman.com/pax/**

Self-sealing tyres

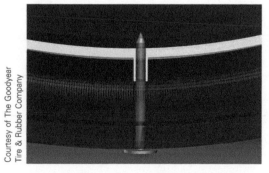

Courtesy of The Goodyear Tire & Rubber Company

The action of a self-sealing tyre

An additional lining under the tread area is coated with a sealant that can permanently seal most nail type punctures, up to 4 mm in diameter. The lining seals the area around the puncture and can fill the hole once the object is removed.

Self-supporting tyres

Special bead design:
• Enhanced retention after pressure loss
• Acceptable seating pressure

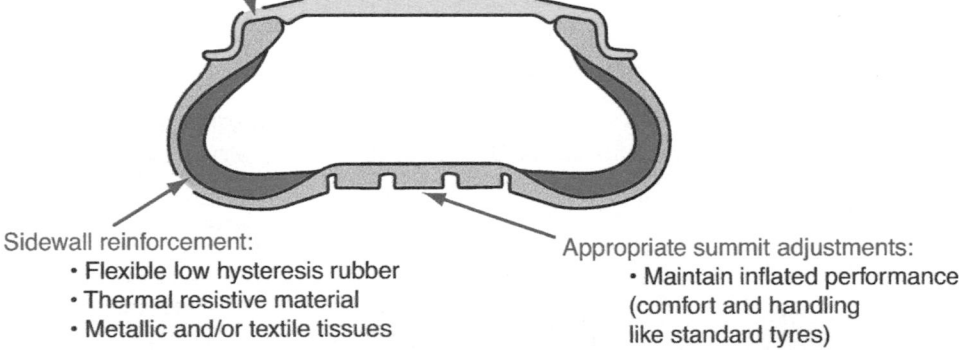

Sidewall reinforcement:
• Flexible low hysteresis rubber
• Thermal resistive material
• Metallic and/or textile tissues

Appropriate summit adjustments:
• Maintain inflated performance
(comfort and handling
like standard tyres)

Self-supporting tyres feature a stiffer internal construction which is capable of temporarily carrying the weight of the vehicle, even after the tyre has lost all air pressure.

Auxiliary supported systems

Auxiliary supported systems combine special wheels and tyres used for original equipment vehicle applications. In these systems the tread rests on a support ring attached to the wheel when the tyre loses pressure.

Courtesy of The Goodyear Tire & Rubber Company

Tyre tread pattern design

The design of a tyre, particularly the tread pattern, is dictated by the type of vehicle it is fitted to and how the vehicle will be used.

State what each of the following design features will affect:

Tread compound: _____

Circumferential grooves: _____

Cross ribs (lateral notches): _____

Blocks: _____

Sipes or blades: _____

_____

_____

State the type of tread pattern and a typical application for the three different tyres shown below.

_____ _____ _____
_____ _____ _____

The asymmetrical tread is different from one side to the other, the inside half gives good grip when travelling in a straight line, the outside half gives good grip in turns. Most asymmetrical tyres are also directional.

How can a directional tyre be recognized?

_____

Tyre mixing

Cars and vans with single rear wheels must NOT have radial tyres on the front wheels and cross ply tyres on the rear wheels. It is also illegal to have a cross ply tyre on one side of an axle with a radial on the other.

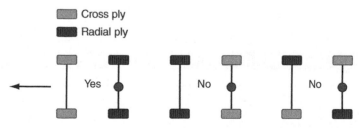

Tyre size markings

The size of a tyre is indicated from markings on the sidewall, for example 185/60 R14 82H.

Complete the table below to show what is meant by each of these markings and indicate on the drawing the positions of the two main dimensions.

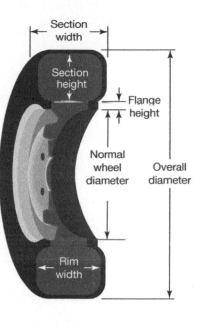

185	_____
60	_____
R	_____
14	_____
82	_____
H	_____

TIP   For tyre markings and much more go to **www.tyres-pneus-online.co.uk**

What does **TWI** stand for on the sidewall of a tyre and what is its purpose?

_____

_____

_____

The load index and tyre speed symbol markings are used in conjunction with tables which will give the weight and speed combination for a given tyre.

Select the weight and speed for a tyre with a load index (LI) of **82** and a speed rating of **H** using information from the tables below.

PART OF A LOAD INDEX TABLE

LI	kg	LI	kg	LI	kg	LI	kg	LI	kg	LI	kg
0	45	20	80	40	140	60	250	80	450	100	800
1	46.2	21	82.5	41	145	61	257	81	462	101	825
2	47.5	22	85	42	150	62	265	82	475	102	850
3	48.7	23	87.5	43	155	63	272	83	487	103	875

TYRE SPEED SYMBOL MARKINGS TABLE

SYMBOL	km/h	mph
Q	160	100
R	170	105
S	180	113
T	190	118
H	210	130
V	240	149
VR	210+	130+
Y	300	186
ZR	240+	150+

How do tyre pressures vary from the start of a long, fast journey to immediately afterwards? What is the reason for this variation?

_____

_____

Tyre profiles

Tyres were designed to a basic circular cross-section, the tyre width being approximately equal to the radial height (100% profile).

Modern vehicles use a wider, lower type of tyre, which is said to have either a low height/width, low profile or low aspect ratio.

The height to width is quoted as a percentage, such as 70%; therefore, for a tyre with a 70% aspect ratio, the height dimension is 70% of the width.

The aspect ratio is normally indicated on the sidewall marking; however, if it is not indicated, it usually means that the tyre is of the now normal aspect ratio which is about 82%.

Tyre having 100% profile ratio

Draw a tyre having a 70% profile ratio

Racing car tyre 35% profile ratio

State three advantages of low profile tyres when compared with 100% profile tyres:

1 _____

2 _____

3 _____

TIP Tyres with a larger or smaller diameter than originally installed could affect the operation of the anti-lock braking system, traction control and the accuracy of the speedometer.

Tyre faults

Certain tyre defects make it illegal to use a vehicle on a public road. Outline the main legal requirements with regard to the tyre faults listed on page 159.

Tread wear

_____

_____

Cuts

_____

_____

Suitability

_____

Inflation

_____

Incorrect pressure can cause uneven tyre wear, this can be either in the centre or on the outside of the tread as shown in the figures below.

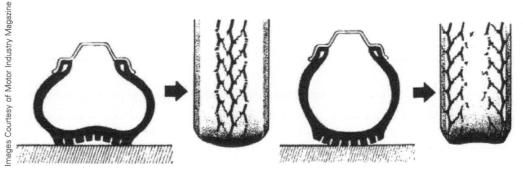

'Under-inflation'   'Over-inflation'

Tread wear can also be caused by incorrect wheel alignment, i.e. 'toe' in or 'toe out' as shown in the figures below.

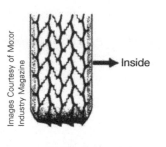
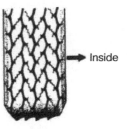

Excessive 'toe in'   Excessive 'toe out'

Steering and suspension geometry can also affect tyre wear. Camber angle is particularly important as shown in the figures below.

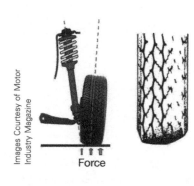
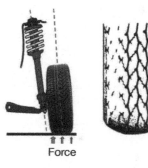

Incorrect positive camber   Incorrect negative camber

Maintenance

Routine tyre maintenance helps to ensure safe, efficient operation and maintains performance and tyre life. List three routine checks and adjustments for tyres and wheels:

1 _____

2 _____

3 _____

Describe with the aid of a sketch, how tyre 'tread depth' is measured:

_____

_____

TIP The manufacturer's specification for original equipment is the best guide when choosing replacement tyres for a vehicle. Changes from this, however, may be possible, when a vehicle is used for particular applications, such as continued high speed driving or cross-country operation. Check the specification.

Multiple choice questions

Choose the correct answer from a), b) or c) and place a tick [✓] after your answer.

1 One function of the tread on a tyre is to:

a) expel water and grease from between the tyre and road surface [ ]

b) stop the tyre casing from being worn badly and perishing [ ]

c) help make the steering lighter and keep the vehicle safe. [ ]

2 Passenger car tyres must have a minimum legal tread depth of:

a) 0.16 mm [ ]

b) 1.6 mm [ ]

c) 1 mm. [ ]

3 A tyre fitted to a car is illegal for road use if:

a) it has not been balanced [ ]

b) it has a cut of 20 mm or above [ ]

c) it is not properly inflated. [ ]

4 A radial and a cross ply tyre cannot legally be fitted on the same axle of a car.

a) False [ ]

b) True. [ ]

5 A tyre has a sidewall marking of 195/65 R 15T. This indicates the tyre is:

a) run at a pressure of 65 psi [ ]

b) for a 195 mm diameter wheel [ ]

c) for a 15 inch diameter wheel. [ ]

6 A tyre has 165/65 R 14H written on the sidewall. The letter 'R' indicates that the tyre is:

a) of a radial ply construction [ ]

b) reinforced for off-road use [ ]

c) only suitable for road use. [ ]

7 A tyre has a sidewall marking of 185/65 R 15H. This indicates the tyre:

a) is for a 0.65 m diameter wheel [ ]

b) has an aspect ratio of 65% [ ]

c) has an aspect ratio of 15%. [ ]

8 Tyre pressures are checked when they are cold because:

a) the air pressure in a tyre increases with heat [ ]

b) it ensures all tyres are the same pressure [ ]

c) the tyre walls are stiff and do not flex. [ ]

SECTION 3

Braking systems

USE THIS SPACE FOR LEARNER NOTES

Learning objectives

After studying this section you should be able to:

- Identify light vehicle braking system components and the key engineering principles that are related to braking systems.
- Describe the construction and operation of light vehicle braking systems and state common terms used.
- Compare key light vehicle braking system components and assemblies against alternatives to identify differences in construction and operation.

Key terms

Drum brake   A brake in which curved shoes press on the inside of a metal drum to produce friction.

Disc brake   A brake in which friction pads grip a rotating disc in order to slow the vehicle down.

Leading shoe   Shoe in a drum brake system which pivots outwards into the drum.

Trailing shoe   Shoe in a drum brake system which is forced away from the drum by its rotation.

Self-servo action   Self-energizing effect which helps to apply a brake shoe to a drum.

Caliper   Housing for the piston(s) which operates the brake pads.

Hygroscopic   Capable of absorbing moisture from the atmosphere.

Master cylinder   A cylinder in the hydraulic circuit which pressurizes fluid.

Brake fade   Loss of friction in the brakes due to overheating.

 www   http://auto.howstuffworks.com/

http://www.fearofphysics.com/Friction/

Remember – when working with braking systems you may be dealing with:

● Brake dust.
● Hydraulic fluid under pressure.
● Chemicals which are carcinogenic or toxic.
● Rotating components.
● Waste disposal.
● Hot components.

BRAKES

State three functions of a braking system on a vehicle:

1 _____

2 _____

3 _____

Braking system principles

A modern light vehicle braking system is operated by fluid and is known as a 'hydraulic' system. This type of system uses brake fluid to transfer pressure from the brake pedal to the pads or shoes. Brake fluid is not compressible, which means that the pressure transmitted in the system is equal throughout the system. If, however, the size of the piston diameter of the pump (master cylinder) and the pistons in the cylinders operating the pads or discs is different, the force going in and the force coming out will also be different.

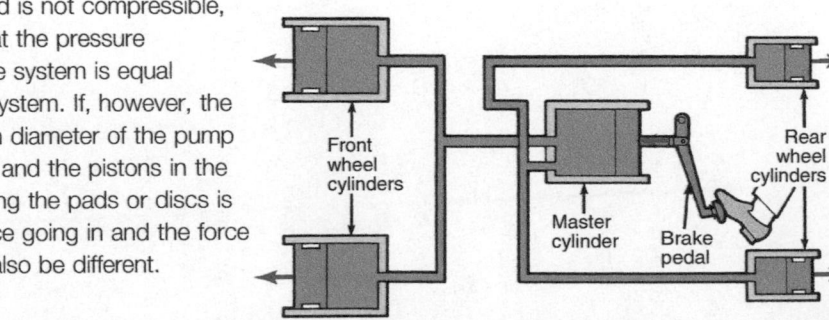

Front wheel cylinders

Rear wheel cylinders

Master cylinder

Brake pedal

Which pistons are normally larger in diameter?

_____

Is there any disadvantage to this?

_____

_____

System layouts – dual braking (split line)

What is the advantage of the diagonally split system when compared with the front/rear split?

_____

_____

_____

The components of the braking system are connected by rigid and flexible pipes. What is the reason for using flexible pipes in part of the hydraulic system?

_____

Hydraulic braking system components

Brake fluid

Brake fluid must conform to the international standards set by the American Society for Automotive Engineers (SAE). It also has a DOT number.

TIP

Many brake fluids are 'hygroscopic' which means they can absorb water from the atmosphere. This lowers the boiling point. DO NOT leave the top off brake fluid containers. DOT standards refer to the boiling point, i.e. DOT 4 and DOT 5.

DOT 4 fluid with a 1 per cent water content boils at approximately 245°C, but with a 7 per cent water content it is down to 135°C.

The main requirements of brake fluid are listed below. State why these are important.

Low viscosity – _____

Compatibility with brake components – _____

_____

Lubricating properties – _____

Resistance to chemical ageing – _____

High boiling point – _____

Resistance to freezing – _____

TIP Brake fluid must not come into contact with paintwork. If it does, wash immediately with water. Use only manufacturer's approved brake fluid. Any other lubricant should **never** be used in a braking system. It may cause the rubber components in the system to swell and disintegrate.

DRUM BRAKES

Single-leading and trailing arrangement

The figure opposite shows a 'double-piston wheel cylinder' acting on a *leading brake shoe* and a *trailing brake shoe*. The leading shoe is pulled harder into the drum by drum rotation whereas the trailing shoe tends to be forced away from the drum by its rotation. A much greater braking effect is therefore obtained from a leading shoe.

Name the parts in the drum brake assembly. Indicate drum rotation and show the leading and the trailing shoe.

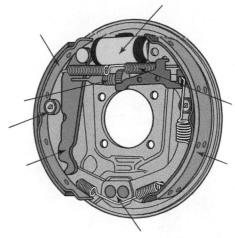

Why does the leading shoe wear at a greater rate than the trailing shoe?

_____

_____

What friction materials may be used for brake linings?

_____

What health hazards are associated with the dust from this friction material?

_____

State the purpose of the components listed below.

Wheel cylinder – _____

Return spring – _____

Adjuster – _____

Parking brake lever – _____

_____

TIP When dismantling brake shoe assemblies it is a good idea to sketch out the position of the springs and shoes. This can make life much easier when replacing the components.

WHEEL CYLINDERS

The wheel cylinder consists of a plain cylinder and piston(s) which, when energized by fluid pressure, actuate the brake shoes. A double-piston type is shown below. Label the drawing.

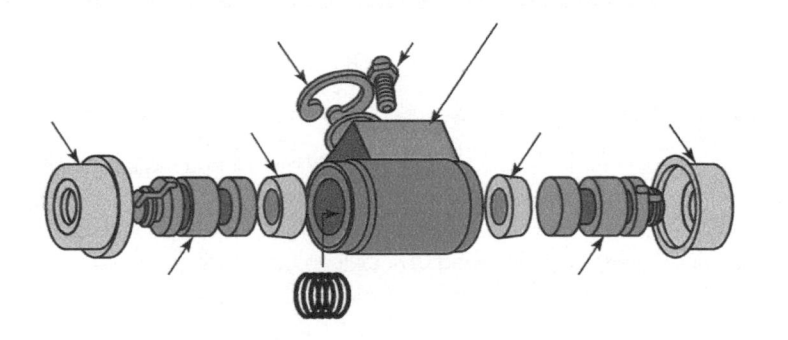

TIP DO NOT press the brake pedal while a brake drum is off. This could cause the pistons in the wheel cylinders to pop out of the cylinders.

DISC BRAKES

The braking surfaces on a disc brake are more easily cooled than on a drum brake. They are self-cleaning and less prone to brake fade.

In the fixed caliper disc brake shown here, a common fluid pressure applied to the two pistons forces the friction pads against the disc. Label the drawing.

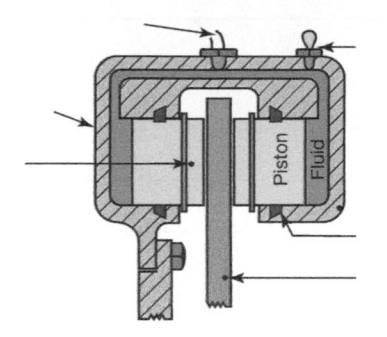

The discs themselves are either solid or ventilated. The ventilated disc has fins cast between the braking surfaces (see figure below).

What are the advantages and disadvantages of this when compared to a solid disc?

_____

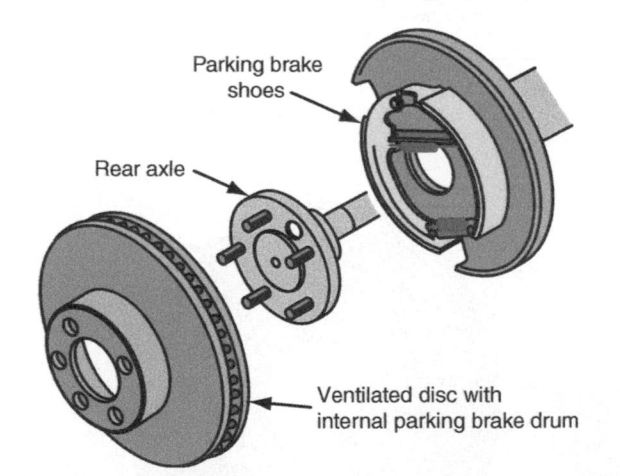

Parking brake shoes

Rear axle

Ventilated disc with internal parking brake drum

The figure above shows a section of a ventilated disc, it also shows a parking brake incorporated into drum cast into the disc.

State briefly what is meant by 'brake fade':

_____

_____

An alternative design of brake caliper is the floating or sliding caliper. With a single piston on one side only it needs less space on the wheel side, it operates cooler and can incorporate a handbrake mechanism. Name the component parts of the floating caliper disc brake assembly shown below.

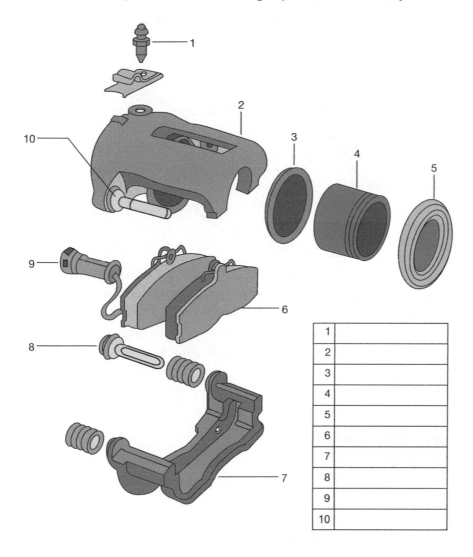

1	
2	
3	
4	
5	
6	
7	
8	
9	
10	

Disc pad wear sensors

Some brake pads have wear sensing indicators. They can be audible as indicated in the figure below where a spring steel tab contacts the disc as the pad wears and squeals when the disc rotates, or an electronic system which uses sensors in the pads. These complete an electronic circuit to illuminate a warning light when the pads wear to beyond a predetermined point.

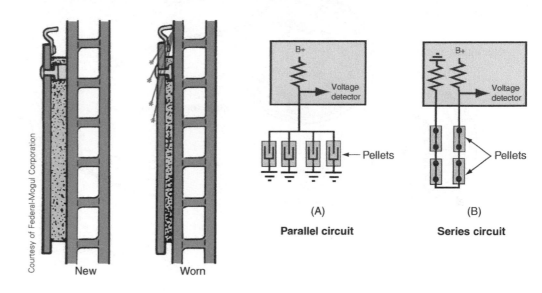

Courtesy of Federal-Mogul Corporation

New   Worn

(A)
Parallel circuit

(B)
Series circuit

TIP   To find out more about disc brakes, visit:
http://auto.howstuffworks.com/auto-parts/brakes/brake-types/disc-brake6.htm

PARKING BRAKE

The parking brake on most cars operates on two wheels only and is normally mechanically actuated by cables.

The handbrake lever (figure opposite) is connected to the cable and locks with a pawl (A) and ratchet (B).

What does the spring (C) do?

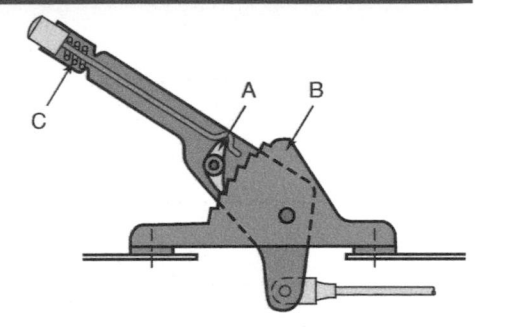

TIP When carrying out a vehicle inspection, the handbrake should only travel for approximately three clicks of the ratchet before it applies. This gives a safe reserve travel.

Label the figure below and state the purpose of the compensator.

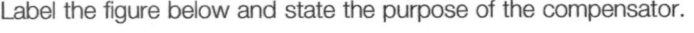

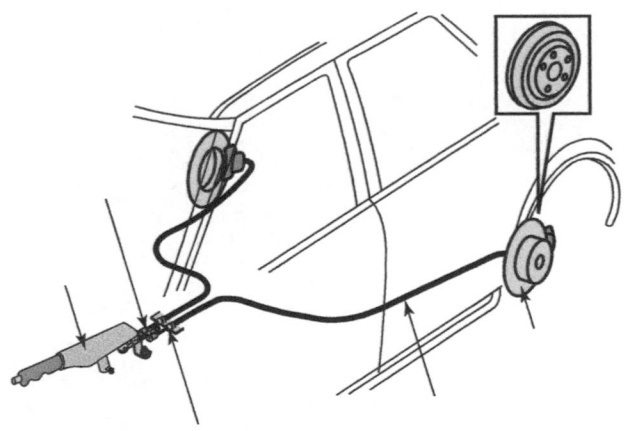

Typical handbrake (parking) sensor

Electronic parking brake

There are two main types of electronic parking brake (EPB) and they are becoming more common. Either a drive unit activates the cable, or an electric motor on each of the two rear-axle brake calipers applies the brake. In the second method, the cable and servo motor are no longer required. In addition to the hydraulic brake line on the brake caliper, only an electric cable that transfers data and power is needed for connection to the separate control unit.

Courtesy of Robert Bosch GmbH, www.bosch-presse.de

Shoe and drum

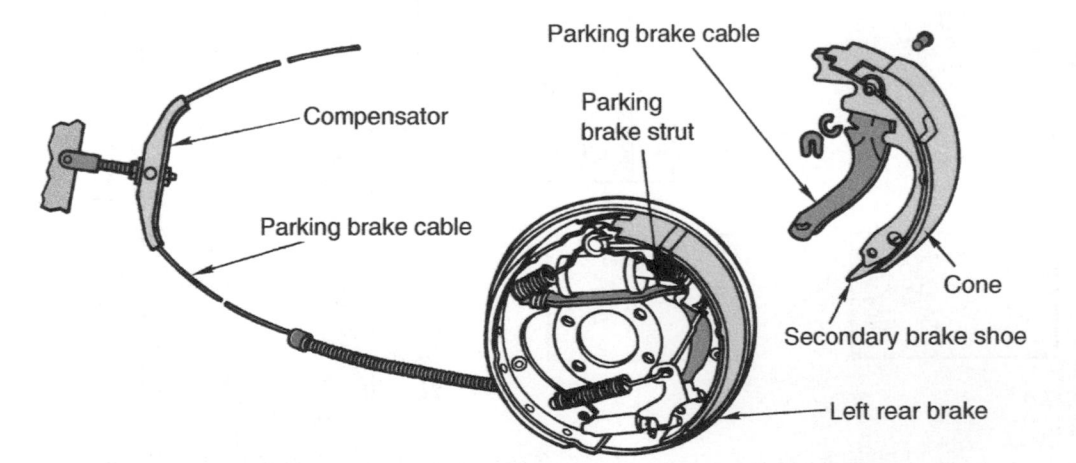

On a typical shoe and drum parking brake system, when the handbrake is applied the compensator and cables exert a balanced pull on the parking brake levers and struts. These move the brake shoes outwards against the drum until the handbrake is released.

Rear disc and auxiliary drum

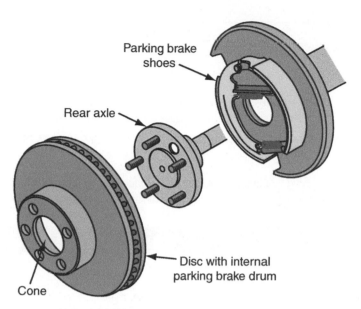

Parking brake shoes

Rear axle

Disc with internal parking brake drum

Cone

This is found on some vehicles with fixed or sliding calipers. The inside of the rear wheel hub and disc assembly is used as the parking brake drum and operates independently of the service brake. They are often manually adjusted through the back plate.

Caliper actuated

This type of design tends to use levers operated by linkage to apply the caliper pistons and pads against the disc. On the design shown, an actuator screw is rotated when the parking brake is applied. This in turn pushes a cone against the inside of the piston which applies the pad to the disc.

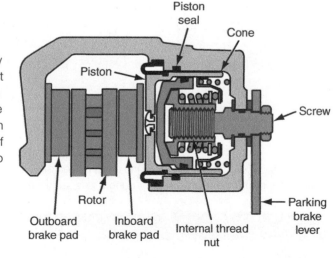

Piston seal

Cone

Piston

Screw

Rotor

Outboard brake pad

Inboard brake pad

Internal thread nut

Parking brake lever

AUTOMATIC BRAKE ADJUSTERS

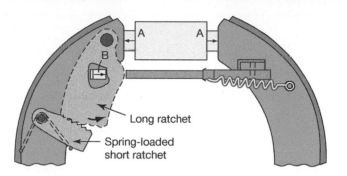

Long ratchet

Spring-loaded short ratchet

As the friction material (brake shoes and pads) wears down, some form of adjustment is required to keep a minimum running clearance between the braking surfaces. Why is this necessary?

_____

This is normally done automatically on modern vehicles. Describe the action of the automatic adjuster shown above, by adding the key words from the list below, to the appropriate gaps in the paragraph.

When the _____ brake is applied, the _____ are pushed _____ at A. If the travel is too _____, the interconnecting _____ will _____ the long _____ at B over the _____ on the short ratchet, so that the shoes will be held further _____ when the footbrake is _____.

apart   great   strut   shoes   out   released   foot   pull   teeth   ratchet

TIP   Some automatic adjusters use adjusting screws. Keep the parts from each side of the vehicle separated as the threads are 'handed' and will not work if put back on the wrong side.

DISC BRAKE ADJUSTMENT

In most disc brake arrangements it is the rubber seal around the piston that maintains the correct clearance between pad and disc. It therefore serves three purposes: as a fluid seal, a retraction spring and an automatic brake adjuster.

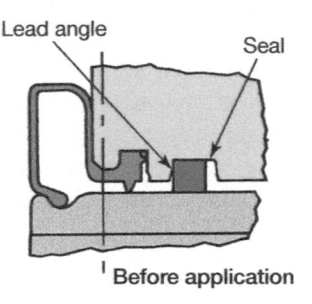
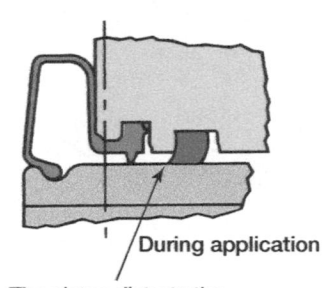
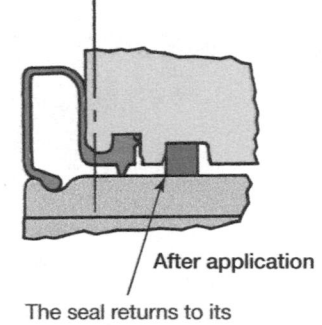

Before application | During application | After application

The piston distorts the seal upon application, it only then moves slightly through the seal, as pad wear takes place

The seal returns to its original shape pulling the pistion back with it, maintaining the correct pad to disc clearance

MASTER CYLINDERS

The master cylinder is a hydraulic cylinder (pump), with a fluid reservoir which is operated by the brake pedal lever and push rod. Irrespective of the type of master cylinder used there must be provision for:

1 Adding additional fluid to the system (fluid compensation) as the pads and shoes wear.

2 Isolation of the fluid reservoir from the main cylinder pressure chamber to prevent pressurized fluid from entering the reservoir when the brakes are applied.

Owing to the use of 'split' braking systems, twin piston (tandem) master cylinders are found on modern vehicles. Why is this?

_____

_____

Tandem master cylinder

The basic components of a tandem master cylinder, as used in 'split line' braking systems, are shown below. Study the diagram.

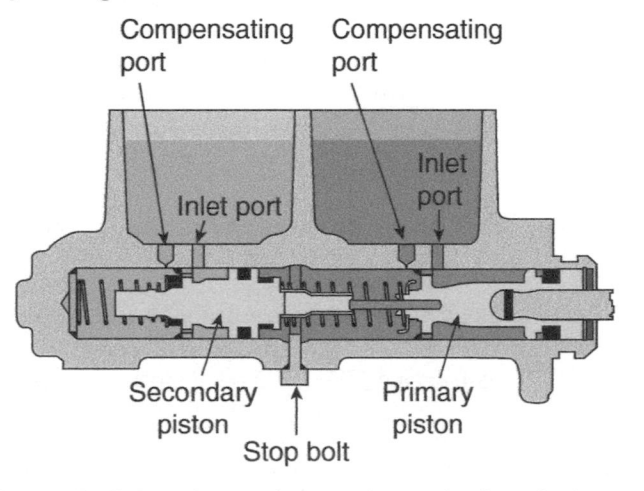

Assuming a front/rear split, if the primary piston activates the front brakes and the secondary piston activates the rear, describe briefly how the system would work if there was fluid loss in:

The front brake system – _____

_____

The rear brake system – _____

_____

What role do the 'compensating' or 'bypass ports' (P) play in the operation of the master cylinder?

_____

BRAKE VACUUM SERVO UNITS

The function of a brake servo unit is to boost the driver's pedal effort, thereby keeping the leverage required (and hence pedal travel) to a minimum.

The necessary pedal assistance is obtained by creating a 'pressure difference' across a large-diameter diaphragm or piston. When applied to a hydraulic piston, the force increases the brake line pressure.

To obtain an air pressure difference, a vacuum is created within the servo. How is this achieved in vehicles fitted with a spark ignition engine?

_____

How is it done with a compression ignition (diesel) engine?

_____

The two diagrams below show the basic principle of operation of a 'direct acting servo'.

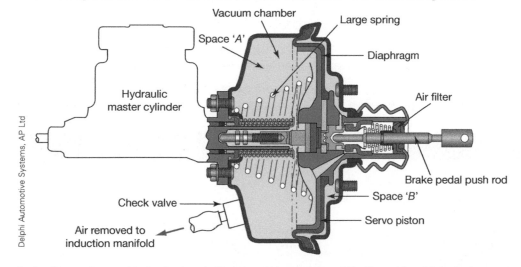

In the figure above with the brakes 'off', space 'A' and space 'B' either side of the diaphragm are subjected to a vacuum. Pressure is equal and there is no movement of the servo piston or the master cylinder.

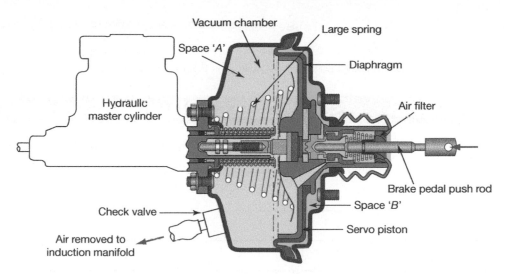

In the figure above the brake pedal is applied and air rushes in through the air filter to space 'B', which is sealed from space 'A' (still under vacuum). The difference in pressure assists the driver's effort on the push rod. The servo piston moves forward and the increased force is felt at the master cylinder, operating the brake.

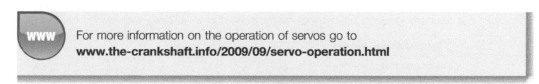

WWW For more information on the operation of servos go to
www.the-crankshaft.info/2009/09/servo-operation.html

INDICATING DEVICES

A number of warning devices are used in the braking system. Some of these are shown in the figure below. Complete the wiring circuits for the diagrams.

Stop lamp actuation

Pad/brake shoe wear indication

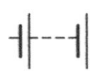

Fluid level

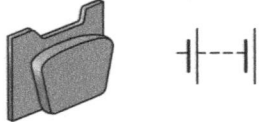

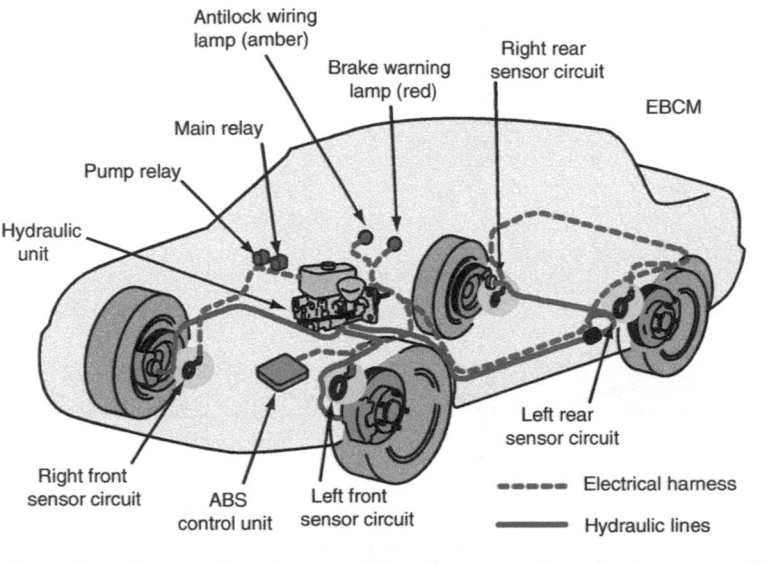

BRAKE TESTING

A brake test is carried out on a vehicle to determine the 'brake efficiency'. The efficiency is an indication of the ability of the braking system to stop the vehicle.

The brake efficiency of a vehicle is usually determined by putting the vehicle on a set of brake rollers and measuring the braking force on each wheel.

TIP A 'Tapley' meter may be used to measure vehicle retardation when the vehicle is driven on a road test. This may still be used for vehicles with transmission handbrakes.

For more information on brake testing go to **www.ukmot.com**

ANTI-LOCK BRAKING SYSTEMS (ABS)

To achieve maximum retardation and maintain steering control during braking, the road wheels must be on the point of locking or skidding. Anti-lock braking systems (ABS) control the braking torque at the road wheels to the maximum permissible without locking-up the wheels. The brake is automatically released (momentarily) to prevent a lock-up occurring.

The system shown has three main sub-assemblies, these are the wheel sensors, the hydraulic unit or modulator and the ABS control unit (ECU).

Describe the function of the following:

Wheel sensors – _____

_____

Modulator – _____

_____

WWW For more on anti-lock brakes go to **www.absbrakes.co.uk**

TRACTION CONTROL SYSTEM (TCS)

Controlling wheel spin during a standing start or when cornering allows maximum usage of the traction available to the tyres.

The TCS shown on page 171 is a development from ABS.

In the figure below wheel spin is controlled by a combination of brake and throttle control. The two key components in the system are a combined ECU and hydraulic modulator and a throttle control intervention motor. Indicate and name the major parts on the drawing.

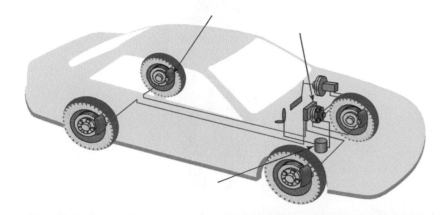

BRAKING SYSTEM FAULTS

Drum brakes

Whenever drums are removed they should be checked for wear, scoring, ovality, cracks and distortion. Linings should be checked for wear, security and contamination. Any oil or brake fluid leaks must be rectified. Also check for seized wheel cylinders and adjusters.

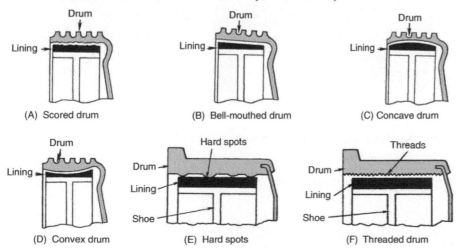

(A) Scored drum

(B) Bell-mouthed drum

(C) Concave drum

(D) Convex drum

(E) Hard spots

(F) Threaded drum

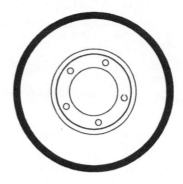

Show on the diagram above where a drum should be measured to check for ovality.

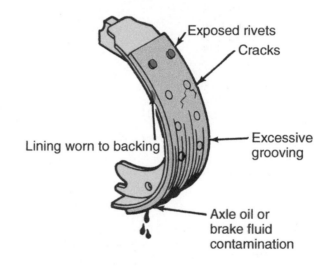

Exposed rivets

Cracks

Lining worn to backing

Excessive grooving

Axle oil or brake fluid contamination

TIP Hydraulic system parts should not be allowed to come into contact with oil or grease. They should not be handled with greasy hands.

Disc brakes

When servicing disc brakes, check the discs for rust, thickness, wear, cracks, blue spots, parallelism and run out.

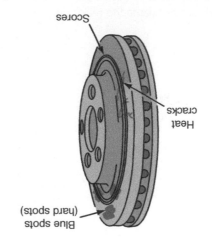

Pads should be checked for wear and seizure in the caliper. Calipers should be checked for leaks and seizure (if applicable).

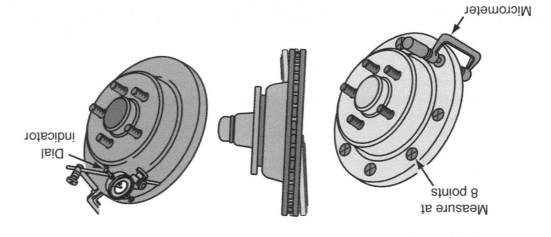

Multiple choice questions

Choose the correct answer from a), b) or c) and place a tick [✓] after your answer.

1 The number of hydraulic pistons used in a conventional sliding caliper disc brake is:

a) Three [ ]
b) Two [ ]
c) One. [ ]

2 'Vapour lock', in a hydraulic braking system, is caused by:

a) a seized master cylinder [ ]
b) boiling brake fluid [ ]
c) freezing brake fluid. [ ]

3 Hydraulically operated braking systems are self-compensating because:

a) the lining wear does not affect pedal travel [ ]
b) the shoes are automatically adjusted [ ]
c) the fluid pressure is equal throughout the system. [ ]

4 The leading shoe in a drum brake is the shoe which:

a) has a 'self-servo' action [ ]
b) has the lowest rate of wear [ ]
c) is fitted at the front only. [ ]

5 Brake fade is caused by the:

a) contraction of brake linings due to overheating [ ]
b) expansion of brake drum due to overheating [ ]
c) reduction of friction due to overheating. [ ]

6 It is recommended that brake fluid is changed periodically because:

a) it has the ability to absorb moisture [ ]
b) it becomes dirty and less compressible [ ]
c) ages and therefore leaks more readily. [ ]

7 If air becomes trapped in a hydraulic brake circuit, the result at the pedal would be:

a) a hard feel to the pedal [ ]
b) increased, 'spongy' travel [ ]
c) no movement whatsoever. [ ]

8 Under emergency braking, the brake pedal on a vehicle fitted with ABS pulses. This indicates that:

a) the system is functioning correctly [ ]
b) the system is not working at all [ ]
c) there is air in the hydraulic system. [ ]

SECTION 4

Suspension systems

USE THIS SPACE FOR LEARNER NOTES

_____

_____

_____

_____

_____

_____

_____

Learning objectives

After studying this section you should be able to:

- Identify and describe the construction and operation of suspension systems and components.
- Compare suspension system components and assemblies against alternatives to identify differences in construction and operation.
- State common terms used in suspension system design.

Key terms

IFS Independent front suspension.
IRS Independent rear suspension.
Live axle Driven axle.
Dead axle Non-driven axle (usually on the rear).
Beam axle Rigid axle.
Shock absorber A suspension device designed to absorb road shocks, i.e. a spring.
Damper A device which dampens the oscillations or vibrations of the road springs.
Bump Upward movement of suspension.
Rebound Downward movement of suspension.
Pitch Forward and backward rocking motion of the vehicle.
Roll Sideways sway or 'heel-over' of a vehicle on corners.
Panhard rod A rod mounted between the body or chassis and the axle, to control the lateral (sideways) movement of the axle.

Remember – when working with suspension, you may be dealing with:

- Jacking and supporting vehicles.
- High fluid or air pressures.
- Loaded components under tension or compression.
- Waste disposal.

SUSPENSION

One purpose of the suspension system on a vehicle is to minimize the effect of road surface irregularities on passengers, vehicle and load.

State three other reasons for the suspension system on a vehicle:

1 _____

2 _____

3 _____

Many different forms of suspension are used on road vehicles. Some are very simple and relatively inexpensive and others are highly sophisticated and expensive. Suspension systems fall into one of two main categories: independent and non-independent systems.

Describe with the aid of simple diagrams the essential difference between the two categories:

Leaf spring suspension

The leaf spring may still be found on some older car rear suspensions and light commercial vehicles. Used in conjunction with a beam axle the leaf spring is a simple non-independent

suspension system. The length, width, thickness and number of leaves vary according to the load requirement. An advantage of the leaf spring is that it can provide total axle location in addition to its springing properties, as well as carrying heavy loads.

Label the multi-leaf spring suspension below:

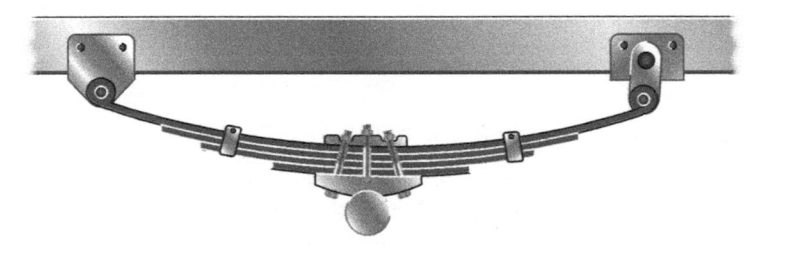

Complete the following description by adding the correct words from the list below.

flexes   shackle   'U'   slipper   centre bolt

The leaf spring is positively located on the axle by a _____ and secured to the axle by _____ bolts . As the spring _____, the distance between the spring eyes, that is, the length of the spring, varies. This is accommodated by the swinging _____ or, in some cases, an open-ended spring operates in a _____ bracket.

In relation to suspension, state the meaning of:

Sprung weight – _____

Unsprung weight – _____

_____

SUSPENSION DAMPERS

Contrary to popular belief, the function of the suspension damper is not to increase the resistance to spring deflection but to control the oscillation (bounce) of the spring. The basic principle of hydraulic dampers is that of converting the energy of the deflecting spring into heat. This is achieved quite simply by pumping oil through a small hole.

The figure below shows the oscillations of an 'undamped' spring as it dissipates energy following an initial deflection.

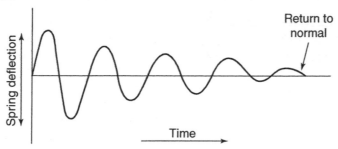

Draw a graph below to illustrate the effect of a damper on the oscillation of a spring:

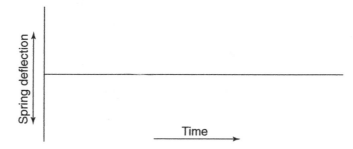

Direct-acting hydraulic damper

The telescopic type, as this type is known, is located directly between the frame and suspension unit or axle, thus eliminating the need for levers or links. The principle of operation is the transfer of fluid through small holes from one side of the piston to the other. Complete the sketch opposite by adding a sectioned view of the interior of the damper. Label the damper shown.

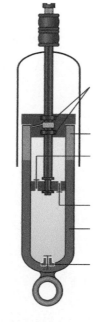

Add the bump and rebound valves to the enlarged valve assembly shown above.

In groups, examine a direct-acting hydraulic damper. Discuss its operation and make brief notes on the purpose of the fluid reservoir.

TIP Dampers are normally replaced in pairs. Before fitting to the vehicle, they should be primed by holding them upright and operating the damper through its full movement up and down.

Gas charged dampers

With ordinary dampers under really arduous conditions, such as when travelling quickly over very rough roads, the oil tends to foam as it passes rapidly from chamber to chamber and mixes with air. This reduces the damping effect. Gas pressurized dampers overcome this problem. Although they operate in a similar way to conventional oil-filled dampers, Nitrogen is used to keep the oil pressurized, to minimize oil foaming and increase efficiency. The damper shown below is of the 'monoshock' design. There is no reservoir for the fluid to move back and forth into. The valves are in the main piston.

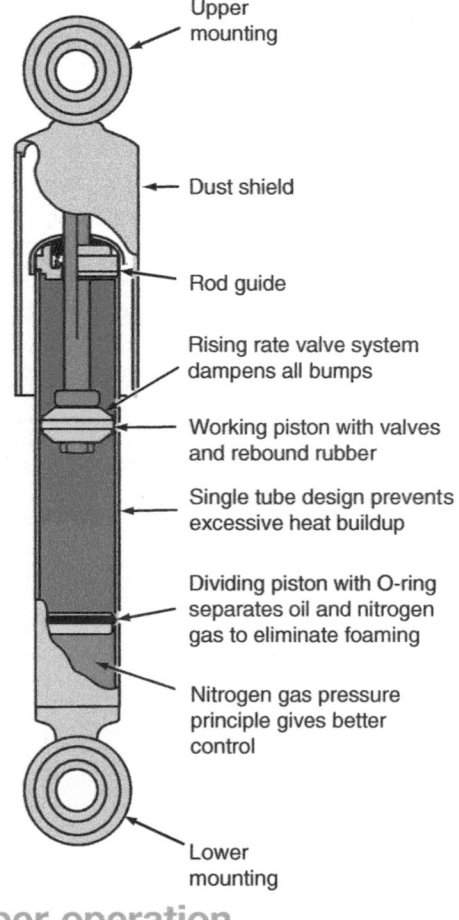

Upper mounting

Dust shield

Rod guide

Rising rate valve system dampens all bumps

Working piston with valves and rebound rubber

Single tube design prevents excessive heat buildup

Dividing piston with O-ring separates oil and nitrogen gas to eliminate foaming

Nitrogen gas pressure principle gives better control

Lower mounting

Checking damper operation

Bounce test

This involves pushing down on the car body at each corner and noting the oscillations of the vehicle before it again becomes stationary in its normal level position. The 'rule of thumb' guidance (Munroe) is that a vehicle taking more than one and a half oscillations has ineffective damping of the spring.

Other methods involve the use of equipment which will produce a graphical record of vehicle oscillation.

COIL SPRING SUSPENSION

Fill in the missing words in the following paragraph:

The coil spring is used on many car front and rear suspension systems. When compared with the leaf spring the coil spring is _____ in weight, provides a ride which is _____ and wheel deflection can be _____.

TIP Some suspension ball joints are subjected to direct loading as a result of vehicle weight. It is important to realize this when checking a suspension system and remove the load when checking for wear.

Show the position of the coil spring. on the three independent front suspension arrangements below. Add arrows to the drawings to show the vehicle weight load path through the suspension and identify the loaded and unloaded ball joints.

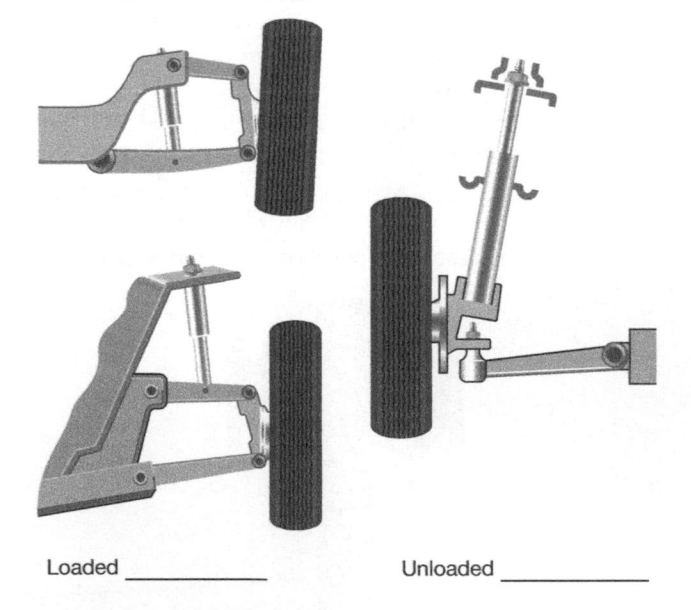

Loaded _____   Unloaded _____

Name the popular type of suspension shown below. _____

State the purpose of the convoluted rubber cover inside the spring.

_____

It can be seen that the spring and its seating are angled relative to the strut. State the reason for this arrangement.

_____

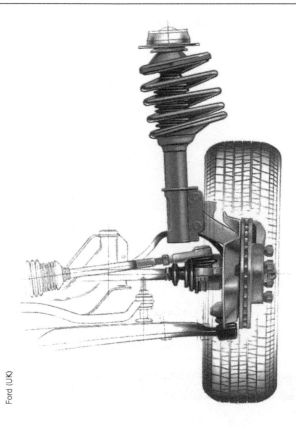

Ford (UK)

RUBBER SPRINGS

The version of rubber suspension shown is one which was used extensively on a popular small car. The use of rubber in suspension systems on modern vehicles tends to be in the form of 'bump stops' which stiffen the suspension spring at maximum deflection. One of the advantages of rubber springs is that, for small wheel movements, the ride is soft and becomes harder as wheel deflection increases. They are also small and light, and tend to give out less energy in rebound than they receive in deflection.

The deflection of the rubber spring in the suspension shown below is small when compared with the actual rise and fall of the wheel. State how this is achieved.

_____

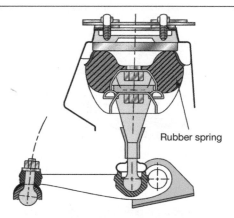

Rubber spring

AIR SUSPENSION

Air springs usually consist of reinforced rubber bags situated between the axle and chassis. Air suspension is a truly progressive suspension providing a soft, cushioned ride. It tends to be used on some larger luxury vehicles and can be controlled electronically. An important feature is the use of levelling valves in the system, which maintains a constant vehicle ride height irrespective of vehicle loading. The ride height can also be adjusted.

HYDRO-PNEUMATIC SUSPENSION

As with air suspension, the spring is pneumatic. The main difference is that a fixed quantity of gas (nitrogen) is contained in a variable-sized chamber; liquid is used to transmit the force from the suspension link to the gas (via an intervening diaphragm).

Complete the drawing below to illustrate the principle of operation of a hydro-pneumatic suspension system:

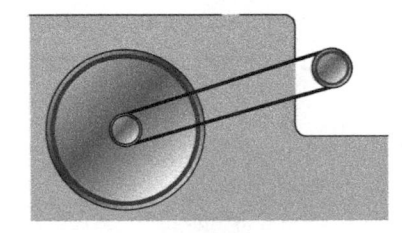

One system of hydro-pneumatic suspension incorporates a levelling valve and a hydraulic accumulator. In this system the vehicle maintains a constant ground clearance irrespective of load. The driver can also adjust the vehicle height or ground clearance to any one of three positions.

Give one example of a vehicle that uses this form of suspension: _____

The simple layout shows a suspension system with a levelling valve and an accumulator.

Add arrows to the hydraulic circuit to indicate the direction of fluid flow. Label the drawing and describe briefly how the system operates:

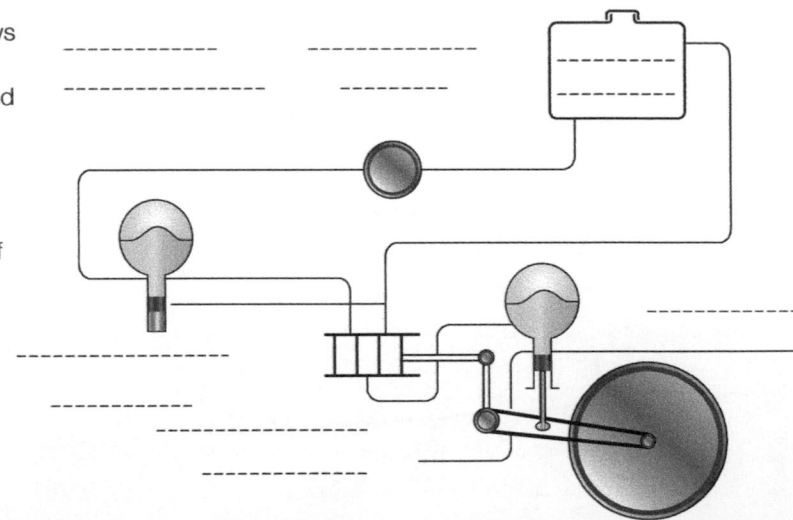

- _____

- _____

- _____

_____

BUMP AND REBOUND STOPS

On the drawing below show a rubber bump stop and a strap to limit rebound in a suspension system.

State the purpose of these fitments.

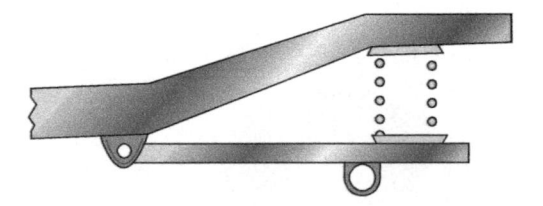

_____

_____

_____

TRIM HEIGHT

What is meant by 'trim height' (or standing height) and how is it measured?

_____

_____

_____

INFLUENCES OF LOAD VARIATIONS ON WHEEL GEOMETRY AND ALIGNMENT

The wheel alignment and suspension geometry can vary as the wheels rise and fall to follow the road contour, and can be affected by load. The degree to which alignment and geometry do alter depends largely on the actual suspension design and the amount of suspension movement involved.

REAR WHEEL 'SUSPENSION STEER'

The semi-trailing link IRS (independent rear suspension) shown below produces a steering effect when the vehicle is cornering. Describe this:

_____

_____

_____

AXLE LOCATION

Coil, torsion bar, rubber and air springs serve in general merely to support and control the vertical load and do not locate the axle or wheel assembly in any way.

Additional arms or links must be provided to locate the wheels both fore and aft and laterally. These also resist the forces due to drive and brake torque reaction.

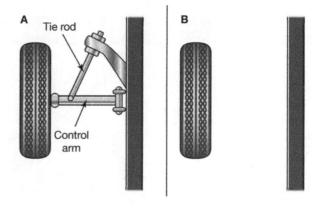

Axle location (independent front suspension)

The tie rod shown at A, above, locates the wheel assembly to prevent rearward movement of the wheel when the brakes are applied, when the vehicle accelerates or when the wheel strikes a bump.

Sketch, at B, an alternative method of providing this location. On some suspension systems the anti-roll bar serves also as a tie rod.

The torsion beam is now widely used on the rear of front wheel drive cars.

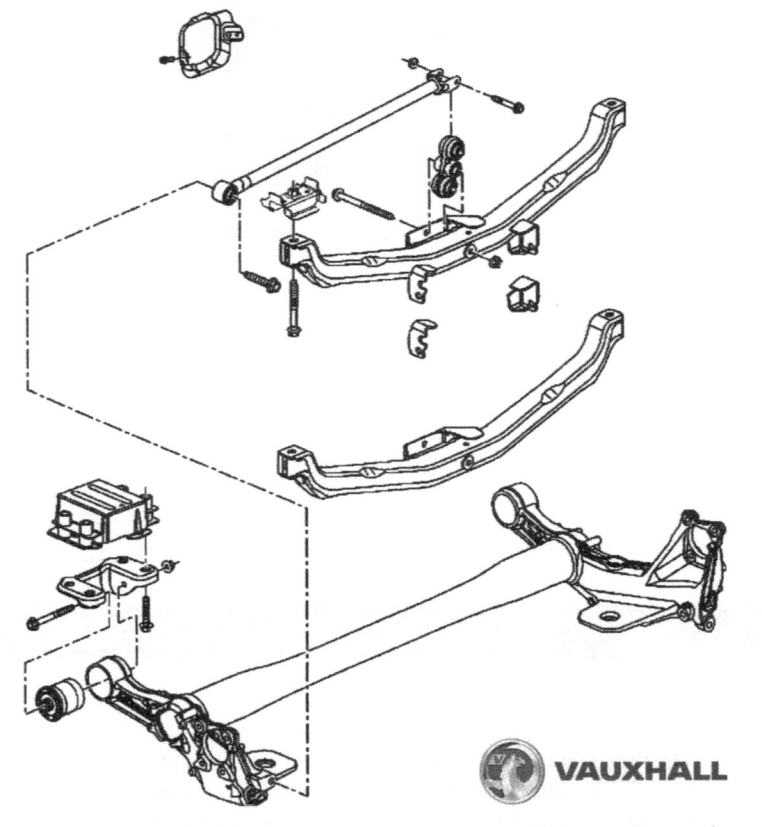

VAUXHALL

Complete the following description about torsion beam suspension using the correct words provided in the list below.

beam	bending	location	high	anti-roll
constant	section	tilt	twist	independently

Because the _____ is V or H _____, it will _____ relatively easily and yet have a _____ resistance to _____. It is used to provide good wheel _____ with stiff _____ properties. The wheels can move up and down _____ without the _____ associated with a beam axle and _____ wheel alignment is maintained.

The suspension sub-frame assembly is attached to the chassis/body through bonded rubber mountings.

State why a sub-frame is considered desirable.

_____

_____

Name the rear suspension system shown in the figure below. _____

State the purpose of component 'B'. _____

When replacing the damper with this layout, would it be necessary to use a spring compressor, and if not, why not?

_____

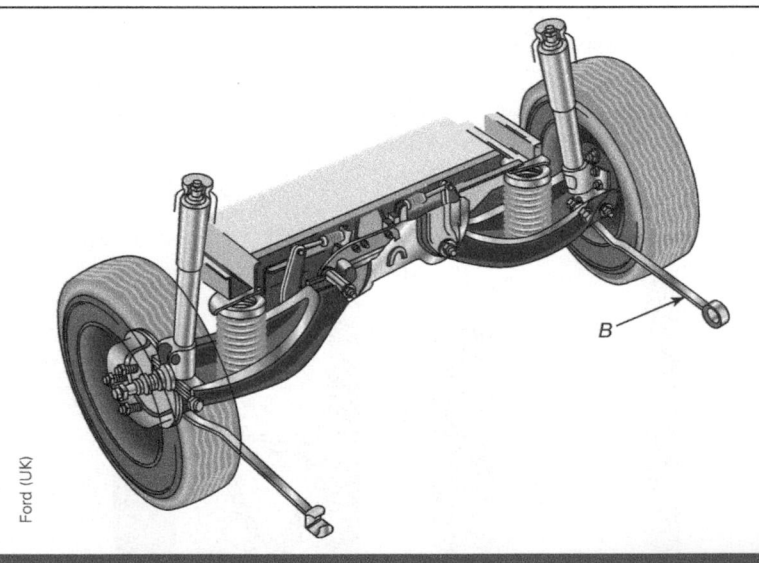

Ford (UK)

DE-DION AXLE

The De-Dion axle is a compromise between the live beam axle and independent suspension. The final drive and differential are mounted on the chassis and the road wheel assemblies are carried on a beam type dead axle. Coil springs are used and the axle is located by trailing links and a Panhard rod. In some systems a Watts linkage provides lateral location rather than a Panhard rod.

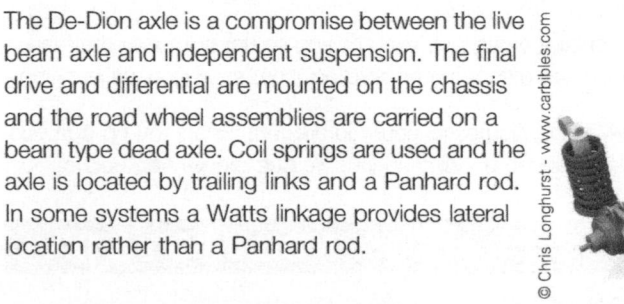

© Chris Longhurst - www.carbibles.com

ADVANCED ADAPTIVE (ACTIVE) SUSPENSION CONTROL

In an ideal suspension system, the spring and damper rate would be automatically adjusted during vehicle operation to ensure that the ride and handling standards are optimum under all operating conditions.

Some modern systems use sensors which can detect a number of variables under all driving conditions; this information is fed to an ECU. The ECU controls double-acting hydraulic cylinders mounted at each wheel which can react very quickly and independently to changing conditions. These act as both damper and spring. Some systems are pneumatic using a number of different configurations.

 Find three models of vehicle which are fitted with active suspension. Discuss these within the whole group.

The figure below shows a vehicle fitted with an active suspension system.

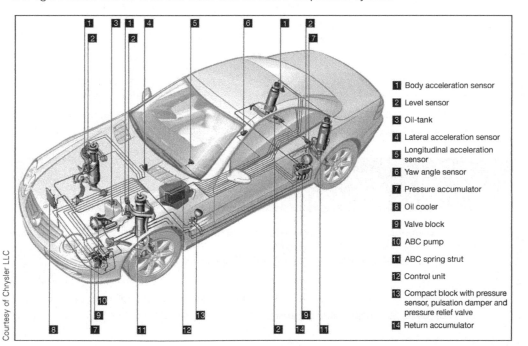

1 Body acceleration sensor
2 Level sensor
3 Oil-tank
4 Lateral acceleration sensor
5 Longitudinal acceleration sensor
6 Yaw angle sensor
7 Pressure accumulator
8 Oil cooler
9 Valve block
10 ABC pump
11 ABC spring strut
12 Control unit
13 Compact block with pressure sensor, pulsation damper and pressure relief valve
14 Return accumulator

Courtesy of Chrysler LLC

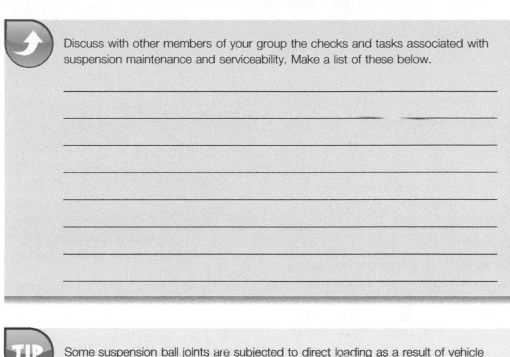 Discuss with other members of your group the checks and tasks associated with suspension maintenance and serviceability. Make a list of these below.

_____

_____

_____

_____

_____

_____

_____

TIP Some suspension ball joints are subjected to direct loading as a result of vehicle weight. It is important to realize this when checking a suspension system and remove the load when checking for wear.

CHECKING SUSPENSION ALIGNMENT AND GEOMETRY

Describe briefly the suspension alignment and geometry checks normally carried out on the rear suspension.

_____

_____

_____

There are many makes of four-wheel alignment equipment which often use lasers to show alignment and make accurate measurements.

WWW For more information on alignment go to **www.alignmycar.co.uk**

DIAGNOSTICS: SUSPENSION SYSTEM – SYMPTOMS, FAULTS AND CAUSES

State a likely cause for each symptom and system fault listed below. Each cause will suggest any corrective action required. This is not an exhaustive list of faults and causes.

Symptoms	Faults	Probable causes
Pulling to one side; vibration; poor roadholding	Abnormal, excessive tyre wear	
Ride height low at one corner; axle misalignment	Broken road spring	
Axle misalignment; crabbing	Loose 'U' bolts	
Poor road holding	Worn dampers	
Incorrect trim height	Corroded spring mounting; 'tired' springs	
Steering pull; noise	Worn radius arm brushes	
Abnormal tyre wear; heavy steering	Worn ball joints	
Vehicle instability under full load	Missing bump stop	

Multiple choice questions

Choose the correct answer from a), b) or c) and place a tick [✓] after your answer.

1 When the wheel of the vehicle is moving upwards towards the body it is said to be on:
a) bump [ ]
b) rebound [ ]
c) pitch. [ ]

2 The link which controls lateral location of an axle is called a:
a) torsion bar [ ]
b) anti-tramp bar [ ]
c) panhard rod. [ ]

3 The type of spring used on a MacPherson strut suspension is a:
a) leaf [ ]
b) torsion [ ]
c) coil. [ ]

4 One function of the vehicle's suspension system is to:
a) keep the road wheels in contact with the road's surface [ ]
b) allow the body to move up and down [ ]
c) make the body follow the road contour. [ ]

5 Which part of the vehicle's suspension absorbs road shocks?
a) sub-frame mounting [ ]
b) damper [ ]
c) spring. [ ]

6 A 'transverse link' system is a type of:
a) independent suspension [ ]
b) non-independent suspension [ ]
c) beam axle arrangement. [ ]

7 One advantage of a coil spring over a leaf spring, for the rear suspension of a vehicle, is:
a) it can transmit driving thrust [ ]
b) it is lighter in weight [ ]
c) it dampens its own oscillations. [ ]

8 A rear axle assembly may be located in position by:
a) coil springs [ ]
b) radius arms [ ]
c) hydraulic dampers. [ ]

SECTION 5

Steering systems

USE THIS SPACE FOR LEARNER NOTES

Learning objectives

After studying this section you should be able to:

● Identify light vehicle steering system components.

● Identify and describe the construction and operation of vehicle steering systems and their components.

● Compare key components and assemblies against alternatives to identify differences in construction and operation.

● State common terms used in steering system design.

Key terms

Rack Toothed bar.
Pinion Gear wheel which engages with a rack.
Tie rod Connecting rod or bar, usually under tension.
Track rod Bar connecting the steering arms.
Drop arm Connects steering box to drag link of steering system.
Idler arm Similar to the drop arm but having a guiding function only.
Steering box Changes rotary movement into linear movement.
Drag link Connects drop arm to first steering arm.
Ackerman linkage Form of steering arranged to give true rolling motion round corners.
Toe in or toe out Inward or outward inclination of the leading edge of the front wheels.

Remember – when working with steering, you may be dealing with:

● Jacking and supporting vehicles.

● High fluid pressures.

● Loaded components under tension or compression.

● Rotating components.

● Airbags.

● Waste disposal.

http://www.aa1car.com/index_alphabetical.htm

www.carbibles.com/steering_bible.html

http://auto.howstuffworks.com/steering.htm

STEERING SYSTEM LAYOUTS

The steering system provides a means of changing or maintaining the direction of a vehicle in a controlled manner. List three other requirements of the system:

1 _____

2 _____

3 _____

Beam axle, single track rod

This system may be found on light commercial vehicles which use a beam front axle. Add the steering linkage to the layout below.

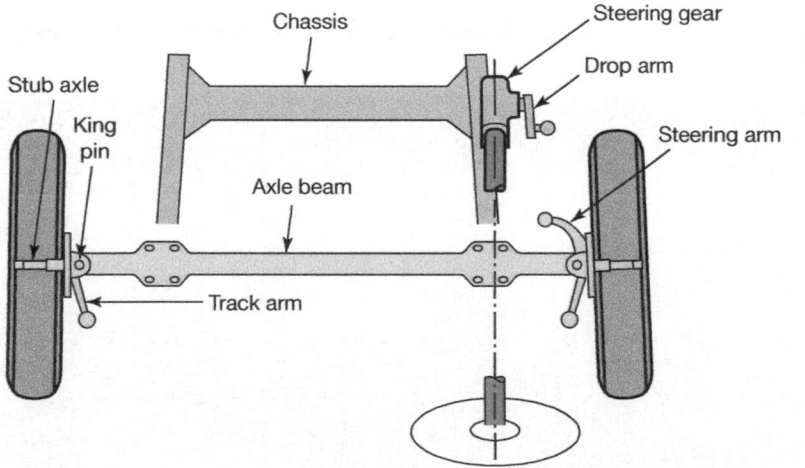

With this system, as the axle tilts owing to suspension movement, the single track rod moves correspondingly and the steering is unaffected. It is unacceptable to employ a single track rod linkage with independent front suspension (IFS). Why is this?

_____

_____

Divided track rod car steering system

Virtually all cars and many light goods vehicles have independent front suspension. The steering linkage for these must therefore be designed to accommodate the up and down movement of either steered wheel without affecting the other steered wheel. In most systems two short track rods which pivot in a similar arc to the suspension links are connected to the stub axles.

One such system incorporates a steering gearbox with 'idler' and three track rod layout. Complete the drawing of this type of system by adding the linkage. Label the parts using the terms listed below.

| idler | steering box | drop arm | steering arm |

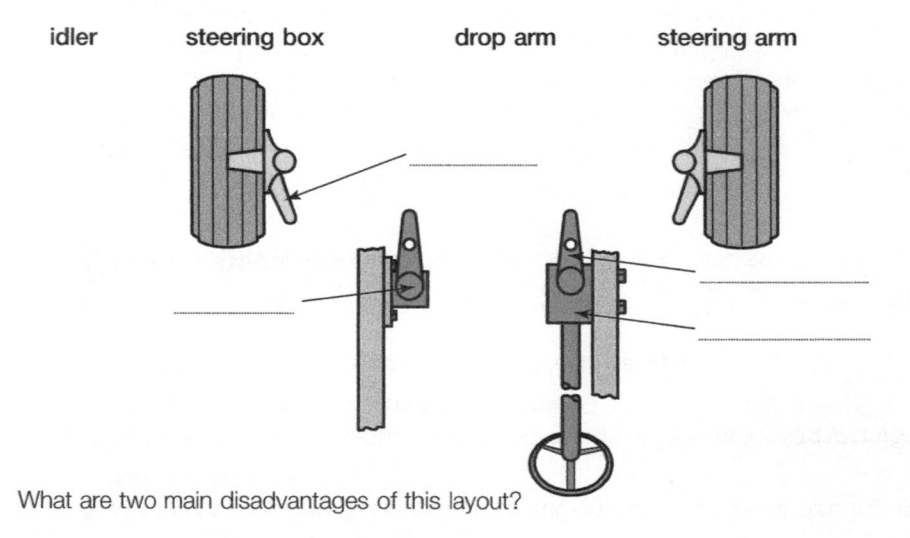

What are two main disadvantages of this layout?

1 _____

2 _____

These steering systems use a steering gearbox. A popular type is known as the re-circulating ball box and can be manual or power assisted.
The figure opposite shows a section through this type of box. A line of individual ball bearings are used to create a 'thread' inside the ball nut which runs around the worm shaft. The balls run in their own circuit via the internal thread and external guide tubes. This makes for a relatively friction free operation.

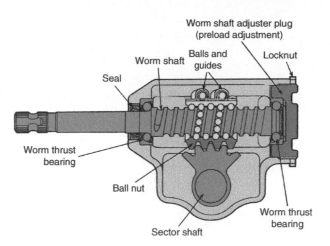

Worm shaft adjuster plug (preload adjustment)
Balls and guides
Locknut
Worm shaft
Seal
Worm thrust bearing
Ball nut
Sector shaft
Worm thrust bearing

The figure below shows an exploded view of a rack and pinion system.

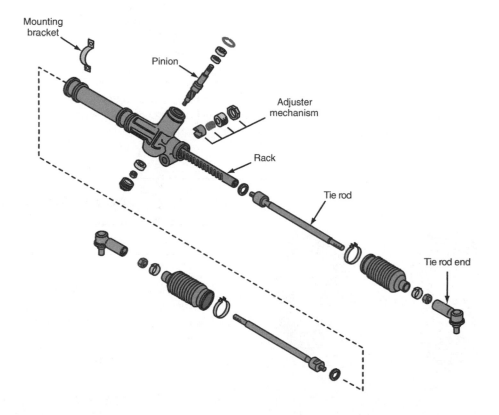

Mounting bracket
Pinion
Adjuster mechanism
Rack
Tie rod
Tie rod end

Complete the following paragraph on the system by filling in the correct missing words from the list below.

rack	pinion	wheel	turned
meshed	sideways	swivel	steering
tie rods	worm	base	suspension

The _____ is a toothed bar contained in a metal housing. Its _____ movement is transmitted to the _____, to change _____ direction. The _____ is a toothed or _____ gear mounted at the _____ of the _____ column, where it is moved by the steering wheel. The pinion is _____ with the teeth in the rack, which is moved as the steering wheel is _____. The tie rods _____ in a similar arc to the _____.

Name the numbered components on the rack and pinion assembly shown below:

1

5

2

4

3

1 _____

2 _____

3 _____

4 _____

5 _____

One slight drawback of a rack and pinion steering gear is its reversibility. Explain what is meant by this.

_____

_____

Linkage ball joints

The joints shown use a ball-pin and spring-loaded bearing socket. State the purpose of the spring and label the drawings.

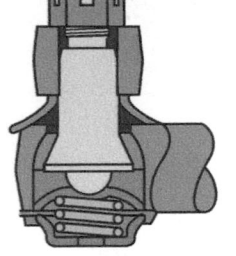
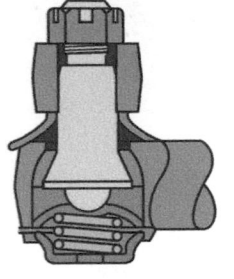

_____

_____

State the purpose of part A.

_____

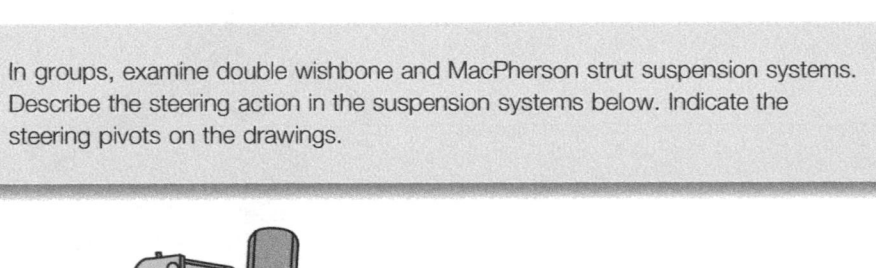

In groups, examine double wishbone and MacPherson strut suspension systems. Describe the steering action in the suspension systems below. Indicate the steering pivots on the drawings.

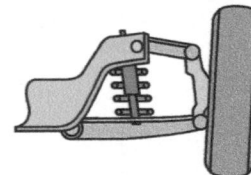

Double wishbone

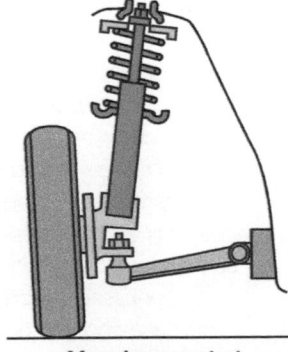

Macpherson strut

FRONT HUB

Examine a wheel hub which is not driven and complete the drawing below to include bearings and grease seal. Label all parts.

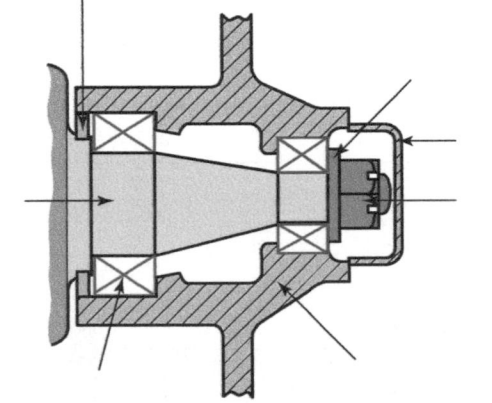

Which type of bearings are used?

_____

A popular type of grease seal used in front hubs is the spring-loaded rubber seal.

State three reasons why grease can leak past a serviceable seal:

1 _____

2 _____

3 _____

True rolling motion

When a vehicle travels on a curved path during cornering, true rolling is obtained only when the wheels roll on arcs which have a 'common centre' or common axis. Show the common centre on the drawing below:

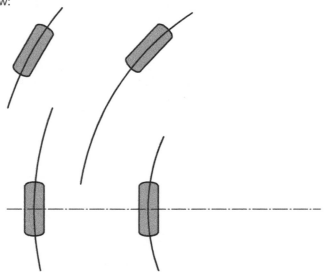

State why true rolling of the wheel is necessary.

_____

_____

ACKERMAN SYSTEM

Complete the following description by using the correct words selected from the list below.

front	arms	outwards	cornering	inwards
distance	longer	steering	shorter	
swivel	centres	rolling	track	

The difference in _____ lock angles, to give true _____ of the wheels while _____, can be achieved by making the track rod _____ than the _____ between the _____-pin centres when the _____ rod is behind the axle. That is, the steering _____ are usually inclined _____. If the track rod is in _____ of the axle, as on many cars, then it is _____ than the distance between the swivel-pin _____ and the steering arms must be inclined _____.

Add a track rod and steering arms to the drawing below and extend lines through the steering arms to show the point of intersection on the vehicle centre line if the linkage is to satisfy conditions for true rolling.

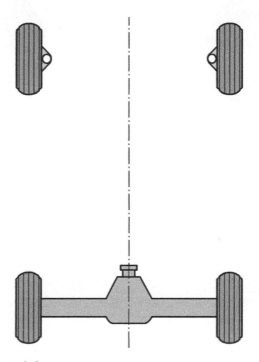

This type of steering linkage is known as: _____

The figure below shows the position taken up by the track rod when the wheels are on lock. Indicate on the drawing a typical angle for the inner wheel for the angle of lock shown on the outer (left-hand) wheel.

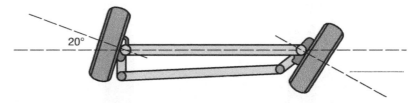

The difference in steering angles shown above is known as: _____

FRONT WHEEL ALIGNMENT

The wheels shown below are _____, that is, dimension A is slightly less than dimension B. When dimension B is slightly less than dimension A, the wheels would be said to have _____.

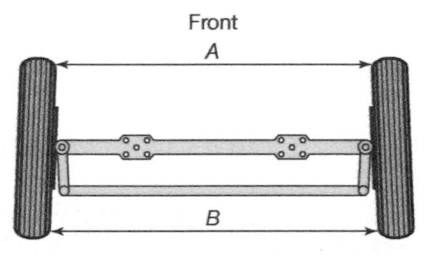

Front
A

B

State the purpose of toe in or toe out:

_____

_____

_____

State the main effects of incorrect wheel alignment.

_____

_____

TIP On independent front suspension systems, equal adjustments should be made to the two outer rods in order to maintain the correct toe out on turns and to keep the correct steering wheel position.

Toe out on turns

The inner steered wheel turns through a greater angle than the outer wheel when cornering (Ackerman) to give an amount of 'toe out on turn'.

Describe with the aid of the figure shown below, a method of checking toe out on turns, and state a typical setting for this.

_____

_____

Lock stops

Lock stops in the steering mechanism limit maximum lock to avoid tyres fouling the chassis or bodywork and prevent extreme travel of the steering gear.

CASTOR ANGLE

One of the desirable features of a steering system is the ability of the road wheels to self-centre after turning a corner. This self-centering effect can be achieved by designing the wheel and steering swivel assembly so that the wheel centre trails behind the swivel axis.

Make a simple sketch on the right to show how the castor is applied to the vehicle and indicate the castor angle.

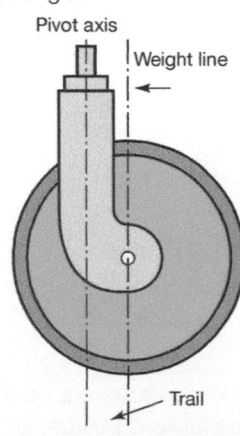

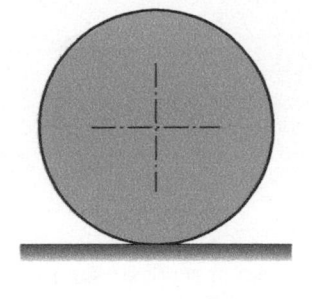

Pivot axis

Weight line

Trail

Principle of castor action

Castor applied to vehicle

Complete the table below for three vehicles, showing the castor angle for each.

Vehicle make	Model	Castor angle

CAMBER ANGLE AND SWIVEL AXIS INCLINATION

Name the angles shown in the diagram and give typical values for each.

Camber angle is the amount that the road wheel is tilted out of the vertical.

Swivel axis inclination is the amount that the swivel axis is inclined from the vertical.

State the purpose of the two angles shown above.

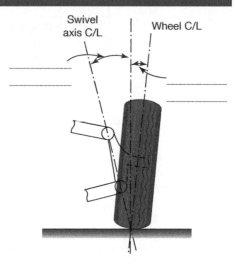

Swivel
axis C/L   Wheel C/L

Camber

_____

_____

Swivel axis inclination

_____

_____

Centre-point steering

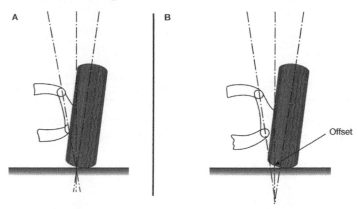

A   B

Offset

The drawing at A illustrates true centre-point steering (the wheel centre line and swivel axis centre line converge at ground level). Many vehicles, however, adopt the arrangement shown at B. This shows the wheel and swivel axis centre lines converging below ground level to give a 'positive offset'; this gives a better steering action.

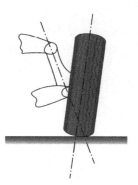

What steering feature is illustrated above?

_____

Give reasons for using this arrangement.

_____

_____

POWER-ASSISTED STEERING (PAS)

Power assistance incorporated into the steering mechanism keeps the steering wheel load to an acceptable level, without increasing the number of turns required from the steering wheel, to turn the steered wheels from lock to lock.

State two advantages of power-assisted steering on the modern vehicle.

1 _____

2 _____

Power assistance can be provided by hydraulic pressure or electrically. It must provide road feel and be positive. It is important to understand that it is only 'assistance' and that should this fail there will still be a functional steering system, albeit requiring more effort from the driver.

The two most popular hydraulic systems used on light vehicles are shown in the figure below. These operate 'integrally' within the steering box or rack. The main components of the system are the hydraulic fluid and reservoir, pump, drive belt, control valve, hoses and piston assembly.

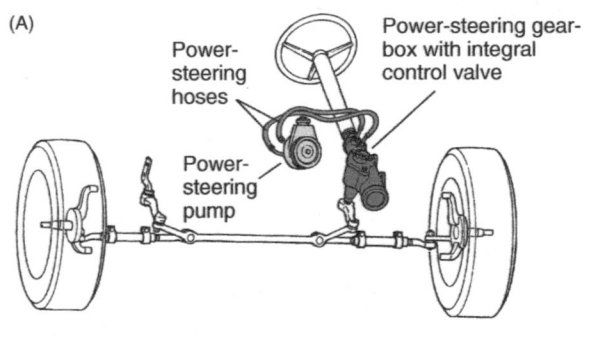

(A)

Power-steering hoses

Power-steering gear-box with integral control valve

Power-steering pump

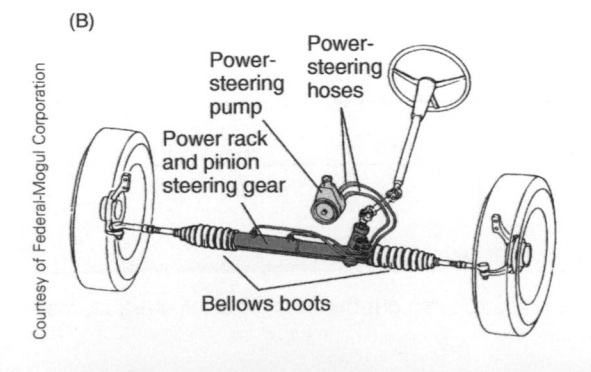

(B)

Power-steering pump

Power-steering hoses

Power rack and pinion steering gear

Bellows boots

Courtesy of Federal-Mogul Corporation

The most popular hydraulic systems used for this purpose are ones which operate on the continuous flow principle.

Integral power assisted steering (PAS) box

A power-assisted recirculating ball-type steering box is shown in the figure below.

Label the simplified arrangement shown below and fill in the missing words using those from the list below to describe briefly how power assistance is achieved.

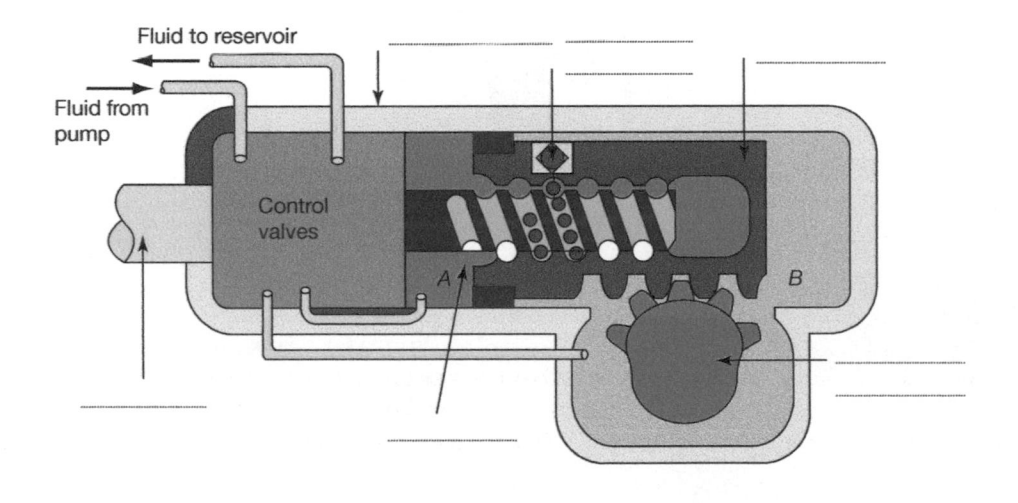

Fluid to reservoir

Fluid from pump

Control valves

A

B

piston   valve   worm   left   nut   recirculating   pressurized   directed

The _____ ball type steering gearbox has a _____ which acts as a hydraulic _____ or ram. When left lock is applied the control _____ directs _____ fluid to the area 'B', assisting the _____ in moving the piston (nut) to the _____. On right lock the pressurized fluid is _____ to area 'A', doing the opposite.

Rack and pinion power assisted steering (PAS)

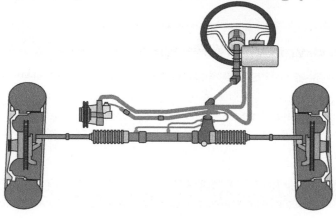

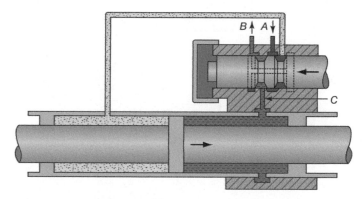

Name the main parts of the system layout shown in the figure above.

Describe the operation of the system shown above.

_____

_____

_____

_____

_____

In addition, a control valve spring or torsion bar provides sensitivity or driver feel. It is also desirable to increase the required driver effort in proportion to the power assistance demanded. This gives a degree of natural feel to the system.

How is this normally achieved?

_____

_____

Electro-hydraulic power assisted steering (PAS)

Some PAS systems use an electric motor to drive the hydraulic pump rather than a 'V' belt from the crankshaft pulley. One benefit of this arrangement is that it provides a relatively simple and compact installation.

State two other advantages of an electric pump in a PAS system.

1 _____

2 _____

Speed sensitivity can be controlled electronically. In the figure below a microprocessor evaluates speed signals from the electronic speedometer and computes the degree of hydraulic power assistance required. The power assistance varies in proportion to road speed. This provides minimum driver effort at the steering wheel when parking but at high speeds the steering action is almost as direct as a mechanical steering gear, providing more precise and responsive steering.

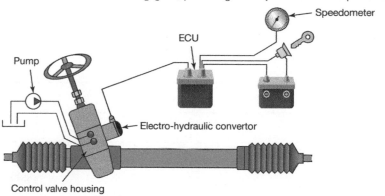

The level of hydraulic reaction transmitted to the hydraulic control valve is varied by the electro-hydraulic converter, which directly affects the actuating force required at the steering wheel.

Electrical power assisted steering (PAS)

In electrical PAS an electric motor is used to assist in the actuation of the steering system, rather than a hydraulic power cylinder. One electrical PAS arrangement is shown in the figure below.

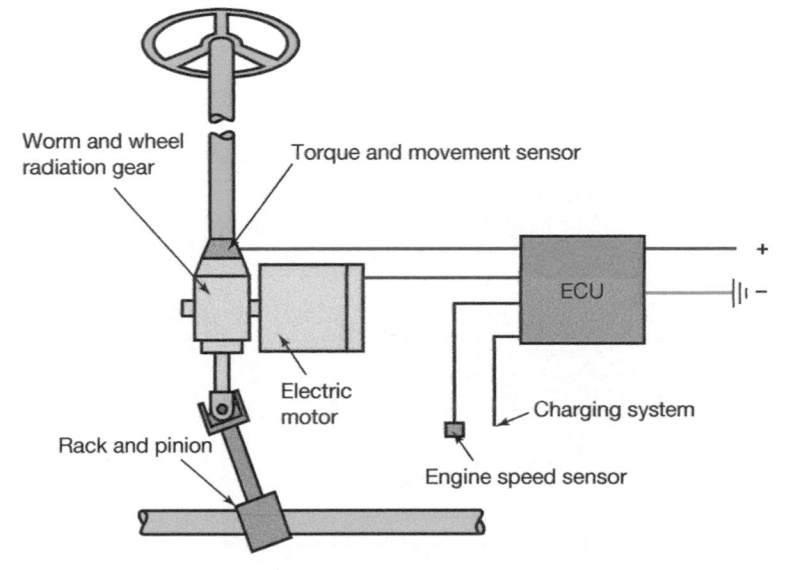

Worm and wheel radiation gear

Torque and movement sensor

ECU

Electric motor

Charging system

Engine speed sensor

Rack and pinion

STEERING SYSTEM MAINTENANCE

In small groups discuss steering system maintenance. Make a list of typical preventive maintenance tasks and checks for the Department for Transport (VOSA-MOT) checks associated with the steering system. The websites below will help.

http://www.ukmot.com/MOT%20test/Steering_Suspension.asp

http://www.motester.co.uk/CarOwnersGuidetoTheMOT/WhatIsTested.aspx

List three of the maintenance tasks and three MOT checks in the spaces below:

Maintenance tasks

1 _____

2 _____

3 _____

MOT checks

1 _____

2 _____

3 _____

TIP Always use the correct hydraulic fluid when topping up PAS reservoirs. Mixing of fluid types can lead to seal failure and leaks.

State the effects of air in a hydraulic PAS system and describe the procedure for bleeding air out of the system.

_____

_____

_____

State the effects of a sticking pressure relief valve in a PAS pump assembly.

_____

_____

The figure below shows two tests being carried out on a PAS system. Label the items indicated.

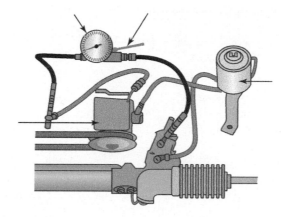

The tests being carried out are:

Pressure balance test. With the test cock open, gently turn the steering to left and right lock. Gauge pressure should be the same (balanced), a typical pressure being approximately 60 bar.

Pump pressure test. With the engine running, temporarily close the test cock (for no more than 5 seconds) and observe the gauge reading. This should be to the manufacturer's specification.

Carry out a check on steering linkage and suspension ball joints. Describe briefly how this is done.

Steering linkage – _____

_____

Suspension ball joints – _____

_____

_____

DIAGNOSTICS: STEERING SYSTEM – SYMPTOMS, FAULTS AND CAUSES

State a likely cause for each symptom/system fault listed below. Each cause will suggest any corrective action required.

Symptoms	Faults	Probable causes
Heavy steering; poor self-centring	Partially seized king pins or ball joints	
Excessive free play (backlash); misalignment of wheels; abnormal tyre wear	Worn track rod end	
Pulling to one side; abnormal tyre wear	Loose tie bar	
Misalignment of steering wheel	Incorrectly fitted drop arm. Incorrectly fitted ball joints	
Loss of power assistance	PAS fluid loss	
Noise as steering is operated	Faulty PAS pump	
Intermittent power assistance and squeal	Slack PAS belt	
Excessive play or backlash at steering wheel; wandering	Excessive rocker shaft end float	

Multiple choice questions

Choose the correct answer from a), b) or c) and place a tick [✓] after your answer.

1 'Toe in' or 'toe out' is a measurement taken of the:
a) wheel alignment [ ]
b) camber angle [ ]
c) vehicle ride height. [ ]

2 One advantage of an optical wheel alignment gauge is that:
a) adjustments can be made to steering geometry from the side of the vehicle [ ]
b) the light beam automatically compensates for any irregularities in the wheel mounting [ ]
c) precise effects of adjustments can be measured whilst making them. [ ]

3 If the front road wheel is set to tilt inwards at the top it has:
a) negative camber [ ]
b) positive castor [ ]
c) positive camber. [ ]

4 The steering angle produced by rearward or forward inclination of the swivel axis is known as:
a) castor angle [ ]
b) swivel axis inclination [ ]
c) camber angle. [ ]

5 What is the name of the steering design which gives true rolling action of the wheels when turning a corner?
a) camber [ ]
b) castor [ ]
c) Ackerman. [ ]

6 When viewed from the front, the upper ball joint of the swivel pin is mounted slightly inboard of the lower ball joint. This is to give:
a) swivel axis inclination [ ]
b) camber angle [ ]
c) castor angle. [ ]

7 On checking the steering system of a vehicle, it is noticed that the wheel can be moved on the stub axle, what could cause this?
a) worn wheel bearing [ ]
b) buckled wheel [ ]
c) broken coil spring. [ ]

8 The steering angle which provides a self-centering force to the road wheels is the:
a) swivel pin inclination angle [ ]
b) castor angle [ ]
c) camber angle [ ]

PART 5

TRANSMISSION

USE THIS SPACE FOR LEARNER NOTES

_____

_____

_____

_____

_____

_____

_____

SECTION 1

Vehicle layouts

USE THIS SPACE FOR LEARNER NOTES

_____

_____

_____

_____

_____

_____

_____

_____

_____

Learning objectives

After studying this section you should be able to:

- Identify the engine arrangements found in a range of light vehicles.
- Identify the driveline configurations for a range of vehicles.
- State advantages and disadvantages of each layout.
- Identify transmission components for front-wheel drive and rear-wheel drive vehicles.
- Explain the difference between all-wheel drive and four-wheel drive.

Key terms

Transverse engine   The engine is fitted across the vehicle.

All-wheel drive (AWD)   A term associated with vehicles that have permanent four-wheel drive.

Four-wheel drive (4WD or 4×4)   This term is usually used where vehicles have selectable four-wheel drive.

Longitudinal (in-line) engine   The engine is positioned in the centre line of the vehicle.

Propeller shaft (propshaft)   Transmits torque from the gearbox to the final drive.

Drive shaft   A shaft designed to take drive from the final drive to the driven wheels.

LAYOUT OF POWER TRAIN AND TRANSMISSION COMPONENTS

Look at the following diagrams and answer the four questions:

1 Describe the engine configuration and drive layout of the following vehicles.

2 State a make and model of vehicle for each layout.

3 Give advantages of each layout using the word bank below:

good weight distribution
easier to fit larger engines
More room in the passenger
    compartment
good traction

better protection in a front end crash
easy engine cooling
good vehicle handling (even weight
    distribution)

4 State the disadvantages of each layout using the following word bank:

Difficult engine cooling
Only two seats
Heavy Steering

Reduced passenger space due to
    transmission tunnel
Long distance for clutch and throttle
    controls

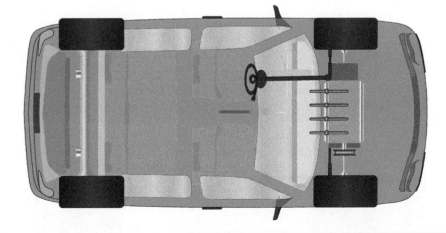

1 _____

2 Make: _____ Model: _____

3 Advantages

- _____
- _____
- _____
- _____

4 Disadvantage

- _____

1 _____

2 Make: _____ Model: _____

3 Advantages

- _____
- _____
- _____
- _____

4 Disadvantage

- _____

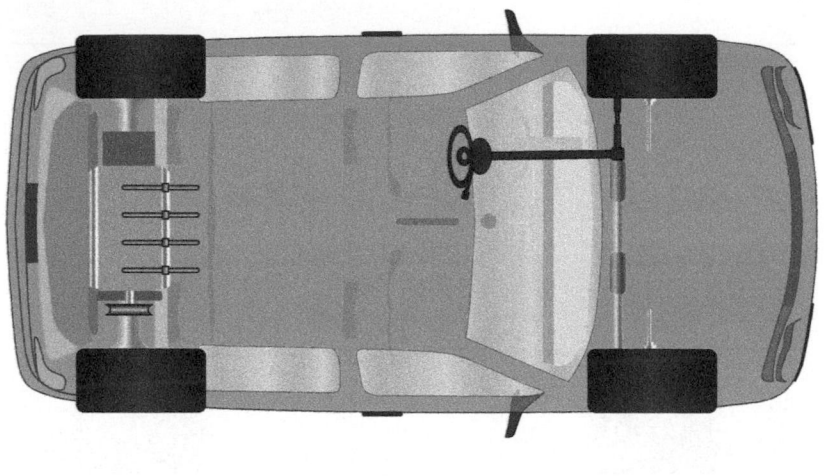

1 _____

2 Make: _____ Model: _____

3 Advantages

   • _____

   • _____

4 Disadvantages

   • _____

   • _____

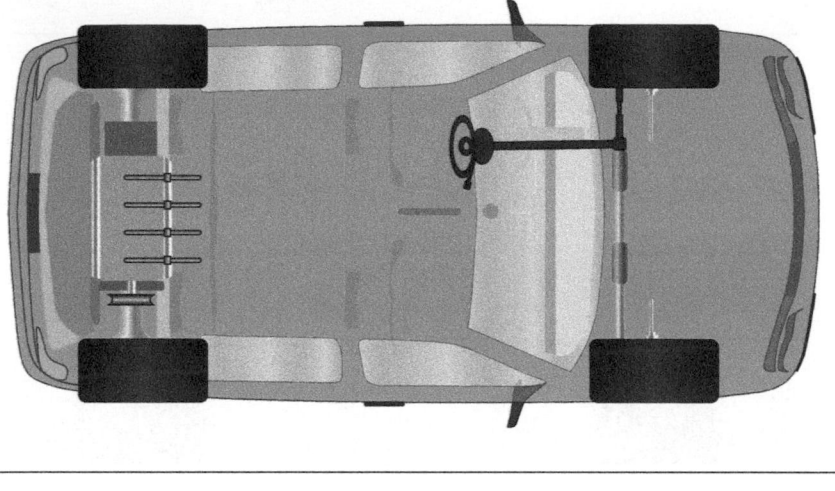

1 _____

2 Make: _____ Model: _____.

3 Advantages

   • _____

   • _____

4 Disadvantage

   • _____

MID-ENGINE TYPES OF LAYOUT

Mid-engine vehicles may have transversely mounted engines.

Or

In-line or longitudinal engines such as the Porsche Boxster.

Describe engine positioning for:

1 Rear-wheel drive – _____

2 Mid-engine – _____

Label the front-wheel drive layout below using the following terms:

Drive shaft (× 2)   **Engine**   **Final drive (differential)**
Clutch   **Driven wheel**   **Gearbox**

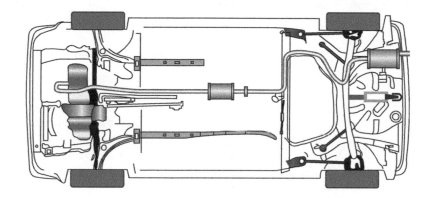

Label the rear-wheel drive layout below using the following terms:

Drive shaft (× 2)   **Engine**   **Gearbox**   **Final drive (differential)**
Clutch   **Driven wheel**   **Propeller shaft**

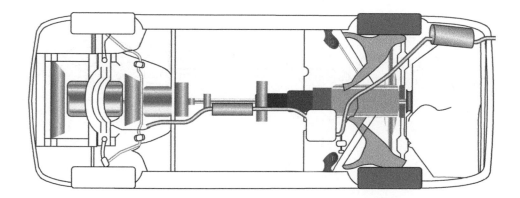

 Arrange with your supervisor in your college or training workshop to have vehicles safely raised on vehicle hoists so you can draw the basic layout of the transmission components of:

- A transverse engine front wheel drive
- A longitudinal mounted front engine rear-wheel drive.

FOUR-WHEEL DRIVE AND ALL-WHEEL DRIVE

Four-wheel drive

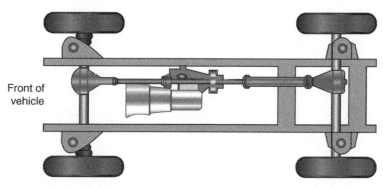

Front of vehicle

This type of drive is mainly found in off-road vehicles. Two or four-wheel drive can be selected by the driver.

Name the numbered components on this four-wheel drive layout:

1 _____   4 _____

2 _____   5 _____

3 _____

All-wheel drive

Permanent four-wheel drive used for performance vehicles. Torque split is often controlled electronically between front and rear axles. These vehicles are often evolved from front-wheel drive vehicles.

Label the drawing with these components:

Front axle differential and final drive
Transfer box
Rear axle differential and final drive
Gearbox
Propeller shaft
Viscous coupling
Engine

Multiple choice questions

Choose the correct answer from a), b) or c) and place a tick [✓] after your answer.

1 **What component allows four-wheel drive to be transmitted to both axles on a four-wheel drive car?**

   a) final drive [ ]

   b) transfer box [ ]

   c) clutch. [ ]

2 **What is a disadvantage of having engines at the rear of a vehicle?**

   a) difficult to cool the engine [ ]

   b) poor traction [ ]

   c) heavy steering. [ ]

3 **Complete the following abbreviations:**

   a) FWD – _____

   b) RWD – _____

   c) AWD – _____

   d) 4WD – _____

SECTION 2

Clutch

USE THIS SPACE FOR LEARNER NOTES

_____

_____

_____

_____

_____

_____

Learning objectives

After studying this section you should be able to:

- Identify clutch components.
- Describe the construction and operation of clutch systems.
- Name and explain the advantages of different types of pressure plate assemblies.
- Identify the key engineering principles that are related to light vehicle clutch systems to include:
  principles of friction
  principle of levers
  torque transmission.
- State common terms used in light vehicle clutch system design.
- Name the different types of clutch linkage.
- Describe common faults and causes that occur with clutches.

Key terms

Pressure plate   Bolted to the flywheel and presses the clutch plate to the flywheel face.
Clutch plate/centre plate/drive plate   Turns the gearbox input shaft.
Diaphragm spring   A dish type spring used in pressure plates.
Clutch release bearing   Thrust bearing that pushes on the diaphragm spring fingers.
Clutch slip   A fault where torque transmission is reduced.
Clutch judder   A fault causing unwanted vibration.
Clutch drag   A fault where the clutch does not release fully.
Release mechanism   Means of the driver temporarily stopping clutch plate rotation.
Coefficient of friction   Ratio of the amount of force needed to cause an object to start to slide.
Slave cylinder   Hydraulic cylinder moving the clutch arm.
Torque   Turning force measured in Newton metres.

WWW   http://auto.howstuffworks.com/clutch.htm
www.luk.com
http://www.automotive-clutches.com

CLUTCHES

Where is the clutch fitted on the majority of light vehicles?

_____

State three functions of the clutch:

1 _____

2 _____

3 _____

Single-plate clutch

There are many different types of clutch in use on road vehicles but by far the most popular is the single-plate friction type. This type of clutch is operated by the driver pressing and releasing a pedal and is used in conjunction with a manual type gearbox.

Label the diagram below.

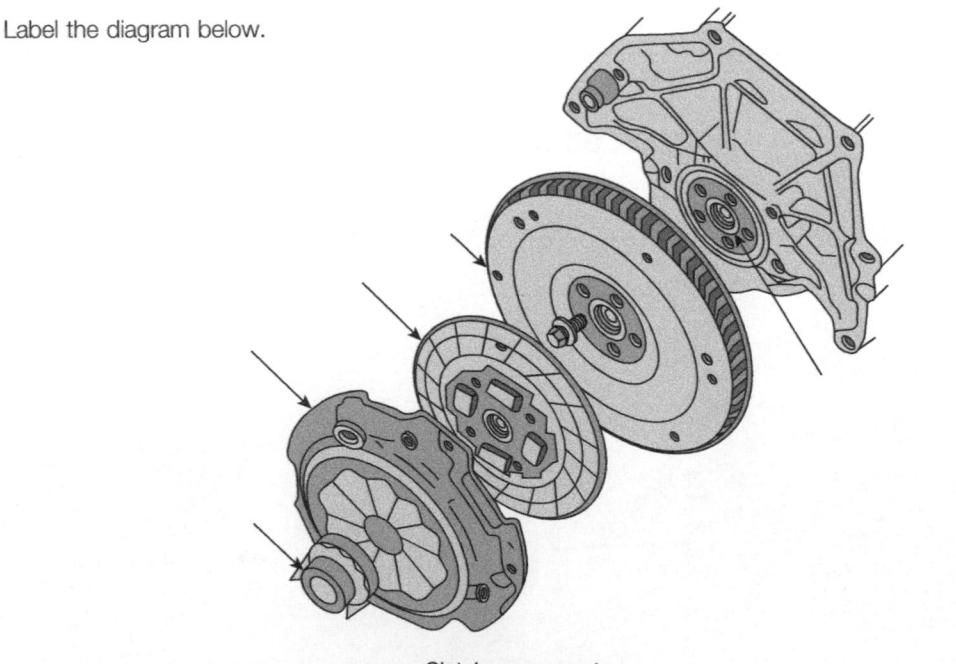

Clutch components

When a vehicle moves off from rest, the clutch must make a connection between the rotating engine crankshaft and the stationary gearbox primary shaft.

Describe how clutch engagement allows the vehicle to move off smoothly.

_____

_____

Clutch operation

Complete the following paragraphs describing the clutch operation when engaged using a selection of the following words:

turning	presses	output	input	flywheel
pulls	clamped	clutch plate	splined	

With the clutch pedal released the diaphragm spring _____ the pressure pad onto the _____ linings, causing the clutch plate to be _____ to the flywheel.

The clutch plate turns with the engine, which transmits the drive to the gearbox _____ shaft through the _____ hub on the clutch plate.

Therefore whenever the engine is running the gearbox input shaft is _____.

Flywheel bolted to the engine crankshaft

Clutch plate

Gearbox input shaft

Engine crankshaft

Release bearing

Diaphragm spring

Pivot points

Pressure pad

Clutch cover

Clutch engaged (drivers foot off the clutch pedal)

Using arrows, show the movement of the release bearing and the pressure pad.

Complete the following explanation of a clutch being disengaged.

When the clutch pedal is pressed the release bearing pushes on the _____, which pivots on the cover and causes the pressure plate to _____ the clutch driven plate.

The clutch plate will then _____ while the flywheel and pressure plate continue to rotate either side.

Why is it necessary to provide some form of adjustment (manual or automatic) in the clutch operating system from the clutch pedal?

_____

_____

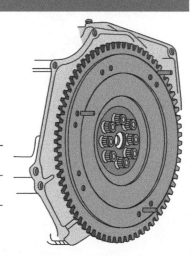

Pivot point

Clutch disengaged (driver holding the clutch pedal down)

FLYWHEEL

The flywheel is bolted to the crankshaft.

The surface of the flywheel is machined to allow the clutch plate linings to engage with the flywheel.

Give two other functions of the flywheel:

1 _____

2 _____

_____

How is the pressure plate located exactly on the flywheel enabling the gearbox input shaft to be positioned centrally?

_____

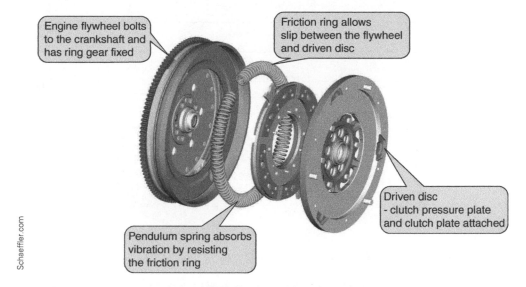

Engine flywheel bolts to the crankshaft and has ring gear fixed

Friction ring allows slip between the flywheel and driven disc

Driven disc - clutch pressure plate and clutch plate attached

Pendulum spring absorbs vibration by resisting the friction ring

Schaeffler.com

Centrifugal pendulum type

This design is commonly used on Diesel engines where, at low speed, engine vibrations occur. The advantages of this design are:

- It increases driving comfort especially at low engine revs
- It absorbs engine vibrations
- It reduces wear and stress on gearbox synchronizer components

 www **www.luk.com**

Name the parts indicated on the diaphragm spring pressure plate.

In this type of clutch the diaphragm spring:

- **provides the clamping force**
- **acts as the release lever.**

Diaphragm spring pressure plate

Examine a diaphragm spring pressure plate and determine:

- the purpose of the drive straps
- where the diaphragm spring pivots
- where the clutch release bearing contacts the diaphragm spring.

The coil spring type is rarely used. It may be found on older commercial vehicles, but it has generally been replaced by the diaphragm spring type.

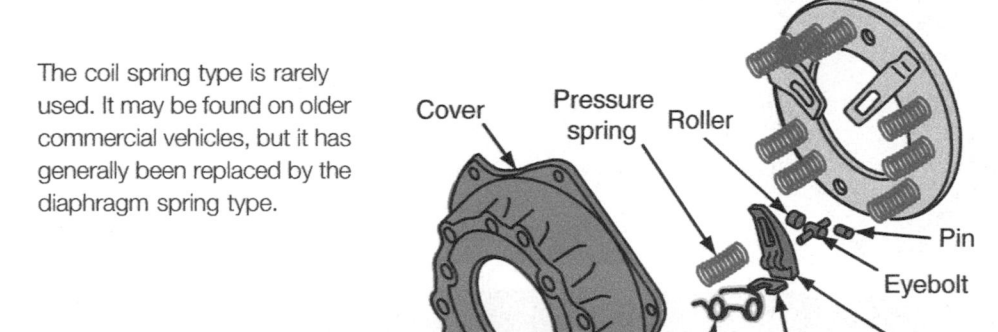

Coil spring pressure plate

A simple coil spring type clutch is illustrated opposite. Label the drawing.

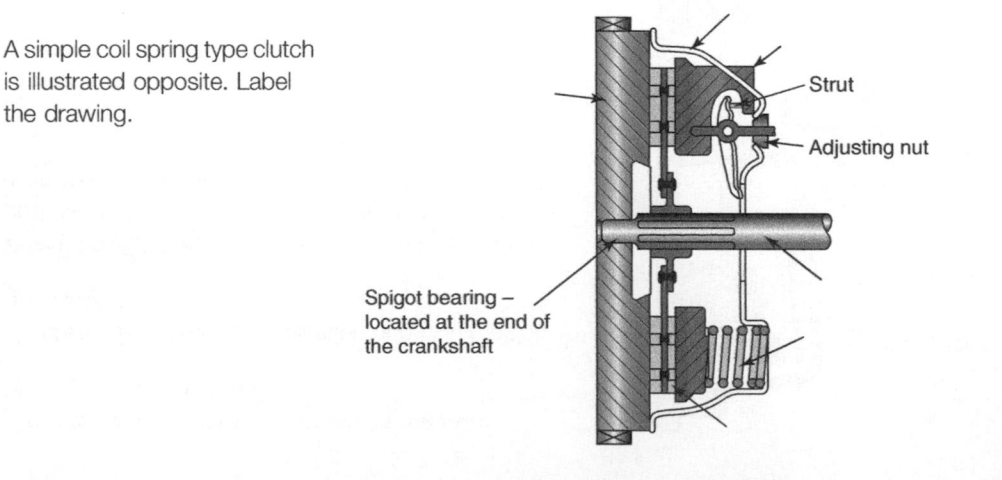

Strut

Adjusting nut

Spigot bearing – located at the end of the crankshaft

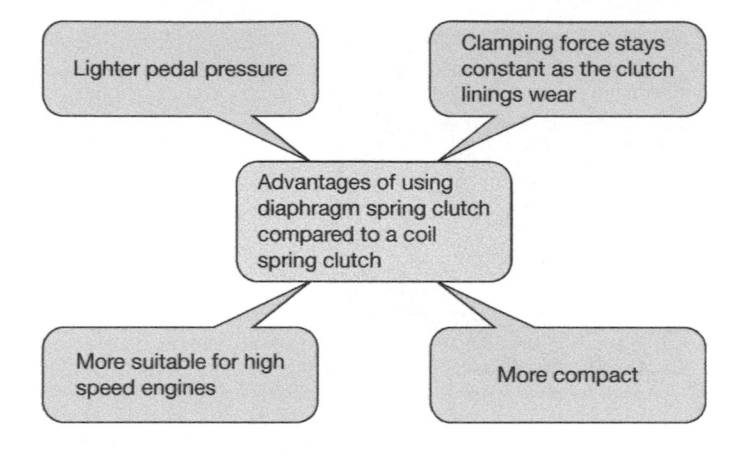

Lighter pedal pressure

Clamping force stays constant as the clutch linings wear

Advantages of using diaphragm spring clutch compared to a coil spring clutch

More suitable for high speed engines

More compact

TIP Many components have more than one name; the clutch plate is one of these!

The clutch plate is also called: **driven plate**, **centre plate** or **clutch disc**.

CLUTCH PLATE CONSTRUCTIONAL FEATURES

The two main types of clutch plate in general use are shown below. Name and describe each one.

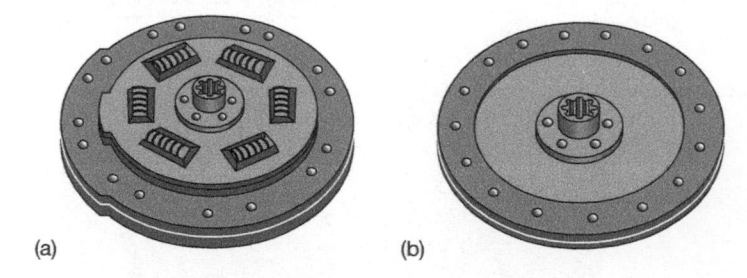

(a)

(b)

Type (a) – _____

_____

Type (b) – _____

_____

Application

The spring centre type is used on most light vehicles. Label the parts of the clutch plate described below:

Clutch plate facings or linings.
Good friction properties

Grooves to promote better cooling

Wave spring allows cooling and smooth take up of drive

Splines which drive the gearbox input shaft

Damper springs, to reduce snatching. Assist in a smooth take up of drive

The spring hub cushions drive take up and absorbs torsional vibration caused by engine torque fluctuations.

Examine several clutch centre plates and note the differences in springs.

Friction materials in the clutch plate

The friction material from which the clutch linings are made must:

● **Maintain its frictional qualities at high temperatures.**
● **Be hard wearing.**
● **Withstand high pressure and centrifugal force.**
● **Offer some resistance to oil impregnation.**

Asbestos bonded with resin has been used in the past, but now non-asbestos materials are used.

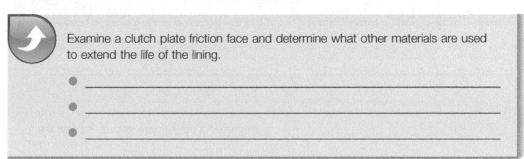
Very old linings could contain asbestos. Be careful when working on classic cars or commercial vehicles. If unsure treat as asbestos.

Materials used for clutch plate facing are now paper-based and ceramic.

Examine a clutch plate friction face and determine what other materials are used to extend the life of the lining.

● _____

● _____

● _____

Factors affecting torque transfer in clutches

The amount of torque that a clutch can transmit depends upon:

● **Mean radius of the clutch plate.**
● **Coefficient of friction of the lining material.**
● **Number of friction surfaces.**
● **Clamping force on the clutch plate.**

What determines the coefficient of friction?

_____

How can the number of friction surfaces be altered?

_____

_____

What determines the clamping force?

_____

Coefficient of friction

The coefficient of friction uses the Greek letter μ (pronounced mew) to describe the friction properties.

Numbers are used to indicate the friction value:

0 = the least amount of friction
1 = the most friction possible.

Typical coefficient of friction figures

- **Clutch lining on cast iron flywheel = 0.35 μ.**
- **Brake lining on cast iron drum = 0.4 μ.**
- **Tyre on normal road surface = 0.6 μ.**
- **Metal on metal (dry) = 0.2 μ.**
- **Metal on metal (lubricated) = 0.1 μ.**

What would happen to the figure if a clutch was contaminated with engine or gearbox oil?

_____

Calculating the torque transmitted by clutches

Torque transmitted in Newton metres (Nm) = μWrn

μ = the coefficient of friction of the lining
W = the force exerted by the pressure plate spring in Newtons (N)
r = the mean radius of the clutch linings in metres (m)
n = the number of friction surfaces

These values are multiplied together to obtain the torque transmitted by the clutch.

What is the mean radius of a clutch plate whose friction rings are 0.125 m from the centre to the outside edge and 0.075 m inside edge radius?

Mean radius (r) = _____

How many contact surfaces have a single-plate clutch? _____

Example calculation

A single-plate clutch has a coefficient of friction of 0.3 and the friction linings have a mean radius of 75 mm. If the total force exerted by the pressure plate spring is 2200 N, calculate the torque transmitted by the clutch.

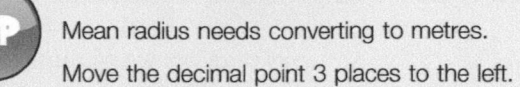

TIP

Mean radius needs converting to metres.

Move the decimal point 3 places to the left.

75 mm = 0.075 m

Torque transmitted = μWrn
Torque transmitted = 0.3 × 2200 × 0.075 × 2
Torque transmitted = 99 Nm

Questions

1 A single-plate clutch has facings of: 115 mm outside radius, 65 mm inside radius and a coefficient of friction of 0.35. The pressure plate spring exerts a force of 2400 N.

   Calculate the maximum torque that could be transmitted by the clutch.

   _____

   _____

   _____

2 A multi-plate clutch has three friction plates of 60 mm mean radius running in oil. The coefficient of friction is 0.15 and the total spring force is 1000 N.

   Calculate the torque transmitted by the clutch.

   _____

   _____

   _____

MULTI-PLATE CLUTCHES

The multi-plate clutch is a clutch assembly which utilizes more than one clutch plate.

State four basic reasons for using multi-plate clutches:

1 _____

2 _____

3 _____

4 _____

CLUTCH RELEASE SYSTEMS

State two methods of clutch operation on light vehicles:

1 _____

2 _____

Label the clutch operating system shown opposite and describe its action during operation:

_____

_____

_____

_____

_____

_____

A typical hydraulic clutch linkage arrangement

In its released position the release bearing is just clear of the release plate or diaphragm fingers or, in some systems, it may be in very light continuous contact. Why is the disengaged position of the release bearing critical with regard to clutch operation?

● _____

● _____

A common fault with this system is fluid leakage from the slave cylinder. What are the effects of this on clutch operation?

_____

_____

Hydraulic operating system

The advantages of a hydraulic operating system are:

● **It isolates noise from the engine to the body when the clutch pedal is a long way from the clutch assembly (rear engine vehicle).**

● **It is self-lubricating and adjusting.**

CABLE OPERATED RELEASE SYSTEM

Label the figure using the word bank below.

locknut   adjusting nut   inner cable   return spring   pedal shaft   clutch housing

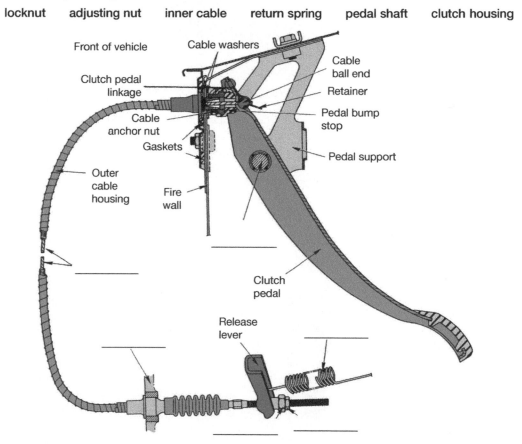

Front of vehicle — Cable washers

Clutch pedal linkage

Cable ball end

Retainer

Pedal bump stop

Cable anchor nut

Gaskets

Pedal support

Outer cable housing

Fire wall

Clutch pedal

Release lever

Cable operated release system

Cable adjustment

The result of pedal travel on friction lining wear is: _____

The result of pedal travel on cable stretch is: _____

Self-adjusting mechanisms

The slave cylinder in a hydraulic clutch operating system can provide automatic adjustment to compensate for friction lining wear. Describe briefly how this is achieved:

_____

An automatic clutch adjustment system is shown in the figure below.

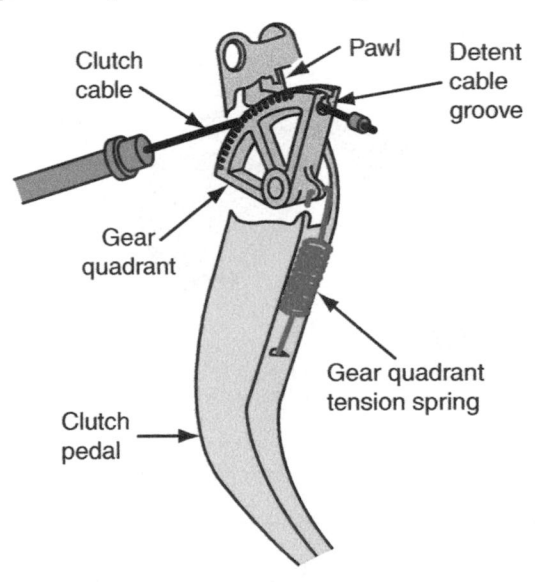

Clutch cable
Pawl
Detent cable groove
Gear quadrant
Gear quadrant tension spring
Clutch pedal

Describe briefly how the system operates:

_____

_____

Release bearing

The release bearing transmits the axial thrust to the release levers, plate or diaphragm or the clutch pressure plate during engagement and disengagement.

The bearing shown is a single row deep groove ball bearing. This type of bearing will withstand the axial thrust imposed during operation. How is this type of release bearing lubricated?

_____

_____

Release bearing carrier

To achieve correct contact with the pressure plate the ball bearing type release bearing moves parallel to the primary shaft. It may be permanently in contact with the pressure plate or held just clear. The release bearing is very often mounted on a carrier which slides on the primary shaft sleeve to provide linear movement of the fork.

Label the diagram of the release bearing carrier below.

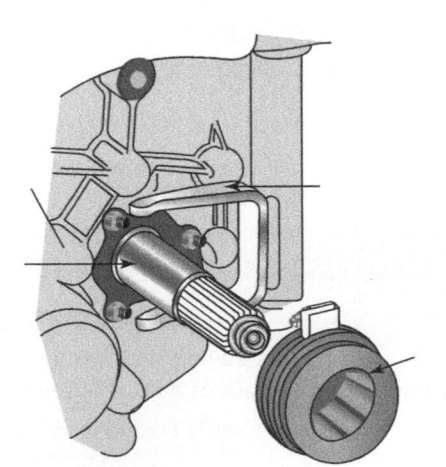

Internal concentric release bearing

This design is now common on hydraulic clutch mechanisms, combining the clutch release bearing with the slave cylinder. The cylinder assembly attaches to the gearbox and has the input shaft passing through it.

A hydraulically operated clutch needs to be bled on certain occasions. When should this clutch type be bled and what equipment is needed?

_____

_____

_____

_____

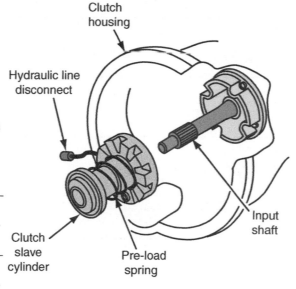

Clutch housing

Hydraulic line disconnect

Clutch slave cylinder

Pre-load spring

Input shaft

CLUTCH COMPONENT CHECKS

Describe the check taking place on the flywheel.

_____

What would be a typically acceptable reading when manufacturer's data is not available?

_____

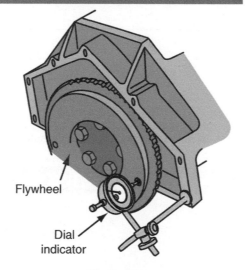

Flywheel

Dial indicator

Flywheel face

Checking the clutch pressure plate

Visual checks on a clutch pressure plate would be for surface cracks, scoring and spring finger breakage. Describe the checks shown below.

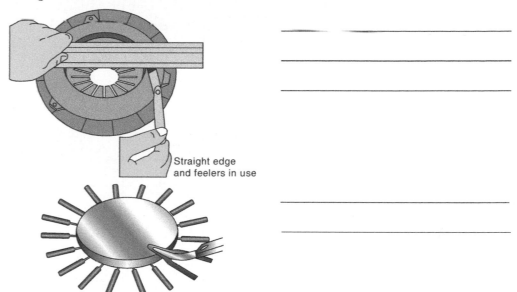

Straight edge and feelers in use

_____

_____

_____

_____

_____

_____

Checking the clutch plate

Describe the checks which must take place on the areas of the clutch plate indicated by the speech bubbles.

Checking the release bearing

Press against the bearing and rotate feeling for roughness.

CLUTCH MAINTENANCE ADJUSTMENTS

Preventive routine maintenance is necessary in order to ensure clutch efficiency, prolong its life and minimize the risk of failure.

List three important maintenance tasks:

1 _____

2 _____

3 _____

Outline the clutch tests to perform when the following problems are identified:

Engagement (slip):

_____

_____

Disengagement (drag):

_____

Abnormal noise:

_____

Abnormal vibration:

_____

Complete the table below which identifies the general rules for efficiency and precautions to be observed during clutch maintenance and repair.

CLUTCH DIAGNOSIS			
Definition	Fault	Symptom	Cause
Engine speed increases without a rise in vehicle speed	_____	Lack of drive _____ _____ _____	1 Oil contaminated clutch plate linings 2 _____ 3 _____ 4 _____
Severe vibration when pulling away from rest	_____	Difficulty pulling away	1 _____ 2 Oil contaminated clutch linings 3 _____ 4 _____
Clutch not clearing	_____ or Spin	Jolt when changing gear whilst vehicle is in motion _____ _____ _____ _____	1 Clutch plate stiff on gearbox splines 2 _____ 3 _____

Definition	Fault	Symptom	Cause
Whirring noise when the clutch is depressed	Noise	Noise and vibration when the clutch is pressed	_____
Sharp clutch	Fierceness or snatch	Difficulty in controlling initial take up drive	Partially seized pedal linkage
			_____

Multiple choice questions

Choose the correct answer from a), b) or c) and place a tick [✓] after your answer.

1 Which of the following components connects to the gearbox input shaft?

a) clutch pressure plate [ ]

b) clutch release bearing [ ]

c) clutch centre plate. [ ]

2 Which of the following faults on a cable operated clutch will cause clutch slip?

a) clutch cable adjusted too tight [ ]

b) clutch cable adjustment too slack [ ]

c) worn clutch release bearing. [ ]

3 What influences the amount of torque a clutch can transmit?

a) the thickness of the flywheel [ ]

b) the thickness of the centre plate [ ]

c) the clamping force of the spring. [ ]

4 Why are clutch plate sprung centre hubs used?

a) to reduce clutch slip [ ]

b) to lower pedal pressure [ ]

c) to assist in the smooth take up of drive. [ ]

5 Which of the following features is associated with a coil spring type of pressure plate?

a) increased pedal pressure [ ]

b) compact and light [ ]

c) consistent spring pressure. [ ]

6 With the engine running and the clutch pedal depressed, which of the following statements are true?

a) flywheel and pressure plate rotating, clutch plate stationary [ ]

b) flywheel, pressure plate and clutch plate all turning [ ]

c) flywheel and clutch plate turning, pressure plate stationary. [ ]

7 With the engine running, the vehicle stationary and the clutch depressed, it is difficult to engage gear. What is the name of this type of fault?

a) clutch slip [ ]

b) clutch judder [ ]

c) clutch drag. [ ]

8 Which of the following faults can cause clutch drag?

a) centre plate friction linings worn [ ]

b) leaking slave cylinder [ ]

c) pressure plate springs weak. [ ]

9 A single plate clutch has a mean radius of 130 mm, with a spring exerting a force of 1800 N and a coefficient of friction of 0.35 μ. Calculate the torque that this clutch can transmit.

a) 163800 Nm [ ]

b) 163.8 Nm [ ]

c) 81.9 Nm [ ]

10 What happens to the release bearing clearance as the clutch linings wear?

a) clearance does not alter [ ]

b) clearance reduces [ ]

c) clearance increases. [ ]

SECTION 3

Gearbox

USE THIS SPACE FOR LEARNER NOTES

_____

_____

_____

_____

_____

_____

_____

Learning objectives

After studying this section you should be able to:

● Identify manual gearbox components.

● Describe the construction and operation of manual gearboxes.

● Compare key gearbox components and assemblies against alternatives to identify differences in construction and operation.

● Calculate gear ratios and speed changes.

● State common gearbox terminology.

Key terms

Synchromesh   Matching gear speeds to allow easy engagement.
Baulk ring   A component that connects with the gear cone to match speeds.
Helical gear   Gear teeth cut at an angle.
Spur gear   'Straight cut' teeth cut at 90° to the gear.
Interlock   Means of preventing two gears being engaged at the same time.
Syncrohub   Splined to the mainshaft and links to a selected gear.
Dog teeth   Teeth on baulk rings and syncrohub assemblies.
Detent   A plunger or ball used to positively locate the selector shaft in gear.

GEARBOX FUNCTION

In a motor vehicle the gearbox serves three purposes:

1 _____

2 _____

3 _____

Under many operating conditions the torque requirement at the driving wheels is far in excess of the torque available from the engine.

State four operating conditions under which the engine torque would need to be multiplied at the gearbox:

1 _____

2 _____

3 _____

4 _____

Types of gearbox

The four types of gearbox are:

1 _____

2 _____

3 _____

4 _____

Which of the four types of gearbox are used in most modern cars?

_____

Originally, gearboxes were nearly all of the sliding mesh type but these are now mainly confined to specialist heavy applications. One interesting exception to this, however, is the use of sliding mesh gears on certain racing cars.

Types of gearing

The spur (straight cut) gear and the helical gear are the two main types of gears used in light vehicle gearboxes.

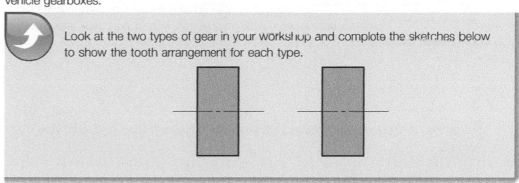

Look at the two types of gear in your workshop and complete the sketches below to show the tooth arrangement for each type.

Complete the table below by listing the advantages and disadvantages of each gear type.

no side load	strong	will not slide into mesh
noisy	slides into mesh easily	more friction due to side load
quiet running	lower tooth load (weak)	

SPUR GEAR	
Advantages	*Disadvantages*
_____	_____
_____	_____

HELICAL GEAR	
Advantages	*Disadvantages*
_____	_____
_____	_____

Torque multiplication

Torque multiplication can be achieved by using different-sized gearwheels. For example, transmitting the drive from a small-diameter gearwheel to a large-diameter gearwheel.

Consider the pair of gears shown below:

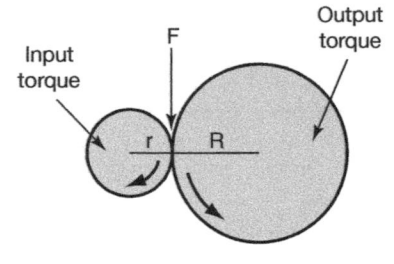

The input torque produces a force (F) which is transmitted from the small gear to the large gear.

Describe how the input torque is multiplied.

_____

_____

_____

Gear ratios

When two gearwheels are meshed to form a 'simple' gear train, the gear ratio can be expressed as a ratio of gear teeth or a ratio of gearwheel speeds.

To increase torque, a small gear drives a larger gear which will cause output speed to drop. In order to calculate the gear ratio:

1 **You need to know the number of teeth on each gear.**

2 **You must determine which is driving and which is being driven.**

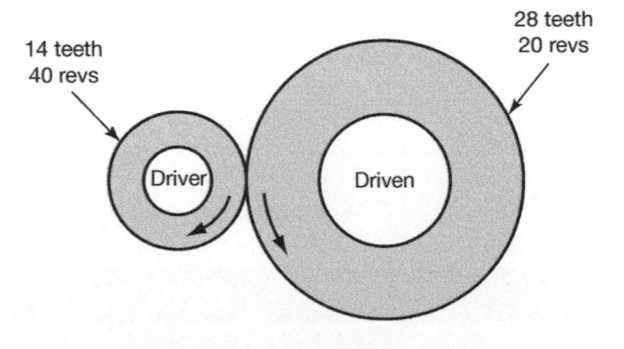

The following formula can be used to calculate gear ratio:

$$\frac{\text{Number of teeth on driven gear}}{\text{Number of teeth on driving gear}} = \text{Gear ratio}$$

e.g. $\frac{28}{14} = 2 : 1$

Alternatively, gear wheel speeds can be used:

$$\frac{\text{Number of revs of driver gear}}{\text{Number of revs of driven gear}} = \text{Gear ratio}$$

e.g. $\frac{40}{20} = 2 : 1$

> **TIP**
>
> Use a calculator for these calculations.
>
> Rounding to two decimal places – if the third decimal is 5 or above, round up the second decimal.
>
> e.g. 4.135 to 2 decimal places = 4.14

Try these.

1 If a gear being driven has 28 teeth and the driving gear has 16 teeth. Calculate the gear ratio.

2 A gear set has 15 teeth driving a gear with 32 teeth. Calculate the gear ratio to two decimal places.

3 A gear having 26 teeth is driving a gear having 21 teeth. Calculate the gear ratio to two decimal places.

When the gearbox multiplies engine torque, a speed reduction occurs between gearbox input and output shafts. When the torque multiplication is high the vehicle speed is _____. The gear in which maximum torque multiplication is obtained is _____.

The torque output from the gearbox is therefore varied according to the speed and load requirement of the vehicle.

Compound gear ratios

This is when there is more than one gear ratio involved. The individual gear ratios are simply multiplied together. For example:

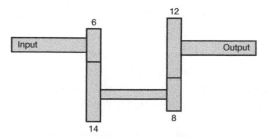

$$\frac{\text{Driven}}{\text{Driver}} \times \frac{\text{Driven}}{\text{Driver}} = \text{compound gear ratio}$$

$$\frac{14}{6} \times \frac{12}{8} = \frac{168}{48} = 3.5 : 1$$

1 Calculate the compound gear ratio to two decimal places.

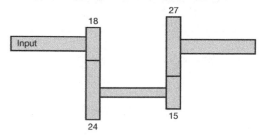

2 Calculate the compound gear ratio to two decimal places.

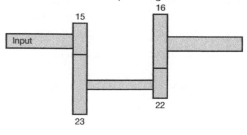

In a compound gear train, how does the direction of rotation of the input shaft compare with that of the output shaft?

Idler gears

Idler gears are used to transfer drive from one gear to another without changing the ratio, but keeping the direction of rotation the same.

Where is this arrangement used in gearboxes?

Shaft speed calculations

The gear ratio is relative to torque changes and speed changes. As torque increases then speed reduces. In which gears is this effect at its maximum?

When considering engine rotation to wheel rotation then both the gear ratio and the final drive ratio are multiplied together. For example:

If first gear has a ratio of 4:1 and the input shaft rotates at 2000 rev/min and delivers a torque of 50 Nm.

The speed of the output shaft will be:

- **Input shaft speed ÷ gear ratio = output shaft speed**
- **2000 ÷ 4 = 500 rev/min**

The torque of the output shaft will be:

- **Gear ratio × torque = output shaft torque**
- **4 × 50 = 200 Nm**

The overall transmission ration, if the final drive ration is 4:1, will be:

- **Gear ratio × final drive ratio = overall ratio**
- **4 × 4 = 16 16:1**

Calculate the following gear speed and ratio problems.

1 A gearbox has a third gear ratio of 1.5:1 and the input shaft rotates at 4000 rev/min, with a torque of 40 Nm:

   a   What is the speed of the output shaft (to 2 decimal places)?

_____

   b   What is the torque delivered by the output shaft?

_____

   c   What is the overall transmission ratio if the final drive ratio is 3.5:1?

_____

2 A gearbox has a fourth gear ratio of 1.2:1 and the input shaft rotates at 3500 rev/min, with a torque of 75 Nm:

   a   What is the speed of the output shaft (to 2 decimal places)?

_____

   b   What is the torque delivered by the output shaft?

_____

   c   What is the overall transmission ratio if the final drive ratio is 4.2:1?

_____

SIMPLE SLIDING-MESH GEARBOX

The figure opposite shows a three-speed sliding-mesh type of gearbox. To obtain the various gear ratios the gearbox layshaft is made up of different sized gearwheels which are connected in turn to the gearwheels on the mainshaft. The drive to the layshaft is via a pair of gearwheels which are permanently meshed.

Complete the missing labels on the diagram below.

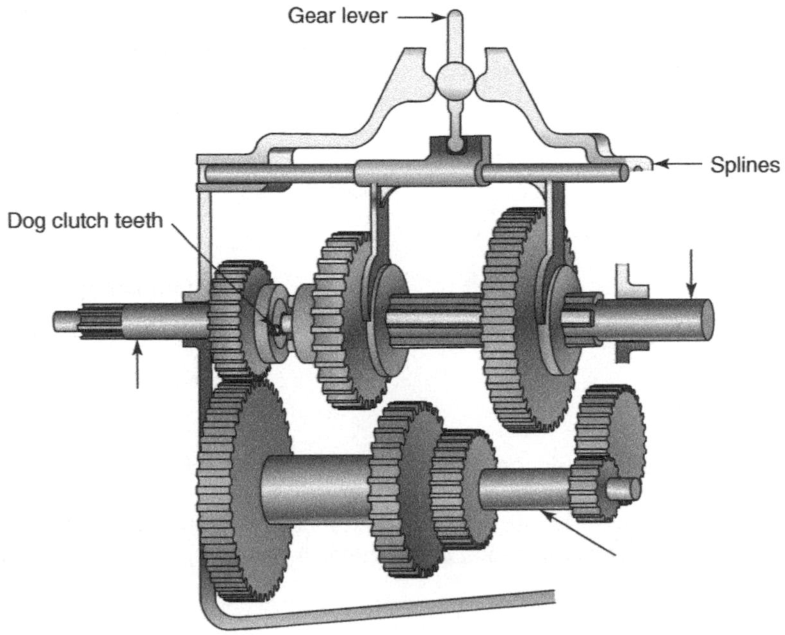

What type of gear is used in a sliding mesh gearbox? _____

How is a gear engaged?

_____

_____

_____

In what position is the gear lever shown in the drawing above? _____

Show the powerflow in first, second and top gear by adding the mainshaft gearwheels and arrows to diagrams (1), (2) and (3) below. On the wheels in (4) show the direction of rotation; view the gear train from the rear of the gearbox.

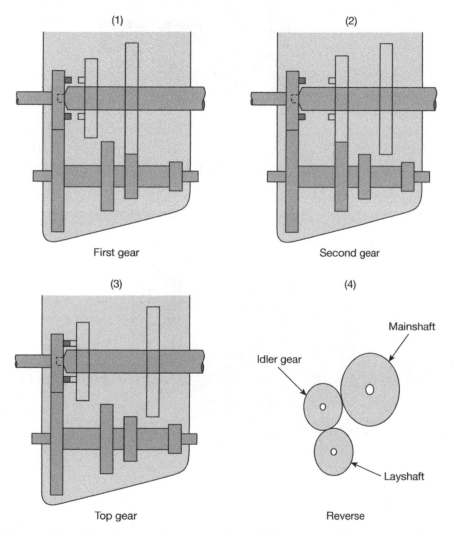

(1)

First gear

(2)

Second gear

(3)

Top gear

(4)

Reverse

The sliding mesh gearbox is a good way to understand how gearboxes are arranged, but is not practical for modern-day motoring as gear changing is difficult and requires a lot of skill.

CONSTANT-MESH GEARBOX

In the constant-mesh gearbox, as the name implies, the gearwheels are permanently in mesh. Constant meshing of gears is achieved by allowing the mainshaft gearwheels to rotate on bushes. Dog clutches, which are normally part of a synchromesh hub, splined to the mainshaft, provide a positive connection when required, to allow the drive to be transmitted to the output shaft.

This operating principle is shown in the diagram below.

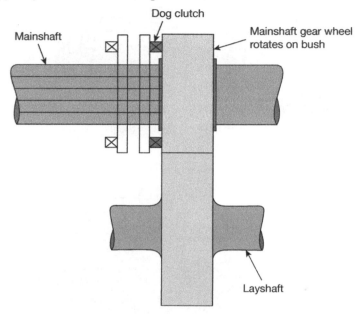

Powerflow

Complete the drawing above to show how a dog clutch member is used to connect the mainshaft gearwheel to the mainshaft. Add arrows to indicate the power flow from the layshaft to the mainshaft.

Constant mesh allows the use of quieter helical gears, but it can be challenging to change gear easily without the use of synchromesh.

To aid easy gear changing most cars and light commercial vehicles use gearboxes which use synchromesh fitted to all forward gears and sometimes reverse.

CONVENTIONAL GEARBOX LAYOUT

Three shaft gearbox

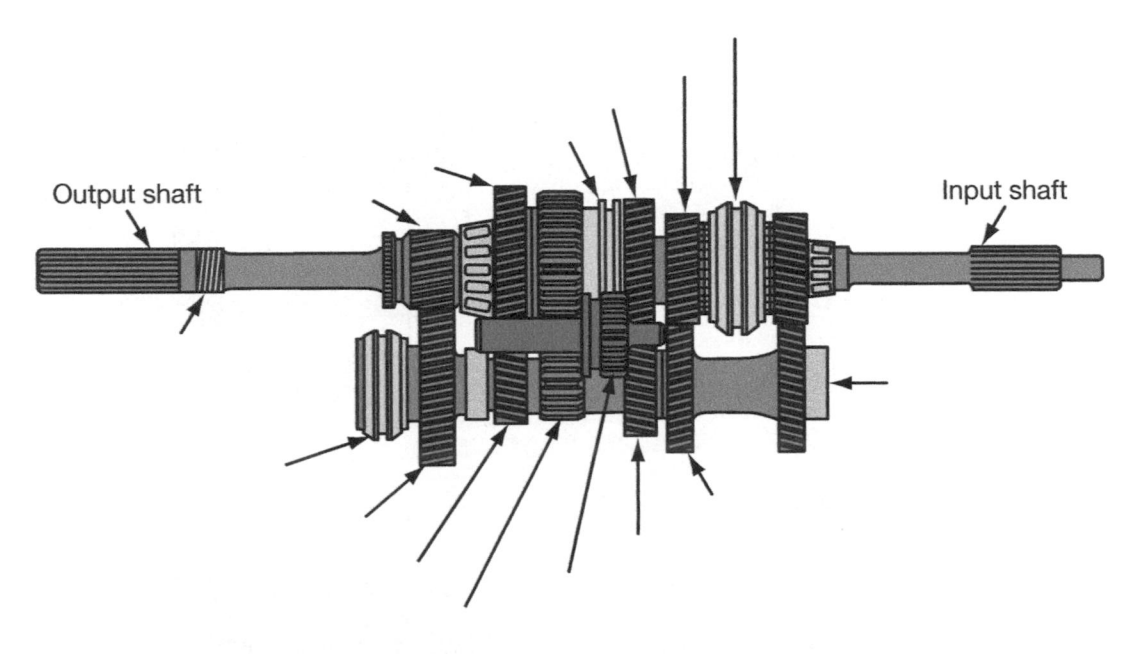

Output shaft

Input shaft

Label the gears and shafts in the 5 speed gearbox above.

Transmission powerflow

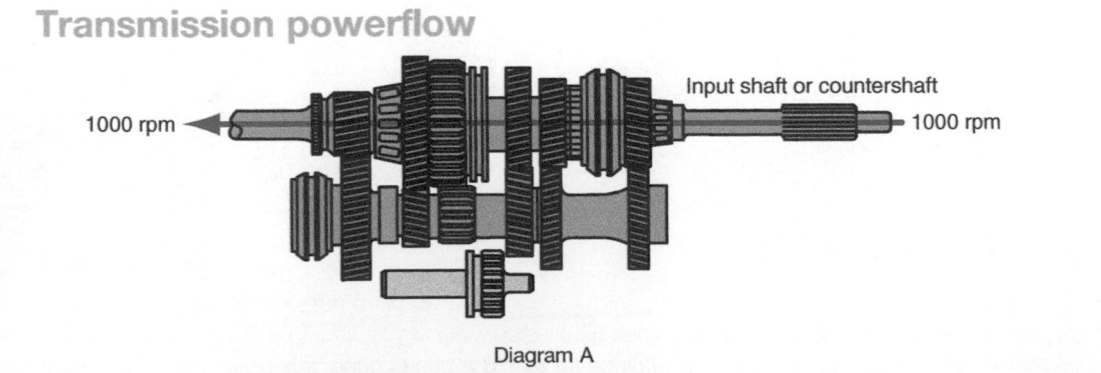

1000 rpm

Input shaft or countershaft

1000 rpm

Diagram A

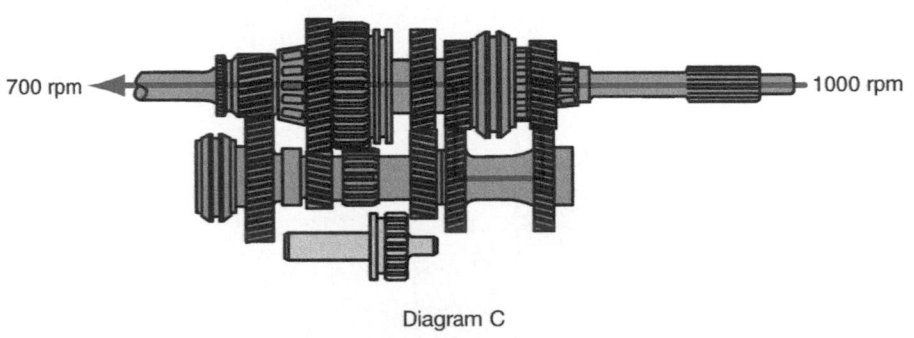

250 rpm

1000 rpm

Diagram B

700 rpm

1000 rpm

Diagram C

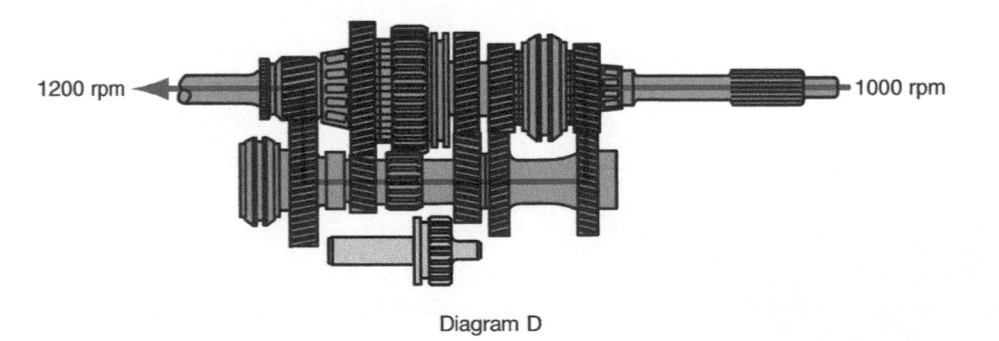

1200 rpm

1000 rpm

Diagram D

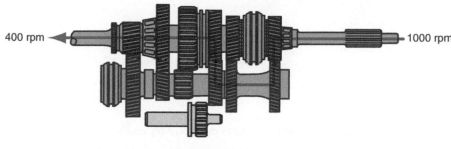

400 rpm ← ← 1000 rpm

Diagram E

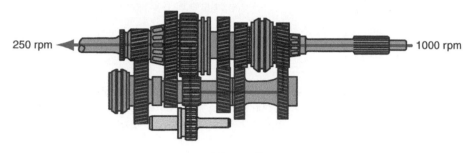

250 rpm ← ← 1000 rpm

Diagram F

On the above powerflow diagrams determine which gear is engaged and calculate the gear ratio.

(A) _____   (D) _____

(B) _____   (E) _____

(C) _____   (F) _____

TRANSAXLE – TWO SHAFT GEARBOX

These gearboxes are widely used in front engine front-wheel drive vehicles and incorporate the final drive and the gearbox in one housing.

How many gearbox shafts are used in this gearbox? _____

How many speeds does the gearbox have? _____

Identify which gear is engaged.

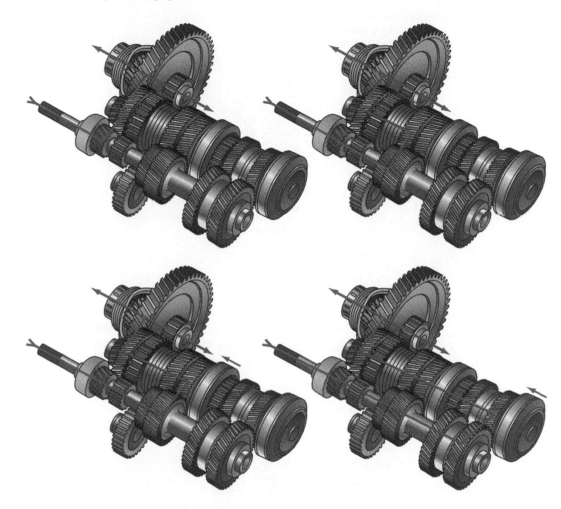

Synchromesh devices

When a gear change is made, the speeds of the meshing dog clutch teeth must be equalized in order to avoid clashing which would result in noise and wear. As the name implies, the synchromesh device synchronizes the speeds of the dog teeth during gear changing.

Baulk ring synchromesh

The characteristics of the baulk ring type are that:

● **The synchronization of the dog clutch speeds allows fast and quiet gear changes.**
● **The dog clutches cannot be engaged until their speeds are equal.**

Name the component parts of the baulk ring synchromesh unit shown in the figure below.

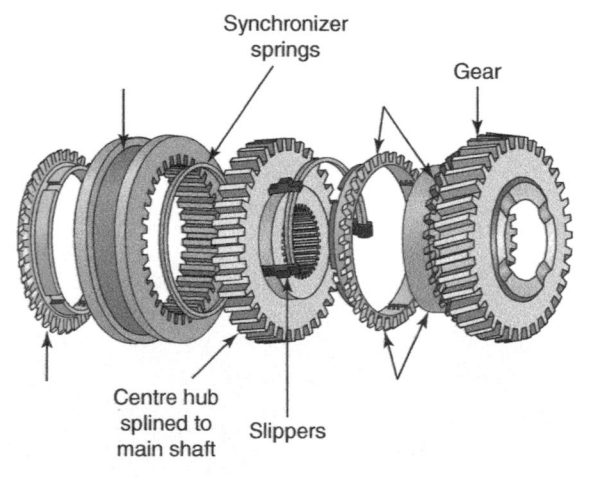

Synchronizer springs

Gear

Centre hub splined to main shaft

Slippers

> Examine a synchromesh unit and observe the limited rotation of the baulk ring. Feel the resistance on the baulk ring as you try to turn and press the baulk onto the gear cone.

Synchromesh gear change sequence

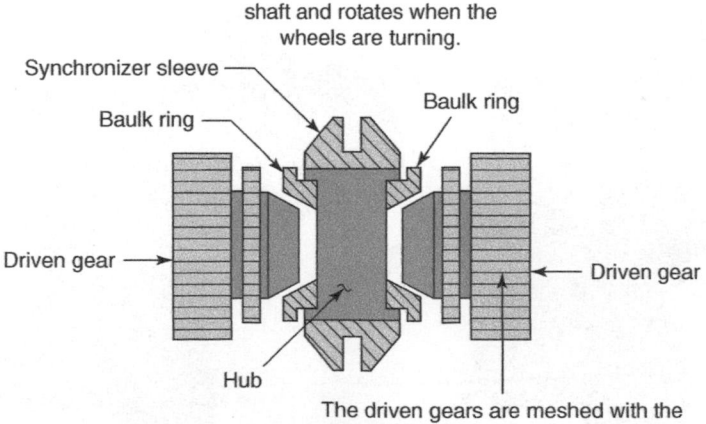

The hub is splined to the output shaft and rotates when the wheels are turning.

Synchronizer sleeve

Baulk ring

Baulk ring

Driven gear

Driven gear

Hub

The driven gears are meshed with the input shaft and rotate when the clutch is engaged with the engine running.

Neutral

Both driven gears have a clearance between the gear cone and the synchronizer sleeve.

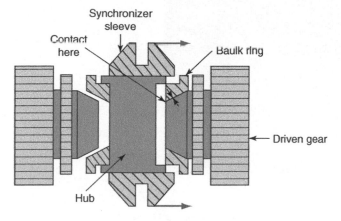

Initial gear lever movement

The synchronizer sleeve pushes the baulk ring against the gear cone causing friction and the gear speeds change until they are matched.

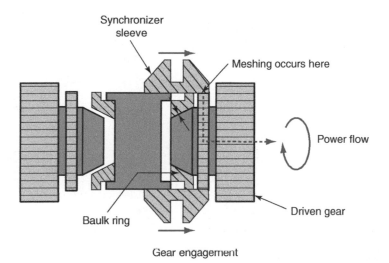

Gear engagement

The outer sleeve moves across locking the dog teeth onto the driven gear, causing power flow from the input shaft to the wheels.

What symptom would be noticed if the baulk ring wears and does not grip onto the gear cone during gear changing?

SELECTOR MECHANISMS

When a gear is selected the movement of the gear lever is transferred to the synchro-hub or gear-wheel through a selector fork. The fork may be fixed to a sliding shaft, or the fork may slide on a fixed shaft. Provision is made for locking the selector fork in the required gear position by a detent and spring.

Examine different types of selector mechanisms and complete the diagram below to include the selector fork and gear lever.

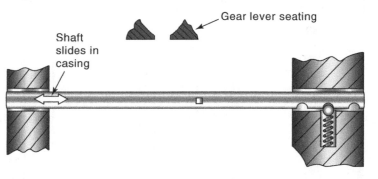

An exploded view of a selector mechanism is shown below. Label the diagram.

1 _____ 4 _____

2 _____ 5 _____

3 _____ 6 _____

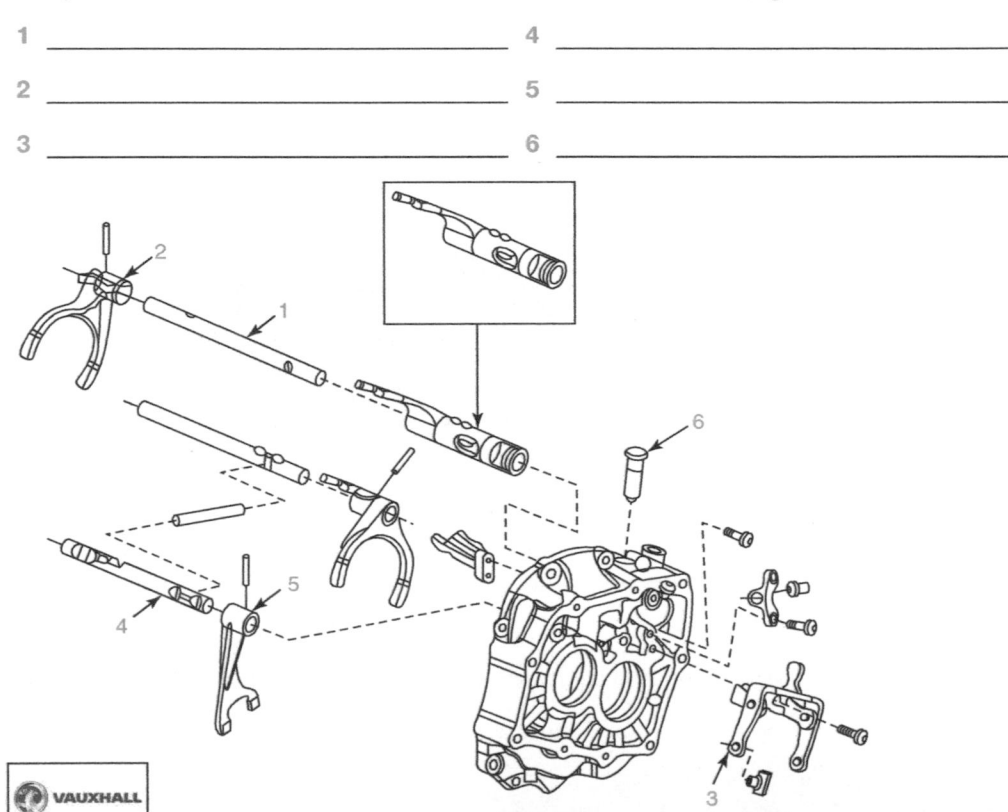

The selector forks fit into the grooves in the synchromesh hub outer sleeves.

Interlock mechanisms

The interlock mechanism prevents the engagement of two gears at once. What would happen if two gears were allowed to be selected at the same time?

_____

The figure below shows a 5 speed interlock mechanism which allows only one shaft to move during gear selection.

5-R rail   3-4 rail   1-2 rail   5-R rail   3-4 rail   1-2 rail   5-R rail   3-4 rail   1-2 rail

Describe the action when each gear is selected on the diagrams above.

First selected: _____

_____

Third selected: _____

_____

Fifth selected: _____

_____

Follower plate interlock mechanism

Explain how this design prevents two gears being selected.

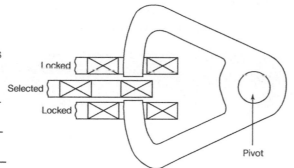

_____

_____

_____

_____

Remote control gearshift systems

The mechanical remote control shown below allows the use of a short gear lever.

Give one reason for using remote control systems.

_____

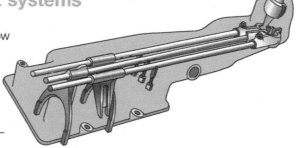

Rod type remote linkage

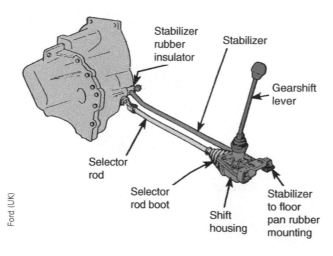

Ford (UK)

Cable type gear change linkage

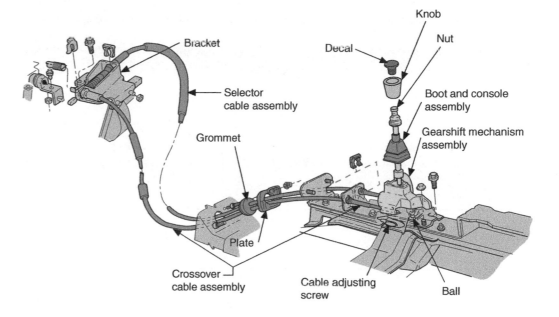

Inspect three vehicles with manual transmission and different types of gear linkage and complete the following table.

Make	Model	Type of gear linkage

Oil seals

Name the two oil seals shown below and give one example of their uses in a gearbox.

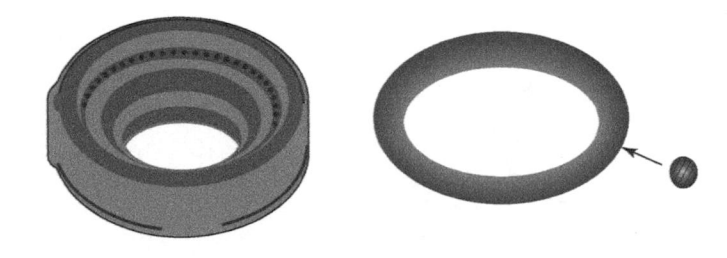

_____   _____

> **TIP** Always lubricate seals before fitment.
>
> Lip type seals must be fitted with the lip facing inwards (towards the source of the oil).

Gearbox breather

A small opening is designed into the top of the gearbox to allow any pressure in the gearbox to escape into the atmosphere.

Why would pressure build up in the gearbox?

_____

If the breather were to become blocked, what could occur?

_____

GEARBOX ROUTINE MAINTENANCE AND LUBRICATION

Regular maintenance and correct lubrication are essential to extend gearbox life, improve efficiency and minimize the risk of failure of components.

Lubrication

The gearbox oil level needs to be checked at routine service intervals as prescribed by the vehicle manufacturer. The level is usually checked by removing the level/filler plug, located about one-third of the way up the gearbox.

Complete the procedure for checking and topping up gearbox oil. There are two distractor words in the following list.

recommended   clean   strip   level   under   plug   bottom   over   top

- Raise vehicle off the ground to enable access to the level/filler _____.
- Ensure the vehicle remains _____.
- _____ the area around the level/filler plug and remove the plug.
- Oil should be level with _____ of threaded hole.
- Top-up, using only the _____ oil.
- Replace the level/filler plug (do not _____ tighten as this could _____ the threads).

The main functions of a gearbox lubricating oil are:

1 _____

2 _____

3 _____

> **TIP** Manual transmissions can have EP oils, engine oils, automatic transmission fluid or manufacturer' specialist oils.
>
> Always check manufacturer's recommendations.

REVERSE LIGHT

When selecting reverse, a white light is emitted warning other road users that the vehicle is reversing.

This is usually integrated into the gearbox and is activated by the reverse selector shaft.

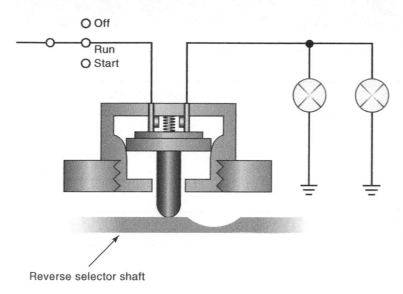

Reverse selector shaft

In the diagram the lights are illuminated. Describe what occurs when reverse is deselected.

_____

WWW http://auto.howstuffworks.com

http://www.carbibles.com/transmission_bible.html

Multiple choice questions

Choose the correct answer from a), b) or c) and place a tick [✓] after your answer.

1 One function of a gearbox is to:

a) reduce engine torque [ ]

b) vary engine speed and torque [ ]

c) multiply engine torque. [ ]

2 The purpose of a gear change interlock is to:

a) prevent engagement of two gears at the same time [ ]

b) prevent the car jumping out of gear [ ]

c) aid smooth gear changes. [ ]

3 The purpose of a ball and spring acting on the selector shaft is to:

a) stop two gears being selected at the same time [ ]

b) hold the selected gear in mesh [ ]

c) prevent accidental selection of reverse. [ ]

4 Gearbox oil level is correct when it is:

a) running out of the filler opening [ ]

b) below the level of the filler opening [ ]

c) level with the bottom of the filler opening. [ ]

5 Which of the following components directly moves the gearbox synchromesh assembly?

a) selector fork [ ]

b) input shaft [ ]

c) interlock mechanism. [ ]

6 How is the overall gear ratio calculated to the wheels?

a) by multiplying the gear ratio with the final drive ratio [ ]

b) by adding the gear ratio to the final drive ratio [ ]

c) by dividing the final drive ratio by the gear ratio. [ ]

7 What is the purpose of using an idler gear between two gears?

a) change the ratio of two gears [ ]

b) to make gears rotate quietly [ ]

c) to enable the two gears to rotate in the same direction. [ ]

8 On a two shaft gearbox, what other term is used for the countershaft?

a) input shaft [ ]

b) layshaft [ ]

c) output shaft. [ ]

9 Why are helical cut gears used in transmission systems?

a) they slide into mesh easily [ ]

b) they are quiet running [ ]

c) they have no side loads. [ ]

10 A two shaft gearbox has 16 teeth driving 28 teeth in third gear, the pinion has 18 teeth and the crown wheel has 72 teeth. If the input torque is 50 Nm what is the output torque to the wheels?

a) 35 Nm [ ]

b) 350 Nm [ ]

c) 200 Nm. [ ]

SECTION 4

Light vehicle final drive systems

USE THIS SPACE FOR LEARNER NOTES

_____

_____

_____

_____

_____

_____

_____

Learning objectives

After studying this section you should be able to:

- Identify final drive components.
- Explain the operation and purpose of the differential.
- Describe the construction and operation of final drive systems.
- Describe the checks and adjustments required on a hypoid final drive.
- Calculate final drive and overall gear ratios including torque and wheel speed changes.
- Describe the construction and operation of the limited slip differential.

Key terms

Hypoid gears Where the pinion is offset from the centre line where it meshes with the crown wheel.
Crown wheel Large gear driven by the pinion.
Pinion Input gear to the final drive.
Sun gear Side gear in the differential, which is splined to the axle shaft.
Differential Series of gears to allow road wheels to rotate at different speeds whilst cornering.

THE FINAL DRIVE

The final drive provides several functions, taking drive primarily from the gearbox and transmitting drive to the wheels.

Final drive location

The final drive assembly may be located in different positions on the vehicle, according to chassis design.

Circle the final drive assembly and state a typical example of a vehicle for each of the following three layouts:

Solid rear axle also known as Hotchkiss drive

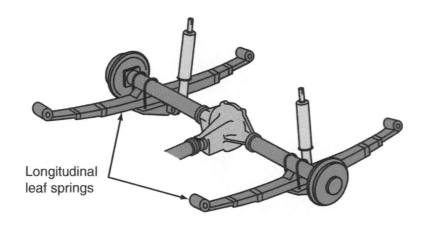

Longitudinal leaf springs

Make – _____   _____

Model – _____   _____

Rear-wheel drive with independent rear suspension

Final drive assembly is mounted to the body using drive shafts with flexible joints transmitting drive to the wheels.

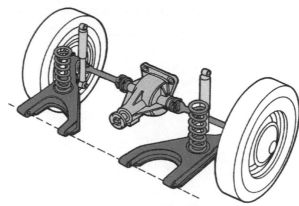

Make – _____ _____ _____

Model – _____ _____ _____

Front engine front-wheel drive

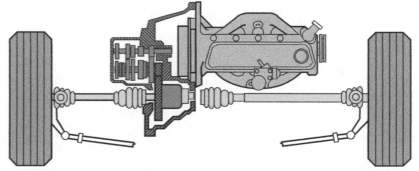

Make – _____   _____

Model – _____   _____

Make – _____   _____

Model – _____   _____

State the function of all types of final drive.

- _____
- _____

In conventional transmission arrangements (front engine with rear live axle) the final drive fulfils another purpose. This is:

_____

The two gears forming the final drive are shown below. Label the diagram.

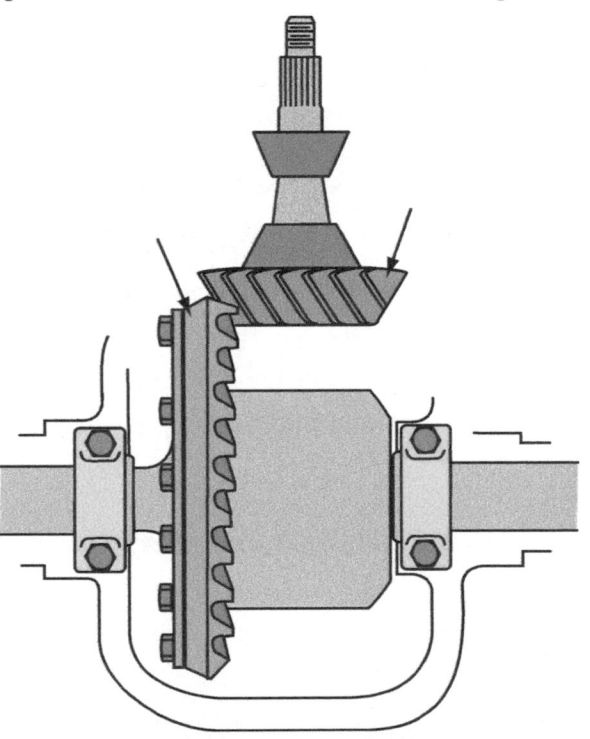

TIP The crown wheel and the pinion are a matched pair which, once meshed and run together, need to always be used as a pair. A replacement pinion must not be used with a used crown wheel.

Examine the following final drive assemblies:

- **Rear-wheel drive (RWD).**
- **Front-wheel drive (FWD).**

Complete the table below for the assemblies examined.

State the formula required to calculate the final gear ratio:

Final gear ratio = _____

Type of layout	Number of teeth on the large gear wheel	Number of teeth on the small gear wheel	Final drive gear ratio
FWD	_____	_____	_____
RWD	_____	_____	_____

FINAL DRIVE BEVEL GEARS

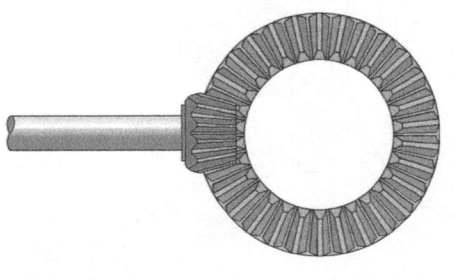

Straight bevel gears

Early vehicles used these bevel gears. What are the disadvantages that caused them to be superseded?

_____

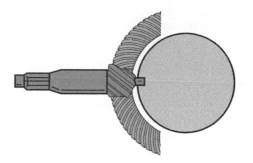

Spiral bevel gears

The spiral bevel gears are stronger and quieter.

Modern vehicles use the bevel gear shown below. Name the type of gear shown below and state why it is now used in preference to the spiral bevel gears.

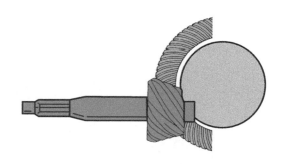

_____

● _____

● _____

● _____

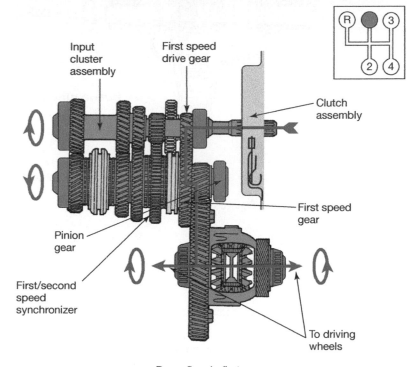

Powerflow in first gear

Due to the positioning of the transaxle layout helical gears are used. Which of the final drive functions is not required?

_____

THE DIFFERENTIAL

When a vehicle is cornering, the inner driven wheel rotates slower than the outer driven wheel. It is the differential which allows this difference in speed to take place while at the same time transmitting an equal driving torque to the road wheels.

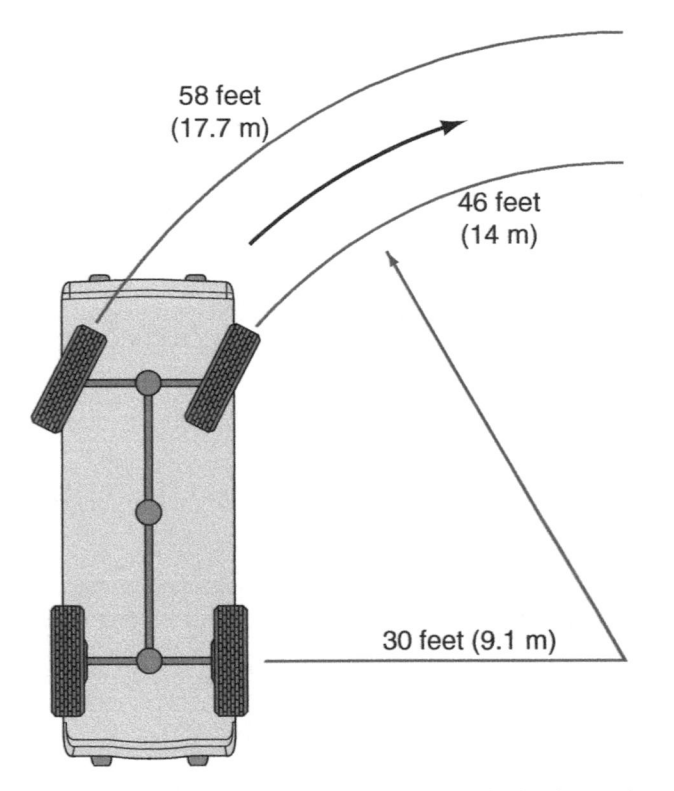

58 feet
(17.7 m)

46 feet
(14 m)

30 feet (9.1 m)

Why does this difference in speed between inner and outer road wheels need to occur?

_____

The differential is normally an assembly of bevel gears housed in a casing (or cage) which is attached to the crown wheel. The essential parts of a differential are:

planet gear	**sun gear**	**pinion**	**planet carrier**
crown wheel	**differential case**	**planet pin**	**half shaft or drive shaft**

Label the diagram below showing the final drive including the differential using the terms provided.

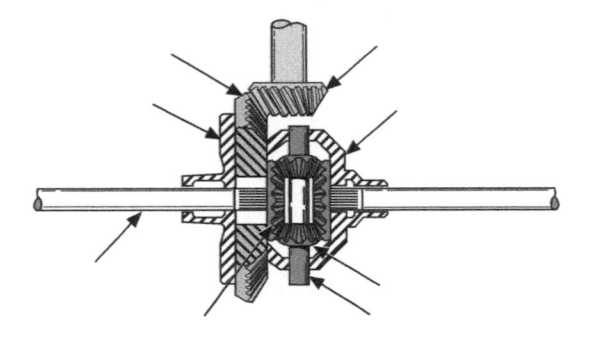

The differential could be described as a torque equalizer. Why is this?

_____

_____

Speed differential

Whatever the inner wheel loses in speed during cornering, the outer wheel will gain in speed.

Complete the speeds for the actions below when the crown wheel speed is 100 rpm.

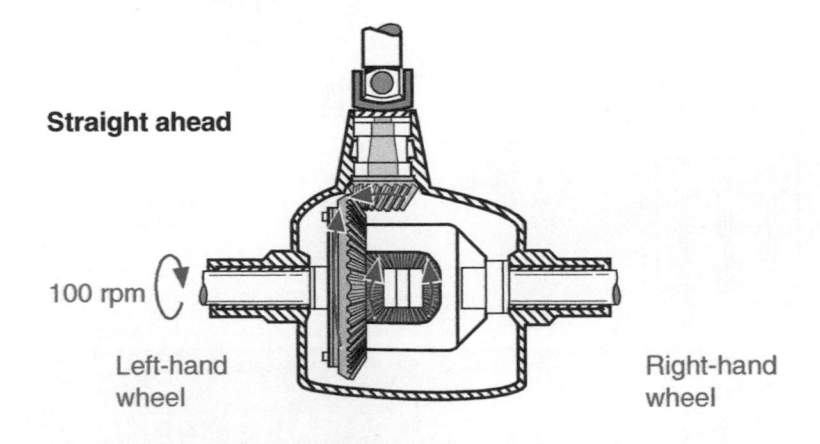

Straight ahead

100 rpm

Left-hand
wheel

Right-hand
wheel

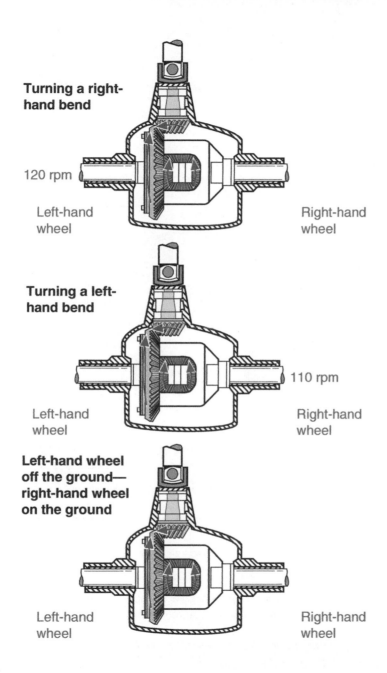

Turning a right-hand bend

120 rpm

Left-hand wheel

Right-hand wheel

Turning a left-hand bend

Left-hand wheel

110 rpm

Right-hand wheel

Left-hand wheel off the ground—right-hand wheel on the ground

Left-hand wheel

Right-hand wheel

During cornering, the inner driven wheel is rotating at 100 rpm and the outer driven wheel is rotating at 200 rev/min. Circle the correct statement which describes what the torque in the half shafts would be:

- Double in the outer.
- Double in the inner.
- The same in each.

Examine a final drive and differential assembly and complete the figure below by adding the main components of the differential. Clearly label each component.

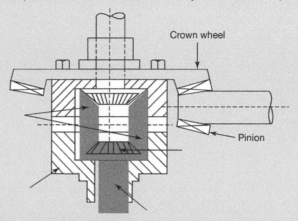

Crown wheel

Pinion

Which gears transmit the drive to the axle shafts? _____

Power is transmitted from the final drive pinion to the axle shafts via the components listed below. Rearrange the list in the correct sequence in accordance with the powerflow through the differential.

Components	Powerflow
Planet (cross) pin	_____
Sun gears	_____
Crown wheel	_____
Planet gears	_____
Axle shafts	_____
Pinion	_____
Differential cage	_____

FINAL DRIVE BEARING ADJUSTMENT

Pinion bearings of the taper-roller type are normally moved towards each other by the adjusting arrangement until they are placed under an initial load which makes them slightly stiff to rotate. This is known as 'pre-load'.

State the reason why such bearings are pre-loaded:

_____

_____

A popular method of obtaining pre-load is to use a collapsible spacer. Show on the drawing below where it would be used and describe how it would be used.

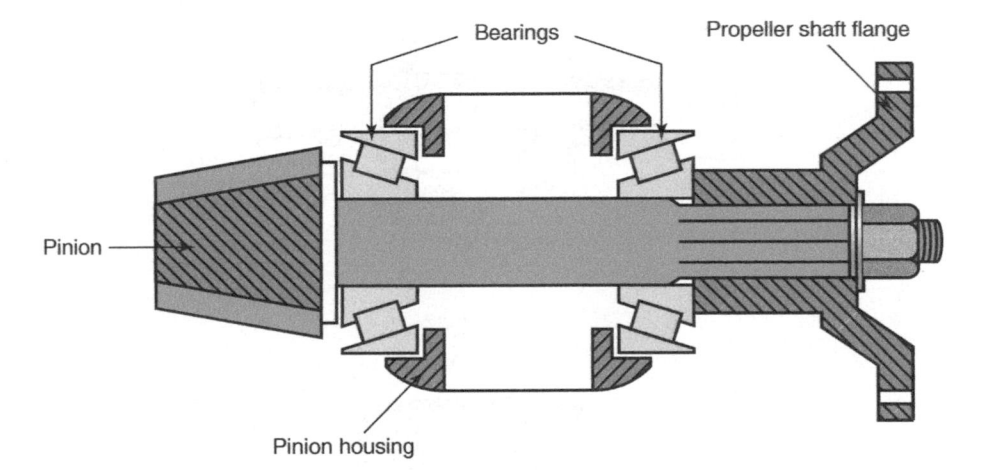

The flange nut is tightened gradually to collapse the spacer applying a load to the pinion bearings. Care must be taken not to over-tighten the nut or a new spacer will be required.

Backlash

Measuring the backlash

A small amount of movement is required between the crown wheel and pinion to allow for thermal expansion and allow lubrication between the gears.

Backlash is measured by holding the pinion still and rocking the crown wheel, measuring with a dial test indicator.

How is the backlash adjusted?

_____

_____

Maintenance

On final drives that are in a separate casing to the gearbox, the oil levels must be checked periodically.

This is achieved by placing a drip tray under the final drive casing and removing the level filler plug. The oil should be level with the bottom of the threads when correct. Top up with the correct oil as specified by the vehicle manufacturer if below the filler hole.

Due to the loads on hypoid final drives they must use an EP oil, such as EP 90.

What does EP stand for? _____

Using websites or oil charts find the oil that is used for three rear-wheel drive final drive assemblies. You may have technical data sources in your workshop. Try the following websites for more information about the types of oil used:

http://www.commaoil.com

http://www.opieoils.co.uk

Vehicle make	Model	Type of oil	Quantity (litres)

OVERALL GEAR RATIO CALCULATION

To calculate the overall ratio from the engine to the road wheels the gearbox ratio must be multiplied by the final drive ratio.

Example

A vehicle has a gearbox ratio of 2.75:1 and a final drive ratio of 3.8:1 and the engine speed is 3000 rpm with an engine torque of 85 Nm.

Calculate the overall ratio, the wheel speed and torque at the wheels.

Speed calculations round to the nearest whole number.

Overall ratio = 2.75 × 3.8 = 10.45:1

Calculation of wheel speed

Engine speed ÷ Overall gear ratio = Wheel speed

3000 ÷ 10.45 = 287 (rounded to whole number)

Calculation of torque at the wheels

Engine torque × Overall ratio = Torque at wheels

85 × 10.45 = 888.25 Nm

Now try these:

1 In first gear, the gear ratio is 3.2:1, with a final drive having a crown wheel with 66 teeth and a pinion with 16 teeth.
   Calculate the overall gear ratio.

   _____

   _____

   If the engine speed is 3500 rpm and torque is 50 Nm.
   Calculate the wheel speed and torque.
   Speed calculations round to the nearest whole number.

   _____

   _____

2 In fifth gear, the gear ratio is 0.85:1, with a final drive having a crown wheel with 72 teeth and a pinion with 20 teeth.
   Calculate the overall gear ratio.

   _____

   _____

   If the engine speed is 2000 rpm and torque is 80 Nm
   Calculate the wheel speed and torque
   Speed calculations round to the nearest whole number.

   _____

   _____

LIMITED SLIP DIFFERENTIALS

What happens to the drive of a vehicle which has a standard differential if one wheel loses traction?

_____

Give four circumstances when this is likely to occur:

1 _____

2 _____

3 _____

4 _____

Sports and racing cars are vehicles with high power-to-weight ratios and as such can, even on good surfaces, cause a driving wheel to spin, for example, during rapid acceleration. Most of the higher quality or top of the range models incorporate a limited slip differential (LSD).

The limited slip device is incorporated in the differential, which automatically applies a brake to the spinning half shaft, thereby maintaining a torque in the other half shaft.

Little torque is required to drive a spinning or slipping road wheel and a normal type differential would transmit the same torque to the non-spinning wheel; thus traction is lost. Limited slip traction control and torque sensing devices all produce 'frictional torque' in the drive to a spinning wheel.

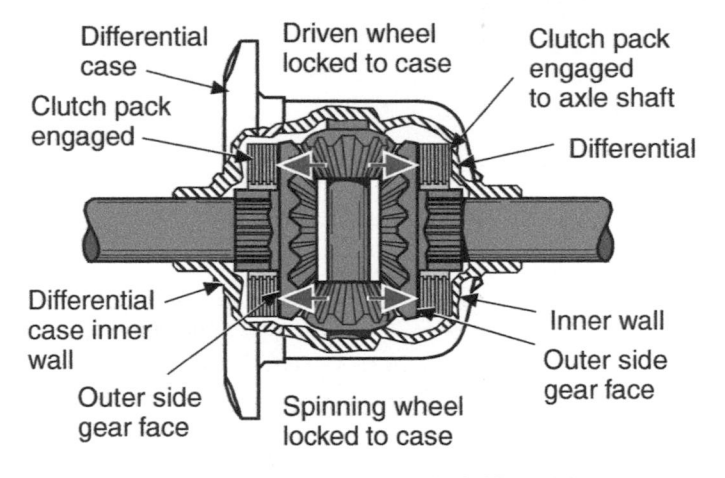

Energized clutches cause locked differential

In the friction clutch type shown above, when the sun gear rotates relative to the differential cage (during wheel spin), the spring-loaded clutch slips generating frictional torque. Owing to the 'torque balancing' action of a differential the same torque is transmitted to the non-spinning wheel.

Describe briefly how to check a limited slip differential for operation:

_____

_____

⚡ Never attempt to drive one wheel while the vehicle is jacked up with the other drive wheel on the ground.

Torsen® differential

Torsen® is also known as torque sensitive limited slip differential, which uses gears to multiply the torque to the wheel that is starting to spin.

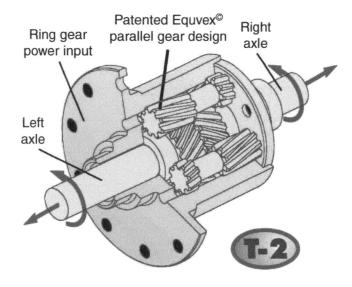

Ring gear power input

Patented Equvex© parallel gear design

Right axle

Left axle

T-2

In small groups use resources available to determine how the torsen limited slip differential operates.

http://www.torsen.com

http://auto.howstuffworks.com

http://www.flashoffroad.com

Multiple choice questions

Choose the correct answer from a), b) or c) and place a tick [✓] after your answer.

1 Which of the following only applies to the final drive on a front engine rear wheel drive vehicle?

a) allows the wheels to turn at different speeds when cornering [ ]

b) final gear reduction [ ]

c) turn the drive through 90°. [ ]

2 What is the advantage of using hypoid bevel gears in the final drive assembly?

a) lower floor line [ ]

b) greater torque increase [ ]

c) less costly to manufacture. [ ]

3 The component which allows the driven wheels to turn at different speeds whilst cornering is called:

a) crown wheel [ ]

b) differential [ ]

c) pinion. [ ]

4 What type of vehicle would be fitted with a limited slip differential?

a) large volume production cars [ ]

b) trucks and buses [ ]

c) high performance cars. [ ]

5 When a vehicle is travelling in a straight line the:

a) sun gears revolve at the same speed [ ]

b) planet gears do not revolve on their shaft [ ]

c) both of the above. [ ]

6 If a vehicle has an overall gear ratio of 7:1, and the input torque to the gearbox is 50 Nm, what is the output torque to the wheels?

a) 35 Nm [ ]

b) 350 Nm [ ]

c) 3500 Nm. [ ]

SECTION 5

Driveline

USE THIS SPACE FOR LEARNER NOTES

_____

_____

_____

_____

_____

_____

_____

_____

_____

_____

Learning objectives

After studying this section you should be able to:

- Identify components and describe the operation of front engine rear-wheel drive driveline components.
- Identify and describe the features of solid axle bearing arrangements.
- Identify components and describe the operation of front-wheel drive driveline components.
- Identify common faults found in light vehicle transmission driveline systems and their causes.

Key terms

Constant velocity   When speed does not vary.
Universal joint (UJ)   Allows small angular changes on shafts.
Half shaft   A shaft enclosed in an axle casing to transmit drive from the final drive to the wheel.
Centre bearing   A bearing that supports long propshafts to reduce vibration and whip.

The figure below shows the layout of a front engine rear-wheel drive arrangement.

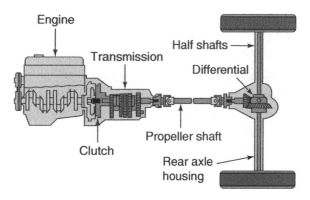

Which of the components above form part of the driveline?

Rear axle layouts

Describe the layouts below and give one advantage of each. List two examples of each layout.

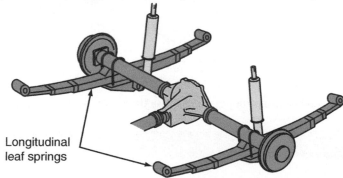

Longitudinal leaf springs

Make   Model

Shock absorber
Coil spring
Universal joint
Differential bolted to body

Arm pivots at right angles to car

Trailing arm

Make   Model

Give three examples of vehicle layouts in which external drive shafts may be used as opposed to conventional type half shafts enclosed in the axle casing.

1

2

3

238

PROPELLER SHAFTS

The propeller shaft transmits the drive from the gearbox to the final drive gear. Name the major parts in the figure below.

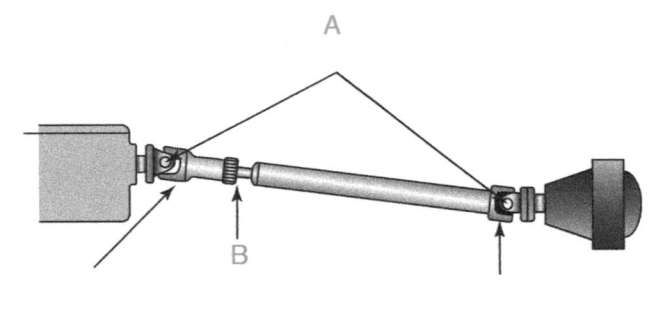

A

B

State the purpose of:

A _____

_____

B _____

_____

Give reasons for using a tubular propeller shaft rather than a solid shaft by completing the following statement using the relevant words from the word bank below.

easier   lighter   heavier   harder   balance   strength   weight

Tubular shafts are much _____ and have similar _____ to a solid shaft of the same diameter, and are _____ to _____.

Universal joints

When drive is being transmitted from one shaft to another through a Hooke-type joint, if one shaft swings through an angle its velocity will vary, that is, the shaft will accelerate and decelerate slightly during every 180° of revolution.

When a single UJ is used the output speed is not constant. The graph shows that, over one full turn, the output accelerates and decelerates twice.

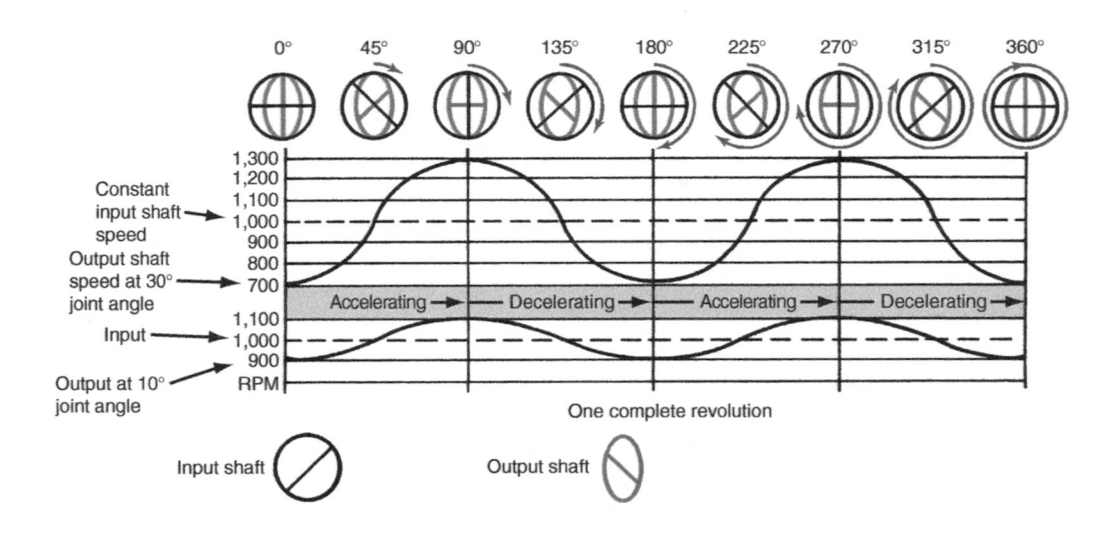

What happens as the joint angle increases?

_____

What vital features are shown on the diagram of the propeller shaft below to overcome the effect of the speed changes mentioned above?

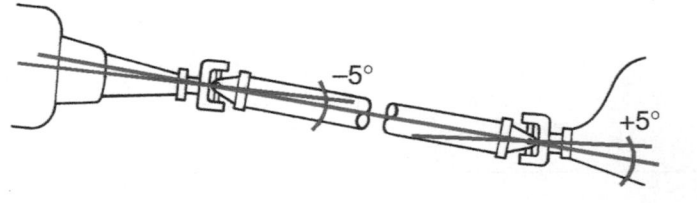

● _____

● _____

A very popular type of universal joint is shown below. The joint is efficient, compact, easy to balance and it will transmit the drive through quite large angles.

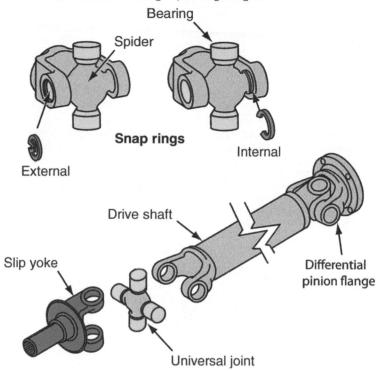

Bearing

Spider

Snap rings

Internal

External

Drive shaft

Differential pinion flange

Slip yoke

Universal joint

Name this type of universal joint. _____.

This rubber hexagonal joint (doughnut type) is sometimes used on transmission shafts. Label the diagram opposite.

Why is this joint used in the driveline shown opposite?

_____

_____

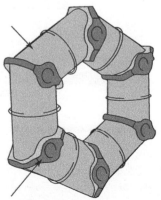

Two-piece propeller shaft

Give reasons for using this arrangement and label the diagram.

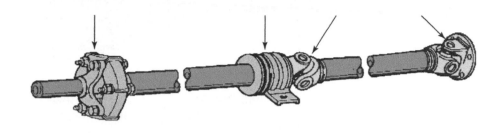

- _____

- _____

- _____

Solid axle hub bearing design

The solid rear axle has half shafts taking drive from the final drive to the wheels. The support on the outer end of the axle is carried out by bearings.

The number, type and location of bearings employed in a hub assembly depend, to a large extent, on the weight of the vehicle.

Three main classifications of hubs are:

1 _____

2 _____

3 _____

On the diagrams of the three bearing types shown in the figure below, determine the type of drive axle and the description number that describes the features from the list.

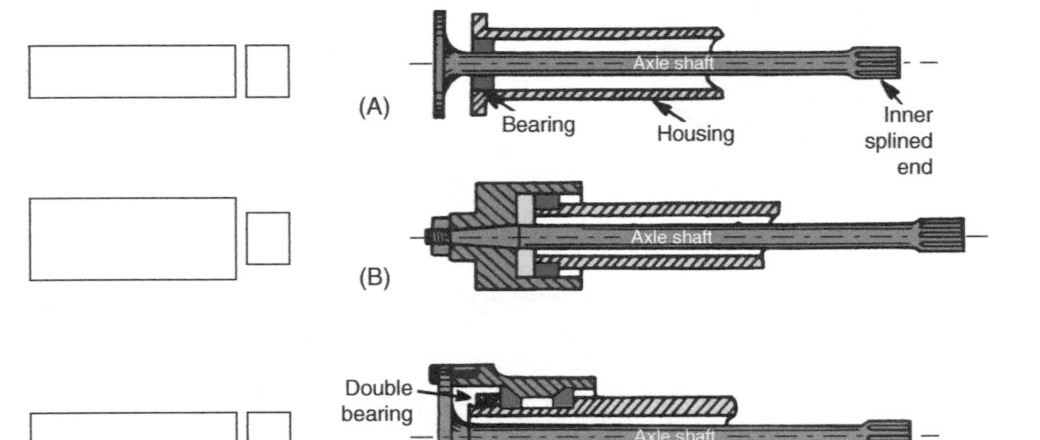

(A) Bearing Housing Inner splined end
Axle shaft

(B) Axle shaft

(C) Double bearing Axle shaft

1. This arrangement is used on commercial vehicles where a large load capacity is required. The half shaft does not bear any weight of the vehicle. The half shaft can be removed without affecting the wheel bearing assembly.

2. Least robust, used only on older cars with solid rear axle.

3. The bearing is positioned on the outside of the axle casing. Used on light commercial vehicles where a higher load bearing capacity is required.

Three-quarter floating   Semi-floating   Fully floating

FRONT-WHEEL DRIVE

Having identified the requirements and operating principles of a rear-wheel driveline, describe the three requirements of a front-wheel drive driveline using the diagram below as a prompt.

1 _____

2 _____

3 _____

Passenger side drive shaft   Up to 40° steering angle
Constant velocity (CV) joints
Driver side drive shaft
Up to 20° operating angle

Front-wheel drive drive shaft

Label the drive shaft components below using the terms:

inboard plunge joint   stub axle   nut
drive shaft   outboard CV joint

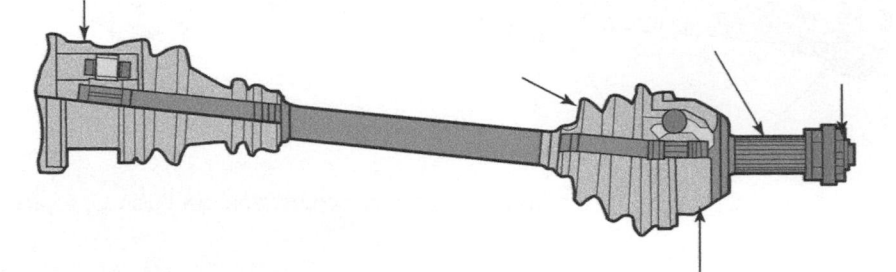

Describe the requirements of the constant velocity (CV) joint.

_____

_____

Describe the requirements of the inboard joint.

_____

_____

Constant velocity joint

The joint used on the outer end of front-wheel drive shafts contains hardened steel balls that transmit the drive using an inner race attached to the drive shaft and an outer race which is part of the stub axle.

Label the parts of the CV joints, selecting the appropriate labels from the list below.

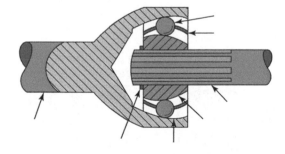

ball bearing	splined housing	inner race
outer race	splines to input shaft	splined input shaft
bearing cage	circlip	stub axle

Match the numbered component parts of the CV joint to the labels below (1 and 11 have been labelled to start you off)

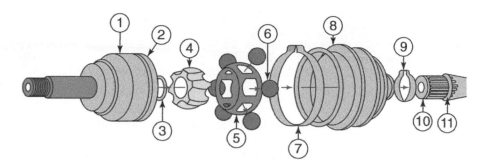

1	Splined housing		Boot clip
	Rubber boot (gaiter)		Cage
	Hardened steel balls		Circlip
11	Locking ring		Inner race
	Circlip groove		Outer race
	Boot clip		

As the name suggests there is no speed variation between the drive shaft and the splined shaft connected to the wheel.

Plunge type CV joint

In which part of the driveline would the joint shown in the figure below be employed, and how does its action compare with the normal type CV joint?

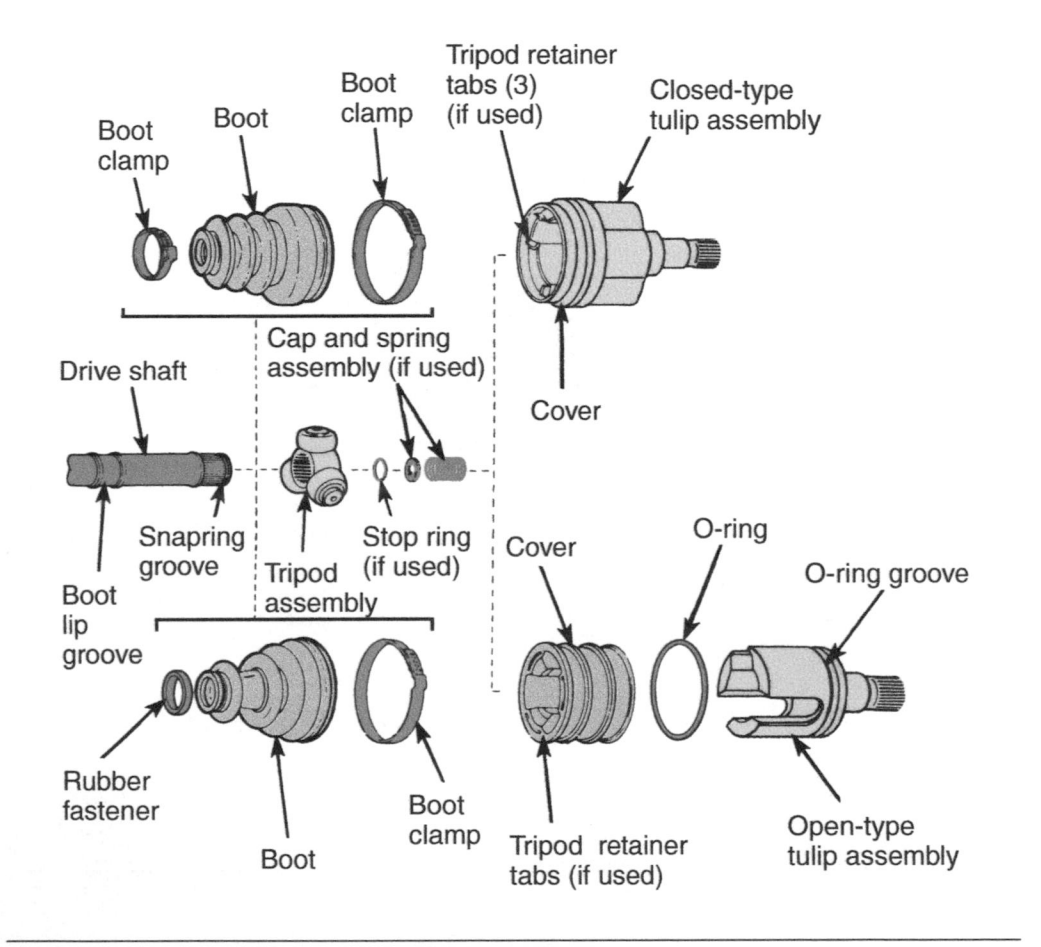

State the purpose of the boot (gaiter).

_____

Front-wheel drive – driveline checks

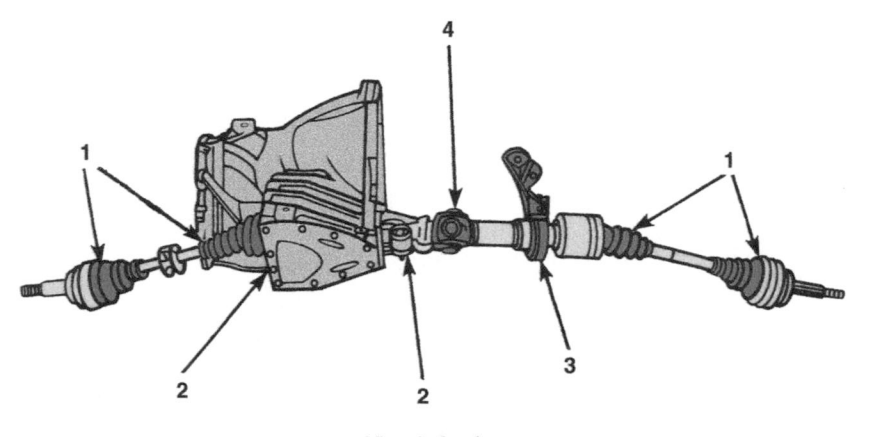

Visual checks

Complete the checks by filling in the blanks using the terms below.

weeping   security   gaiters   play   wear   tightness   splits

1 Check boots (_____) for _____ and clips for _____ and _____.
2 Check differential oil seals for _____.
3 Check the intermediate shaft bearing for excessive _____, _____ and lubrication. (if fitted).
4 Check inboard universal joints for wear and free play (if fitted).

TIP Split boots (gaiters) on CV joints is a MOT failure.

ROAD TEST

When carrying out a road test, check for knocking noises. With the vehicle stationary, turn on full lock, pull away and listen for knocking from the CV joints. Repeat on the other lock.

State the action to be taken if knocking is heard:

- _____

- _____

  _____

- _____

When road testing, check for other road users before pulling away to check for knocking.

How to dismantle a CV joint:

- Dismantle a CV joint after removing it from the axle shaft.
- Remove the first ball bearing from the joint by locating the wide gap on the cage and using a dummy shaft in the inner race to angle to its maximum.
- Remove the ball opposite to the first removed ball.
- Remove all balls.
- Using the wide gaps on the cage, manoeuvre the cage out of the housing.

Never put your fingers inside the joint to push the balls out.

Use a plain screwdriver to prise the balls out.

Fingers have been trapped!

Check the balls and races for signs of wear or overheating.

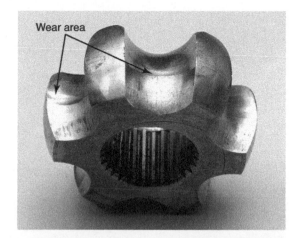

Wear area

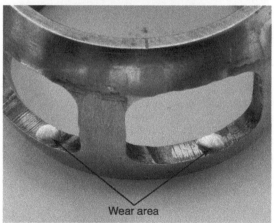

Wear area

Describe the visual signs that would render the joint (bearing) unserviceable.

- _____

- _____

- _____

- _____

243

DIAGNOSTICS: DRIVELINE, SHAFTS AND HUBS – SYMPTOMS, FAULTS AND CAUSES

State a likely cause for each symptom/system fault listed below. Each cause will suggest any corrective action required.

Symptoms	Faults	Probable causes
Backlash on take up of drive and vibration	Worn universal joint trunnions	
Lubricant leakage	Split rubber gaiters	
Transmission vibration	Worn centre bearing	
Knocking noise when turned on lock; vibration	Worn constant velocity joint	
Transmission vibration and noise	Propshaft universal joint worn	

WWW **http://www.gkndriveline.com/drivelinecms/export/sites/driveline/ downloads/brochures/driveshafts_english.pdf**

http://www.gkndriveline.com

http://www.aa1car.com/library/cvjoint2.htm

http://www.aa1car.com

Multiple choice questions

Choose the correct answer from a), b) or c) and place a tick [✓] after your answer.

1 **What type of vehicle is likely to use a solid axle with a fully floating hub?**

   a) small front-wheel drive cars [ ]

   b) high performance sports car [ ]

   c) large truck or bus (commercial vehicles). [ ]

2 **Why is a sliding joint required on a propshaft?**

   a) to enable the propshaft to change length due to suspension movement [ ]

   b) to allow for changes in angular movement [ ]

   c) to allow for engine vibration. [ ]

3 **Which of the following transmission driveline faults would fail an MOT test?**

   a) a knocking CV joint [ ]

   b) balance weight missing from a propshaft [ ]

   c) split CV boot (gaiter). [ ]

4 **On a front engine rear-wheel drive vehicle, a vibration and knocking can be caused by:**

   a) worn propshaft universal joints [ ]

   b) low gearbox oil level [ ]

   c) chipped tooth on the crown wheel. [ ]

5 **What method is used to check constant velocity checks for serviceability?**

   a) drive the vehicle at speed on a straight road [ ]

   b) drive the vehicle over bumpy roads [ ]

   c) pull away on full left and right lock. [ ]

PART 6
ELECTRICAL SYSTEMS

USE THIS SPACE FOR LEARNER NOTES

_____

_____

_____

_____

_____

_____

_____

SECTION 1

Understanding electricity

USE THIS SPACE FOR LEARNER NOTES

_____

_____

_____

_____

_____

_____

Learning objectives

After studying this section you should be able to:

● Explain the basic principles of electricity.

● Identify conductors and insulators and suggest uses of each.

● Use Ohm's Law to determine current, voltage and resistance.

● Determine power using voltage and current.

● Describe conventional and electron flow.

● State differences between Alternating Current (AC) and Direct Current (DC).

● Describe the differences between series and parallel circuits.

Key terms

Voltage   Electrical pressure.
Current   Flow of electrons.
Resistance   Slows the flow of electrons.
Power   Electrical energy.
Conductor   A material that has freely moving electrons which allow current to flow.
Insulator   Material offering considerable resistance to current flow.
Series   Single path for current to follow.
Parallel   Two or more paths for current to follow.
Electron   Negatively charged part of an atom.

www   http://science.howstuffworks.com

http://www.madlab.org/

http://www.electronics-tutorials.ws/

http://www.the-warren.org/quiz/insulators.htm

Electrical principles

```
O Y L V E P Q T X I A T R K M
S P N E A M P S I O R N E L R
W M D S L W I N H R O W M H
F S E Y L L S X I Q C R O Z S
Y W M X J U A S E E O T P W R
C A S K L O W R Q P N C F K V
N E G A T I V E A E D E O W E
R E T E M T L O V P U L H U C
S O W B R Y C I B J C E M G C
R E M W S U T T R E T E M M A
Z U I M R I I H S Q O S E Q I
E I H R S M Y Y I Z R Q T Y L
G O E O E C A E N Z H W E I F
B N P I T S P Y O K U F R U F
T I D R P B E Y C Y S B W R Q
```

AMMETER

AMPS

CONDUCTOR

CURRENT

ELECTRON

INSULATOR

NEGATIVE

OHMMETER

OHMS

PARALLEL

POSITIVE

POWER

SERIES

VOLTMETER

ELECTRICAL/ELECTRONIC PRINCIPLES

Circuit essentials

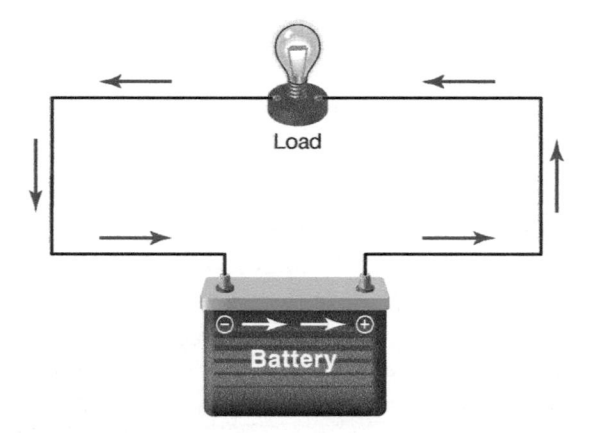

Load

Battery

To allow an electrical current to flow an electric circuit must consist of:

1 A power supply, which will be _____.

2 A consumer which is a device using electrical current such as _____
   _____.

3 Materials that will conduct the electric current from the supply source to the consuming
   device, and then return it to the source such as _____.

Name five different types of devices on a vehicle that consume the current to do useful work:

1 _____

2 _____

3 _____

4 _____

5 _____

Circuit characteristics

Current

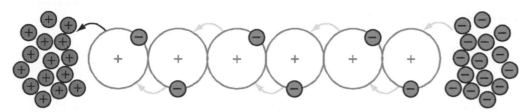

This represents the flow of electrons through a conductor from one atom to another. The greater the flow the more work can be done.

What units are used for current?

_____

Voltage

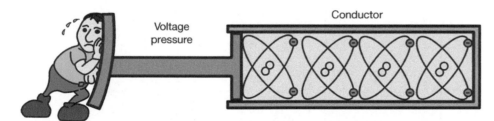

Voltage is the electrical pressure that causes current to flow.

What letter is sometimes used to identify voltage? _____

What units are used for voltage? _____

Which components create this electrical pressure within a vehicle?

● _____

● _____

Other terms used with voltage are:

EMF (Electromotive force) – Voltage available at the battery or generator (no circuit involved).

PD (Potential difference) – Difference in electrical pressure between two points in a circuit.

Resistance Ω

A resistance slows or impairs the flow of electrons.

What are the units of measurement? _____

What is meant by the term 'electrical insulator'?

_____

Draw lines on the figure below to match the material to either conductors or insulators.

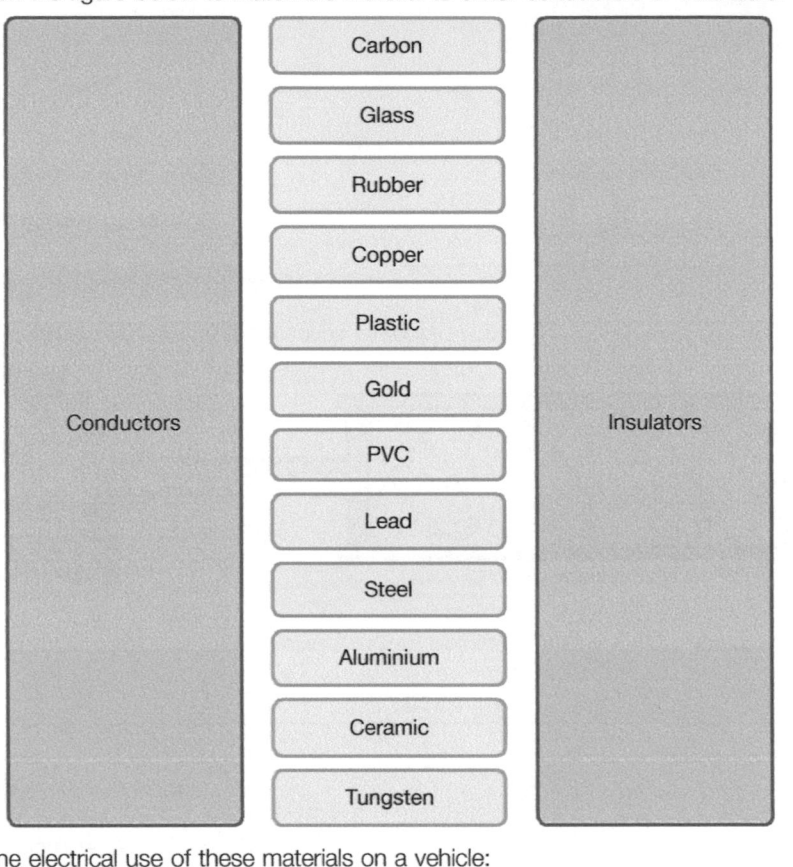

Conductors		Insulators
	Carbon	
	Glass	
	Rubber	
	Copper	
	Plastic	
	Gold	
	PVC	
	Lead	
	Steel	
	Aluminium	
	Ceramic	
	Tungsten	

Describe one electrical use of these materials on a vehicle:

Carbon – _____   Copper – _____

Ceramic – _____   Glass – _____

PVC (polyvinyl chloride) – _____   Lead – _____

Tungsten – _____   Gold – _____

Steel – _____   _____

What common feature applies to most of the conductors mentioned above?

_____

Conventional theory

Conventional theory is used for all diagnostic work and system operation.

Which way does current flow?

_____

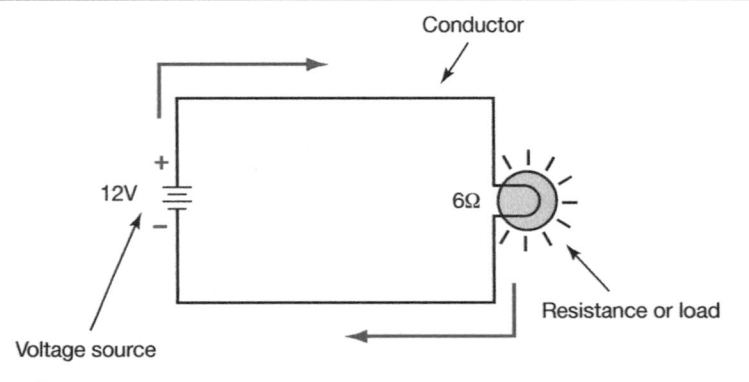

Electron theory

Scientists discovered that electron flow actually goes from negative to positive. This is only used when involved with designing electronics.

TIP Electron theory is not referred to in everyday electrical testing and diagnosis.

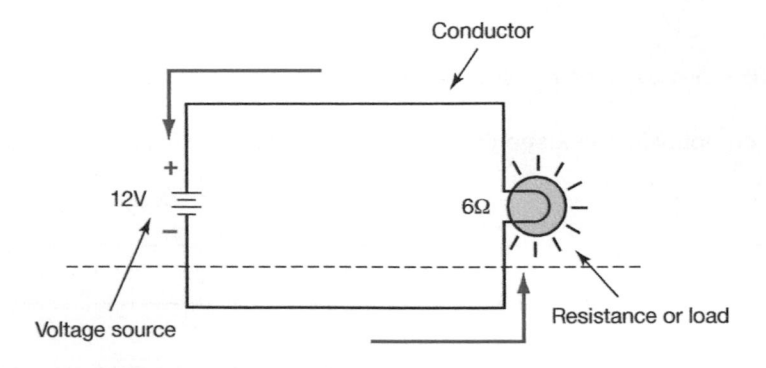

DC and AC theory

DC stands for: _____

AC stands for: _____

DC circuit

This type of circuit has the electrons flowing in one direction only and the voltage stays relatively constant.

What type of vehicle circuit uses DC current?

_____

AC circuit

Alternating current flows back and forth from positive to negative over a set time period. When shown graphically it appears as a sine wave.

What type of vehicle circuits use AC current?

_____

_____

OHM'S LAW: CURRENT, VOLTAGE AND RESISTANCE

Ohm's Law is the expression that relates voltage, current and resistance to each other.

Ohm's Law is used to calculate the third factor if the other two are known.

Cover the unknown factor in the Ohm's law triangle to determine the calculation required.

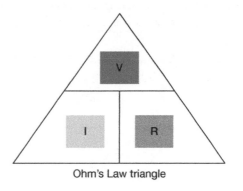

Ohm's Law triangle

When expressed as a formula the electric symbols stand for:

I – _____

R – _____

V – _____

The formula may be transposed to state:

V = _____

R = _____

I = _____

Ohm's Law questions

Example:

Calculate the current flowing in a coil of resistance 4Ω when the electrical pressure is 12V.

Formula: V ÷ R = I
              12 ÷ 4 = 3A

Current flow = 3A

1 Calculate the voltage required to force a current of 2.5A through a resistor of 5Ω.

Formula _____

_____

Voltage = _____

2 An alternator produces a current of 35A when the voltage is 14V. What is the resistance of the alternator?

Formula _____

_____

Resistance = _____

3 Calculate the current flowing in a relay coil of resistance 20Ω when the electrical pressure is 12V.

Formula _____

_____

Current flow = _____

4 What voltage will be required to cause a current flow of 3A through a bulb having a filament resistance of 4.2Ω?

Formula _____

_____

Voltage = _____

5 What will be the total resistance offered by a lighting circuit if a current of 10A flows under a pressure of 13V?

Formula _____

_____

Resistance = _____

6 Two 12V headlamp bulbs each have a resistance of 2.4Ω. Calculate the current flowing in each bulb and the total current flowing in the circuit.

Formula _____

_____

Current flow = _____

ELECTRICAL POWER

Power is the rate of doing work and is measured in _____.

The power triangle is shown in the figure below. Using the same principle as Ohm's Law determine the power formula:

W = _____

the current formula:

I = _____

the voltage formula:

V = _____

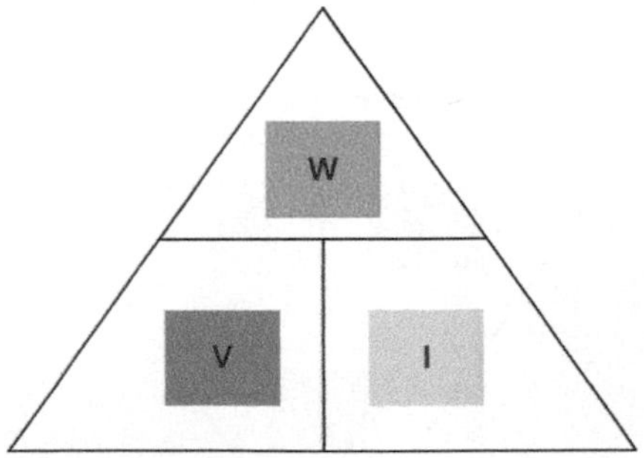

The power consumption in electronic circuits is very small, milliwatts (mW) or microwatts (µW), whereas the power consumed by lamps may be measured in watts (W), and by starter motors in kilowatts (kW).

1　Determine the power in a heated rear screen circuit when the voltage is 12V and the current flow is 3.5A.

Formula _____

_____

Power = _____

2　A starter motor consumes 1320W of power, while cranking the engine with 11V available. Calculate the current flow for this circuit.

Formula _____

_____

Current = _____

3　What is the voltage in a sidelight circuit when the power is 5W and the current is 0.4A?

Formula _____

_____

Voltage = _____

4　A 12V indicator circuit has a current flow of 1.68A. What is the bulb rating in watts?

Formula _____

_____

Current = _____

TIP　In the following problems use Ohm's Law and the power triangles to enable you to complete the calculations.

5　Determine the power in a heated rear screen circuit when the voltage is 12.8V and the resistance is 0.8Ω.

Formula _____

_____

Formula _____

_____

Power = _____

6　What is the resistance in a heated door mirror circuit when the power is 250W and the voltage is 12.5V?

Formula _____

_____

Formula _____

_____

Resistance = _____

SERIES CIRCUITS

This circuit has only one path to follow, from positive through both resistors back to the negative.

What would happen if the 2Ω resistor was to switch to an open circuit?

_____

Why would this type of circuit not be suitable for headlights on a car?

_____

Calculate the total circuit resistance.

_____

The basic laws of series circuits are stated below. Show calculations to obtain values when the battery voltage and resistances are as shown.

The ammeters and voltmeters (see figures below) are shown in their relative testing positions.

Calculate the total circuit resistance.

_____

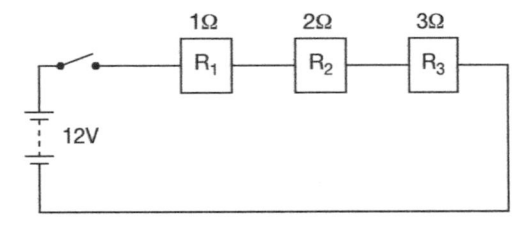

Resistance

The current is the same at any point in a series circuit. Calculate the current.

_____

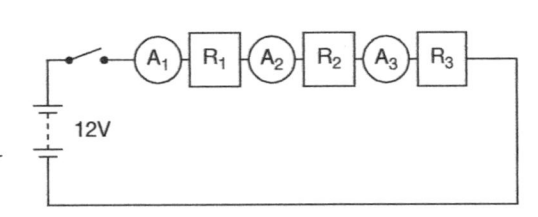

Current

The voltage of a series circuit is the sum of the voltage across each separate resistor.

Calculate the voltage at R1, R2 and R3.

$V = I \times R$

R1 = _____

R2 = _____

R3 = _____

Total voltage = _____

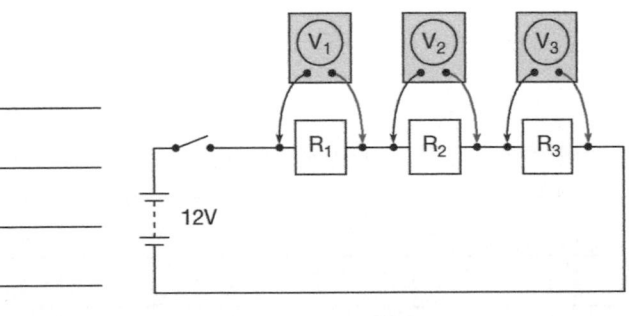

Voltage

PARALLEL CIRCUITS

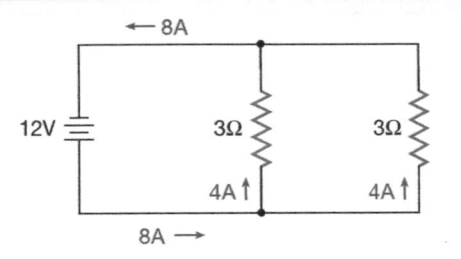

Parallel circuits have more than one path for current to flow. If one resistor went open circuit what would occur in the circuit?

_____

Unlike the series circuit, as more resistances are added to the parallel circuit the total circuit resistance reduces.

To calculate the total resistance with two resistors use the following formula:

R1 × R2 ÷ R1 + R2 = total resistance
(2 × 4) 8 ÷ (2 + 4) 6 = 1.33Ω.

What do you notice about the total resistance?

_____

What would happen to the circuit resistance if another path was introduced?

_____

Current

Current through a parallel circuit is the sum of the current through each separate branch of the circuit.

A1 = 12V ÷ 1Ω = 12A

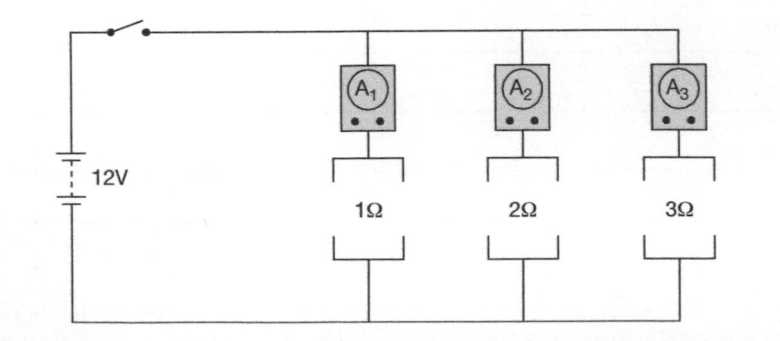

Show calculations for A2 and A3.

A2 = _____

A3 = _____

A1 + A2 + A3 = Total circuit current flow

12 + ___ + ___ = _____.

Voltage

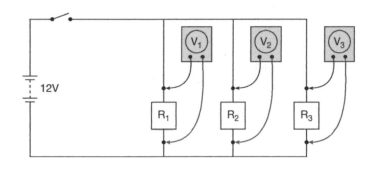

Voltage of a parallel circuit is the same across each separate resistor.

V1 = V2 = V3

 www **http://www.rare-earth-magnets.com/** (magnet university tab)

http://www.ndt-ed.org (Education Resources then Science of NDT)

http://www.school-for-champions.com

http://www.allaboutcircuits.com

SERIES AND PARALLEL CIRCUITS

Build a series circuit as shown in the figure below.

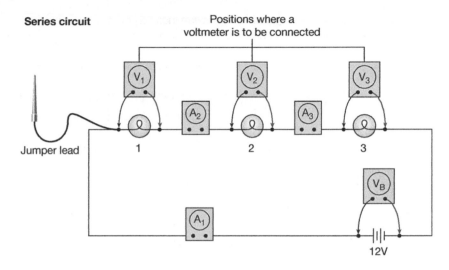

Connect all bulbs in circuit and note the readings in the following table:

Voltage					Current flow		
V₁	*V₂*	*V₃*	*Total*	*V_B*	*A₁*	*A₂*	*A₃*

Why are the bulbs dim? _____

Remove bulb 1, what happens? _____

Why does this occur? _____

 TIP V_B = battery voltage

Place prod of jumper lead to bulb 2 and note the readings in the following table:

Voltage				Current flow	
V_1	V_2	Total	V_B	A_1	A_2

What differences have now occurred and why?

_____

Remove bulb 2, connect prod of jumper lead to bulb 3 and take readings:

Voltage		Current flow
V_3	V_8	A_1

What happens to bulb 3? _____

Build a parallel circuit as shown in the figure below.

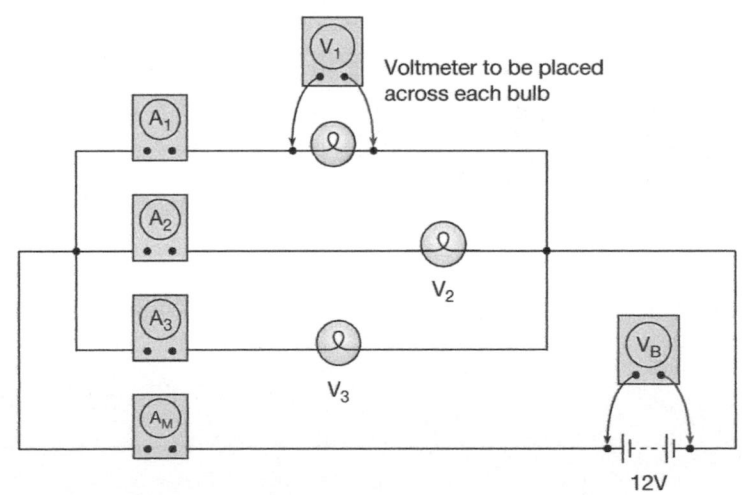

Voltmeter to be placed across each bulb

12V

Connect all the bulbs in the circuit and note the readings in the following table:

Voltage				Current flow				
V_1	V_2	V_3	V_B	A_1	A_2	A_3	Total	A_M

When three bulbs are connected in the parallel circuit, what differences occur with regard to the following, compared with the series circuit? Give reasons.

Voltage _____

_____

Current flow _____

Light intensity _____

Note the current flow when the bulbs are removed in the following table:

	A1	A_2	A_3	Total	A_M
One bulb removed					
Two bulbs removed					

Why didn't the same thing occur as in the series circuit?

_____

BUILDING VARIOUS ELECTRICAL CIRCUITS

Equipment required: Various types of resistances, switches, cables, battery and a digital multimeter.

Assemble the circuits described on page 257. Include in each circuit a switch and battery. In at least two circuits show the correct connection of an ammeter.

You may be provided with either:

(a) motor-vehicle components *or* **(b) a peg board** *or* **(c) a construction kit.**

Sketch the circuit diagrams as required and state the total current flow in each case.

1 Build the circuit below with two resistances in parallel.

State the current flow._____

2 Build a circuit with 4 resistances in series and sketch the circuit.

State the current flow._____

3 Build a circuit with 4 resistances in parallel and sketch the circuit.

State the current flow._____

4 Build the circuit with two resistances in series connected with one resistance in parallel and sketch the circuit.

State the current flow._____

Multiple choice questions

Choose the correct answer from a), b) or c) and place a tick [✓] after your answer.

1 **Current flow is said to be:**

a) the pressure forcing electrons to flow [ ]

b) electron flow multiplied by voltage [ ]

c) the flow of electrons through a conductor. [ ]

2 **Which of the following materials is an electrical conductor:**

a) PVC [ ]

b) carbon [ ]

c) ceramic. [ ]

3 **Which of the following circuits uses alternating current?**

a) wiper circuit [ ]

b) indicators [ ]

c) anti-lock braking system. [ ]

4 **The fog light circuit uses two 21W bulbs if the voltage is 12V, what is the total current flow?**

a) 9A [ ]

b) 1.75A [ ]

c) 3.5A. [ ]

5 **A circuit has three identical bulbs connected in parallel. What would happen to the total circuit resistance and current flow if another bulb was added?**

a) both the resistance and current would increase [ ]

b) the resistance would increase and current decrease [ ]

c) the resistance would decrease and the current increase. [ ]

6 **What unit is used to measure the power of a motor?**

a) Volt [ ]

b) Watt [ ]

c) Amp. [ ]

SECTION 2

Circuit concepts, components and test meter operation

USE THIS SPACE FOR LEARNER NOTES

_____

_____

_____

Learning objectives

After studying this section you should be able to:

- Describe earthing principles and earthing methods.
- Identify the use of different cables used in light vehicle circuits.
- Describe the operation of circuit protection devices and why these are necessary.
- Identify electrical symbols and units found in light vehicle circuits.
- Explain the voltage changes in series circuits.
- Explain the use of potential dividers and calculate voltage values.
- Describe the purpose and operation of electrical relays.
- Describe the features of basic test meters.

Key terms

Relay Electrically operated switch.
Voltage drop Loss of voltage occurring across any part of a circuit which is using current (consuming power).
Polarity conscious A device must be connected to the positive and negative terminals correctly.
Earth return The conductive part of the body is connected to the negative side of the battery.
Fuse Weak link in an electrical circuit to protect from current overloads.
Reluctance When the magnetic field is impaired.

EARTHING PRINCIPLES

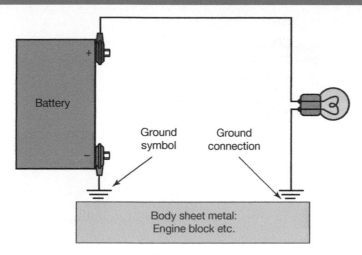

Earth return

On conventional vehicles it is common practice to allow the current, once it has passed through the electrical resistance (consumer) that it has operated, to return to the battery through the body frame instead of by a separate cable. This is known as an earth return system.

Which terminal of the battery is connected to earth on all modern vehicles?

_____

Give four advantages of using an earth return rather than an insulated return.

1 _____   3 _____

2 _____   4 _____

Complete the wiring diagram below incorporating four light bulbs connected in parallel and a switch with a 12V battery. Incorporate the earth return symbol.

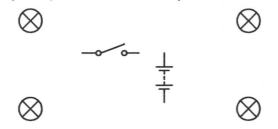

State a type of vehicle that does not use an earth return system:

_____

This type of vehicle uses a system called an insulated return where all the components have an insulated wire from and to the battery. This system is used because there is less chance of a short circuit which could cause a dangerous spark.

VEHICLE WIRING

The selection of cable and type of insulation required for motor vehicle use depends upon two main factors. These are:

1 _____

2 _____

Conductors

The conductive part of the wiring consists of a number of strands of wire wound together.

What material is mainly used? _____

Why are several strands used instead of a single wire?

_____

Insulation

The wire is covered to prevent any chance of the current escaping. This is determined by the voltage requirements of the circuit.

What material is mainly used for general wiring on 12 or 24V circuits?

_____

Ignition circuits have high voltages of up to 75 000V, therefore thick rubber insulation must be used.

Cable resistance

All conductors have a certain amount of resistance which is affected by five factors.

Using the clues, state the five factors that determine the resistance of a wire:

1 _____

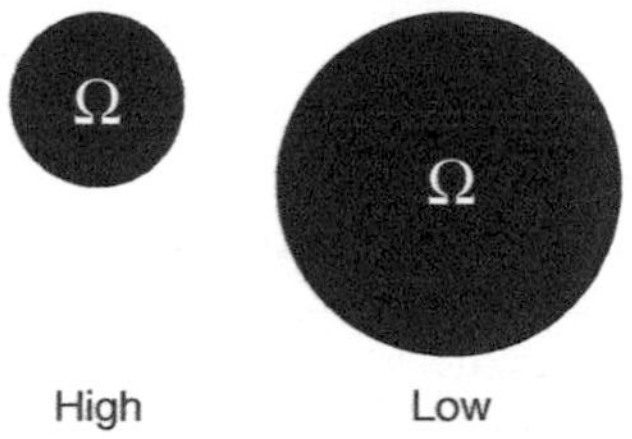

2 _____

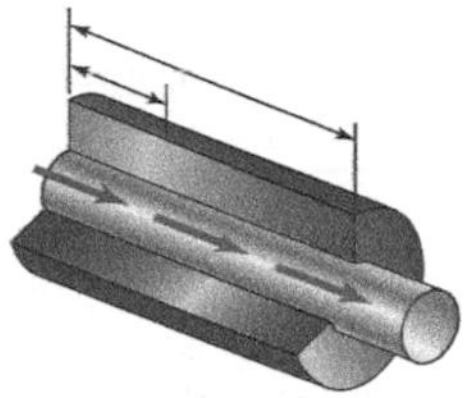

3 _____

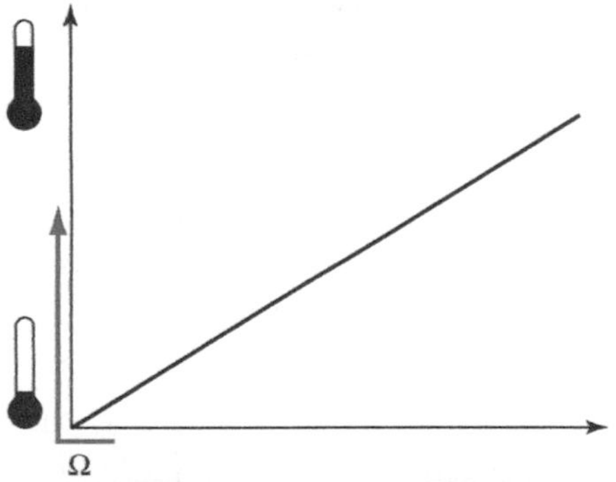

4 _____

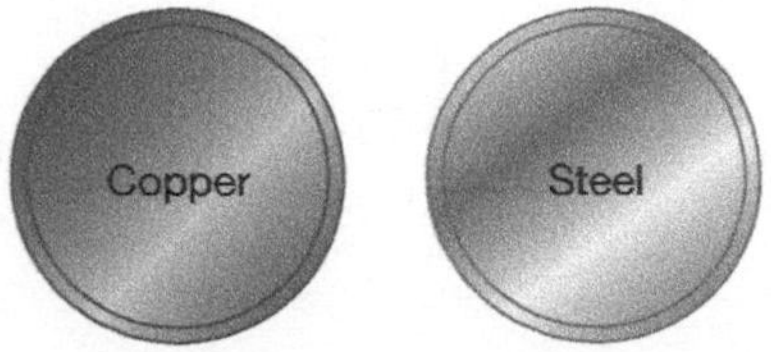

5 _____
_____

High   Low

Copper   Steel

Cable sizing is classified by the number of strands and the size of each strand. This relates to the current carrying capacity.

Number of strands and diameter in mm	Maximum continuous current rating in amps	Typical applications
14/0.25	6.00	Side and tail-lights, instruments, courtesy lights, engine management sensors, in-car entertainment
14/0.30	8.75	
21/0.30	12.75	
28/0.30	17.50	Headlights, heated rear screen, horn, manifold heater, heated seats
35/0.30	21.75	
44/0.30	27.50	
120/0.30	60.00	Alternator
37/0.90	170.00	Starter cable – petrol engines
61/0.90	300.00	Heavy duty starter cable – medium diesel engines
61/1.13	415.00	Heavy duty starter cable – large diesel engines

Select several different sized wires. Measure the wire size and count the number of strands to get an idea of where each wire may be found on a vehicle. Use the table below to record your results.

Strand diameter in mm	Number of strands	Current rating

Cables are often bunched together in a harness or loom. This simplifies fitting and ensures less chance of breakage or short circuits. Each cable within the harness is colour coded. What is the reason for colour coding?

_____

Primary colours: **One particular colour predominates in each circuit.**
Trace colours: **One or more thin lines of different colours are added to the primary colour.**

When checking out a wiring problem always refer to the manufacturer's coding table to determine the letters used for each colour.

British convention		
Colour	**Symbol**	**Destination/Circuit**
Black	B	Earth
Red	R	Sidelight circuits
Orange	O	Wipers
White	W	Ignition system

Colour	Symbol	Destination/Circuit
Brown	N	Main battery lead
Blue	U	Headlight circuit
Green	G	Fused ignition supply
Light green	LG	Instruments
Slate	S	Electric windows
Yellow	Y	Fuel system
Purple	P	Permanent fused supply

This table is for only one convention. Europe and Japan have different codings for some of the colours and circuits.

The main feed cable to a specific switch may have a single colour, whilst beyond the switch the lead has the same main colour but a different coloured trace line passing through it.

Using the coding table for British conventions, complete the following table to show the symbols and colour identifications:

Symbol	Primary colour	Trace colour	Symbol	Primary colour	Trace colour
WR	White	Red	LG____	_____	White
_____	Red	Black	SB	_____	_____
UW	_____	_____	PW	_____	_____
N____	_____	Yellow	_____	Green	Red

CIRCUIT PROTECTION

State the function of a fuse or circuit breaker.

_____

_____

There are three types of fuses used in the automotive industry.

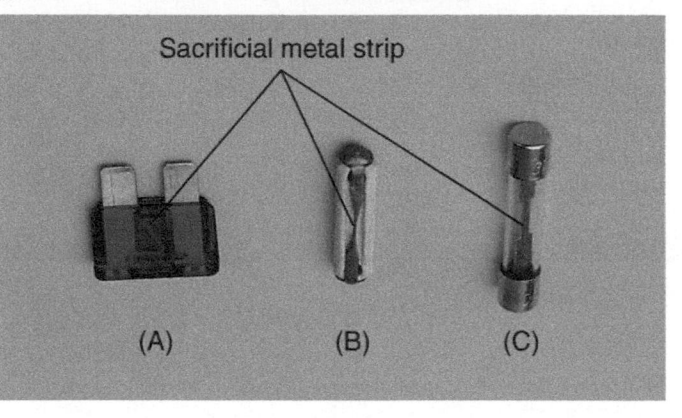

Sacrificial metal strip

(A)   (B)   (C)

State the other two types of fuses shown and where they are most likely to be found:

A   blade type fuse – most commonly found on current vehicles

B   _____

C   _____

The blade type comes in three sizes.

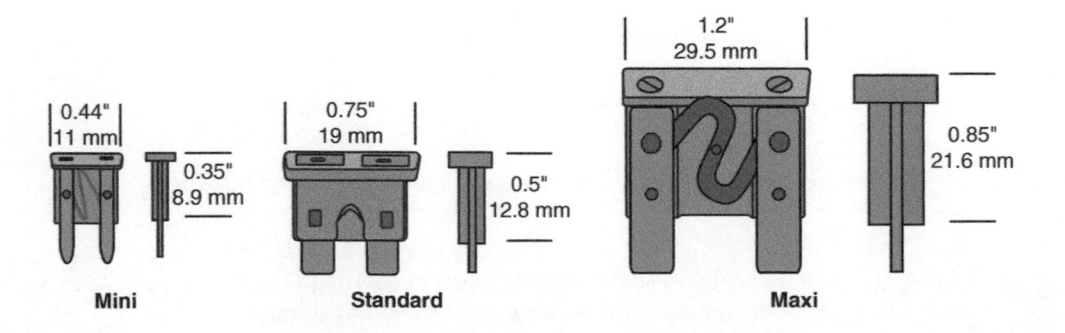

Mini   Standard   Maxi

The body is made from plastic, the colour of which identifies the fuse rating. The fuse rating is usually printed on the top of the plastic body.

What happens if the current becomes greater than the fuse can normally conduct?

_____

Standard blade type fuse ratings		
Identification	Continuous rating	Approximate blow rating
Purple	3 amps	6 amps
Pink	4 amps	8 amps
Orange	5 amps	10 amps
Brown	7.5 amps	15 amps
Red	10 amps	20 amps
Blue	15 amps	30 amps
Yellow	20 amps	40 amps
White	25 amps	50 amps
Green	30 amps	60 amps

Explain the difference between continuous rating and blow rating of fuses.

Continuous rating: _____

_____

Blow rating: _____

_____

The maxi fuse is used as a fusible link with higher ratings. The figure below is a wire type fusible link.

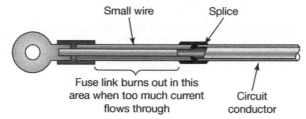

Small wire   Splice

Fuse link burns out in this
area when too much current
flows through   Circuit
conductor

How does a fusible link compare with the fuses shown on page 262 and where would it be used?

_____

_____

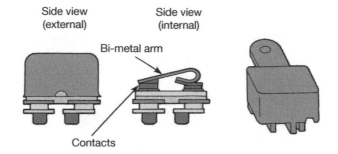

Side view   Side view
(external)   (internal)

Bi-metal arm

Contacts

Bimetal circuit breakers

How does a circuit breaker differ from a fuse?

_____

_____

What type of circuit may use a circuit breaker?

_____

ELECTRICAL SIGNS AND SYMBOLS

Symbols are used in wiring diagrams for ease of understanding and to reduce the size of the drawings.

Label the electrical symbols below using the words from the list.

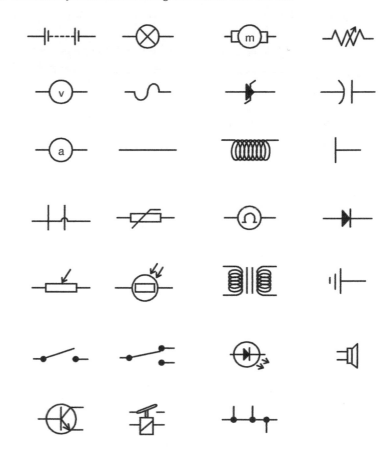

battery	ohmmeter	relay
motor	zener diode	two-way switch
transistor	wires crossing	variable resistor
potentiometer	wires joined	transformer
wire	bulb	light dependent resistor
light-emitting diode	capacitor (condenser)	diode
earth	on–off switch	fuse
ammeter	inductor (coil or solenoid)	loudspeaker
thermistor	voltmeter	earth point

VOLTAGE DROP

Kirchoff's Voltage Law states that the sum of the voltage drop in a circuit will equal the voltage of the battery.

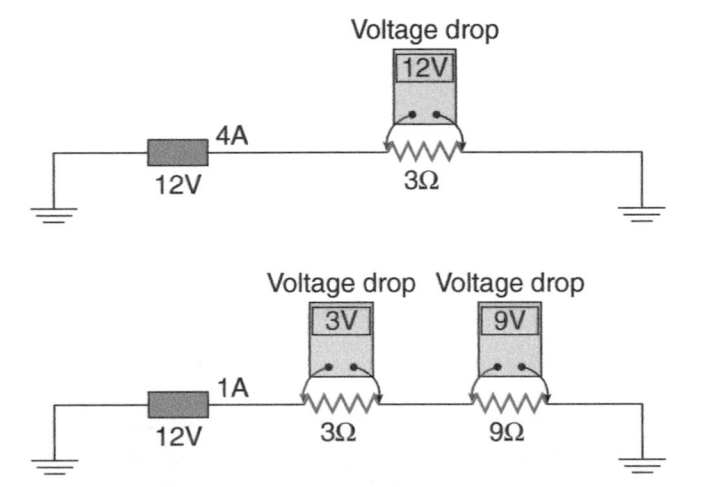

What will be the voltage drop across the 4Ω resistor below? _____

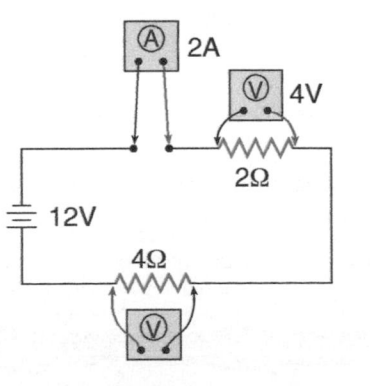

In a series circuit what will happen to the voltage drops as more resistances are introduced?

_____

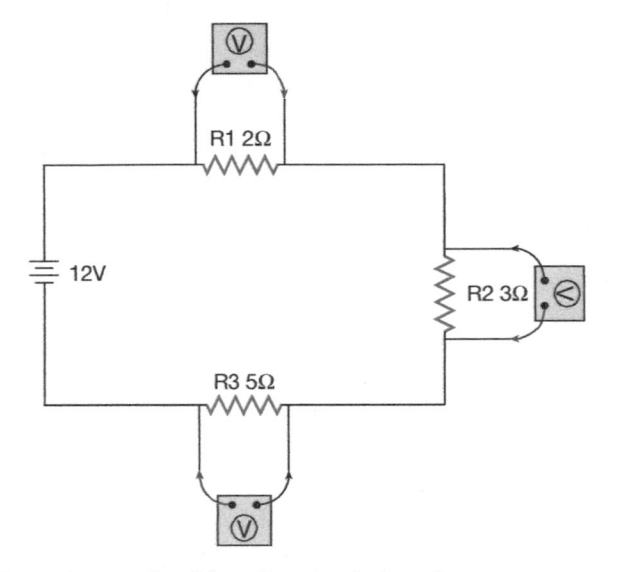

What will be the voltage drop on the 5Ω on the circuit above? _____

Use Ohm's Law to prove your answer.

_____

_____

_____

Parallel voltage drops

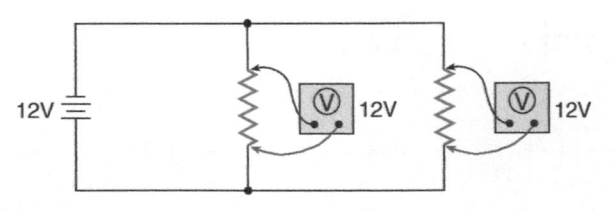

What happens to the voltage drop on each leg of the circuit above?

_____

POTENTIAL DIVIDER

This divides potential (voltage) in a series circuit. If a variable resistor is used, the output voltage can be varied and provide control.

From previous work, if the resistances are equal then voltage at V = half the total voltage. This is known as the output voltage or voltage out (V out). By altering the values of the resistors, this voltage can be varied to give very high or very low values.

On the diagram opposite, voltage out = _____

Voltage out can be calculated using the formula:

$$V \text{ out} = \frac{R_2}{R_1 + R_2} \times \textbf{supply voltage}$$

$$V \text{ out} = \frac{10\Omega}{10\Omega + 10\Omega} = 0.5\Omega \times 12V = 6V$$

Calculate the V out values when:

1 $R_1 = 20$ and $R_2 = 60$

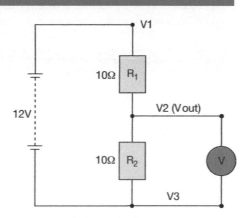

2 $R_1 = 2500$ and $R_2 = 800$ (round to two DP)

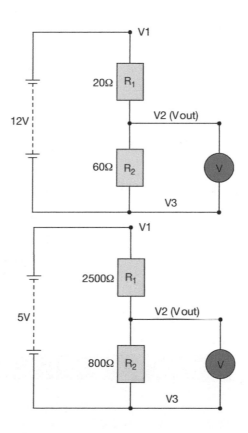

3 $R_1 = 100\Omega$ and $R_2 = 800\Omega$ (round to two DP)

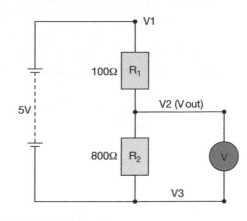

ENGINE TEMPERATURE SENSING

The figure below shows an example of the use of a potential divider.

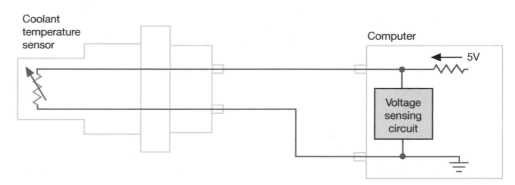

Inside the engine control unit (ECU) computer is a fixed resistance. The coolant temperature sensor is normally a negative temperature coefficient (NTC) type.

Explain the meaning of NTC with regards to resistance.

_____

Questions 2 and 3 above are based on this type of circuit, with question 2 being a cold engine temperature and question 3 a hot engine.

ELECTROMAGNETISM

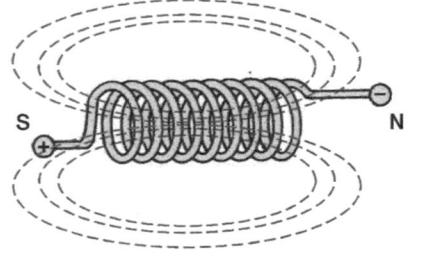

Passing a current through a wire causes a magnetic field which will collapse when current flow stops.

When air is present inside the magnetic field reluctance is caused which is much like resistance in a circuit.

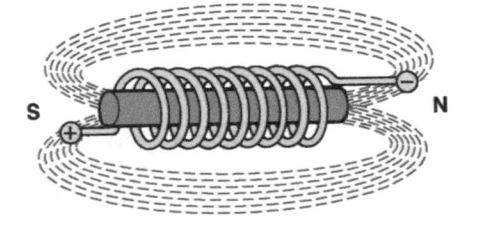

What has been added above to reduce the reluctance and increase the magnetic field?

_____

Give two other ways to increase the magnetic field strength:

1 _____

2 _____

SWITCHES AND RELAYS

A switch in a circuit is used to make or break the flow of current in the circuit. Depending on the application, switches can be simple on–off spring-loaded toggle illustration types or complicated multi-function switches such as the stalk type steering column arrangements.

In many circuits on a motor vehicle the manually operated switches energize relays which in turn carry the main current load for the circuit.

Give two reasons for the use of relays:

1 _____

2 _____

Relay operation

Using the diagram and a selection of words from the list below, complete the statements describing relay operation.

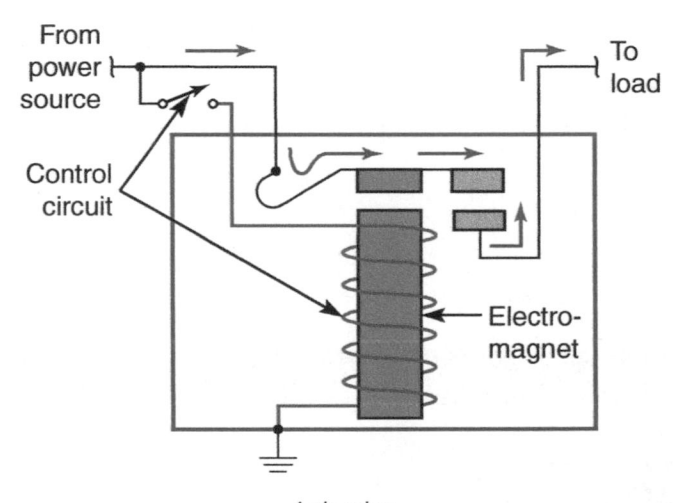

4 pin relay

Note: words might be used more than once in the following paragraphs.

closed	low current	open	power
iron	load	switch	contacts
electromagnet	voltage		

When the control circuit is energized by the _____ switched part of the circuit, the soft _____ core is tuned into an _____, which pulls the high current _____ together connecting the _____ to the _____ circuit.

When the control switch is opened the _____ ceases and the contacts then _____, cutting the power to the load.

Horn circuit

The figure below shows a typical circuit incorporating a 4 pin relay for switching.

Using the following convention for numbering the relay pins, label the drawing with the corresponding pin numbers.

86 – Fused low current control circuit positive feed.
85 – Low current control circuit earth.
30 – Fused high current power source positive feed.
87 – High current feed to the load.

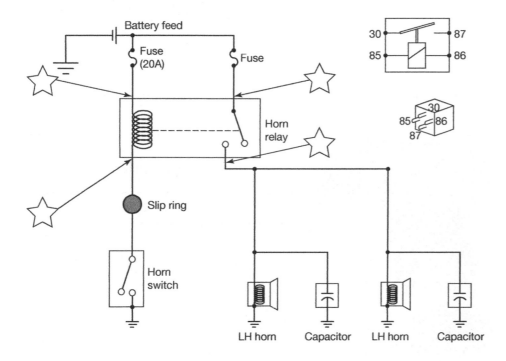

Some interior light circuits employ a delay relay and some heated rear window circuits employ a timer relay.

Give two examples of an intermittent relay:

1 _____

2 _____

5 pin 'either/or' relay

This relay is exactly the same in operation as a 4 pin relay with an extra pin which allows high current to flow from pin 30 to pin 87 constantly when the relay low current control circuit is not activated.

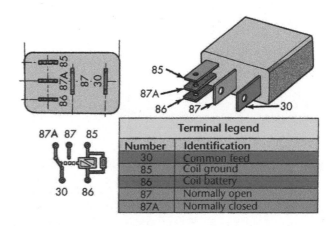

Terminal legend	
Number	**Identification**
30	Common feed
85	Coil ground
86	Coil battery
87	Normally open
87A	Normally closed

Why is it called an 'either/or' relay?

_____

ELECTRICAL TEST EQUIPMENT

Voltmeter

The voltmeter measures electrical pressure in volts. This meter is **polarity conscious**.

Explain the meaning of this term.

_____

_____

How is a voltmeter connected into the circuit?

_____

Pictures supplied by Draper Tools Limited

Draw a voltmeter in position on the circuit below and explain why a voltmeter can be connected in this position.

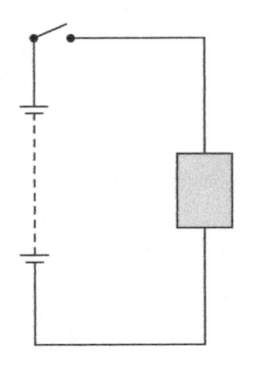

_____

 **TIP** A voltmeter is the best meter for testing purposes as it is safe to use and can operate while the circuit is in operation.

Ammeter

The ammeter measures current flow in amperes. An ammeter is polarity conscious – true or false?

_____

How is an ammeter connected into the circuit? _____

Draw an ammeter in position on the circuit below.

TIP **Caution:** an ammeter must not be connected straight across the battery as it has a low internal resistance allowing large current flow to damage the meter or at least blowing the internal fuse.

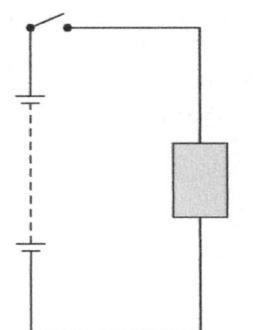

What has to be done to the circuit before connecting the ammeter?

_____

Ohmmeter

The ohmmeter looks similar to the voltmeter and the ammeter. It measures electrical resistance of circuits or the components in circuits.

An ohmmeter is polarity conscious – true or false? _____

What is a feature of the ohmmeter which does not occur with the other types of meter?

_____

_____

What precautions must be taken before checking a component with an ammeter?

_____

 TIP Before checking resistance join the probes and check for a zero reading.

 WWW
http://www.autoshop101.com/forms/hweb1.pdf

http://phet.colorado.edu/en/simulations/category/physics

http://www.allaboutcircuits.com/

http://www.kpsec.freeuk.com/symbol.htm

http://library.thinkquest.org/10784/circuit_symbols.html

Multiple choice questions

Choose the correct answer from a), b) or c) and place a tick [✓] after your answer.

1 **Why is an earth return system used on light vehicles?**

a) for safety reasons [ ]

b) to reduce static electricity [ ]

c) to reduce the amount of cable required. [ ]

2 **If a cable is rated 28/0.30, what does this relate to?**

a) the cable has 30 strands 28mm in diameter [ ]

b) the cable can carry 28 amps for 30 minutes [ ]

c) the cable has 28 strands with a diameter of 0.30 mm. [ ]

3 **Which type of fuse is commonly found on new cars?**

a) blade type [ ]

b) ceramic type [ ]

c) glass cartridge type. [ ]

4 **What does this symbol represent?**

a) potentiometer [ ]

b) capacitor [ ]

c) transistor. [ ]

5 **What will the battery voltage read in this circuit?**

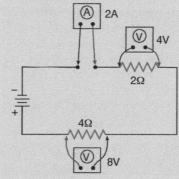

a) 12V [ ]

b) 8V [ ]

c) 6V. [ ]

6 **What occurs in an NTC-type thermistor as temperature rises?**

a) the resistance reduces [ ]

b) the resistance increases [ ]

c) a DC voltage is produced. [ ]

7 **What is the purpose of a relay?**

a) to reduce the current flowing in a circuit [ ]

b) to allow a small current to switch a high current circuit [ ]

c) to reduce the amount of wiring required. [ ]

8 **Which of these meters is connected into a circuit in parallel?**

a) ammeter [ ]

b) ohmmeter [ ]

c) voltmeter. [ ]

9 **Why must an ammeter NOT be connected across the battery terminals?**

a) because it has a low internal resistance [ ]

b) because it has a high internal resistance [ ]

c) because it has its own power supply. [ ]

10 **What should be carried out before using an ohmmeter to test the resistance of a component?**

a) check the internal battery voltage [ ]

b) select the volts DC scale [ ]

c) join the probes and check for zero. [ ]

SECTION 3

Batteries

USE THIS SPACE FOR LEARNER NOTES

_____

_____

_____

_____

_____

_____

_____

_____

_____

Learning objectives

After studying this section you should be able to:

● Identify light vehicle batteries.

● Describe the construction and operation of light vehicle batteries.

● Explain the different ways a battery is rated.

● Describe how to remove and replace batteries.

● Compare light vehicle batteries.

Key terms

Electrolyte   Active fluid in a battery cell.

Battery capacity   How well a battery will perform under certain conditions (e.g. cold cranking and reserve capacity).

Cell   Contains a positive and negative plate surrounded by an electrolyte in a container. Six linked cells make a 12 volt battery.

Lead dioxide   Composition of a positive plate in a fully charged cell.

Lead sulphate   Composition of a positive plate in a fully discharged cell.

Spongy lead   Composition of a negative plate in a fully charged cell.

Plates   A positive and negative plate are in a single cell.

Maintenance-free battery   A sealed battery which requires limited maintenance.

Memory saver   A device connected via the cigarette lighter socket which provides voltage to essential consumers when the vehicle battery is disconnected.

www.varta-automotive.com

www.exide-evolution.com

www.yuasa-battery.co.uk

BASIC CONSTRUCTION

Complete the sentences below using the words in the word bank. Note that there is one distracter word.

electrons   **electrochemical**
electrolyte   **mechanical**
chemical   **electrical**

A battery is a device which converts _____ energy into _____ energy.

Inside a battery there are two plates or electrodes surrounded by an _____.

Chemical reaction in the battery cell produces _____, this effect is called an _____ reaction. A battery stores DC voltage and releases it when connected to a circuit. A number of cells are connected together to make up a battery.

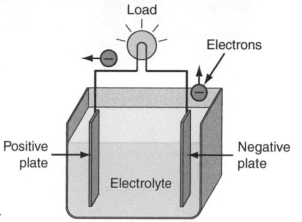

Load

Electrons

Positive plate

Negative plate

Electrolyte

LEAD-ACID BATTERY CONSTRUCTION

A 12V battery container consists of six separate compartments. The picture opposite shows a cutaway lead acid battery. Each compartment contains a set of positive and negative plates and each set is fixed to a bar which rises to form the positive or negative terminal. Between each plate is an insulating separator.

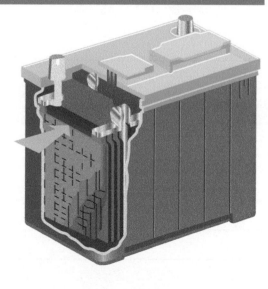

Since each cell has a nominal electrical pressure of 2V, to produce a 12V battery six cells must be joined together in series.

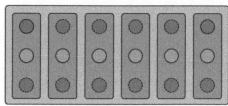

The cells are connected by buss bars (or links), not normally seen on modern batteries. Show where these would be connected in the figure above and indicate the polarity of each cell by symbols.

The battery main components are shown below.

Label the sketch with the following:

Sediment chambers   **Positive plate (3) grid**   **Connecting strap**
Negative plate (4) grid   **Separators**

Describe the following terms and explain their function:

Plate grids: _____

_____

_____

_____

Separators: _____

_____

Container (or case): _____

_____

Electrolyte: _____

_____

_____

Wear the correct PPE when topping up batteries (goggles to protect your eyes, an apron to protect your clothes and gloves to protect your hands).

Battery acid

Safety precautions need to be observed when handling battery acid. Complete the following bullet points using the word bank. Note: there are two extra distracter words.

burn   cold   neutralized   eyewash   goggles   boots   rinse   hot   spilling

- Wear suitable PPE (splash-proof _____, gloves and aprons). It may be necessary to wear a face shield when handling sulphuric acid.
- Do not store acid in _____ locations or in direct sunlight.
- Use extreme care to avoid _____ or splashing the sulphuric acid solution. It can destroy clothing and _____ the eyes and skin.
- Use an _____ station if the sulphuric acid solution is splashed into the eye.
- Spilled or splashed sulphuric acid solution needs to be _____ with a baking soda solution, and _____ the spill area with clean water.

Battery terminals

Name the types of battery connectors shown opposite:

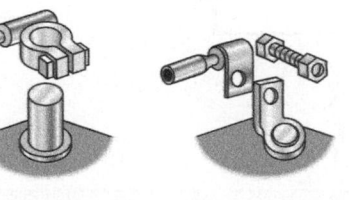

....

BATTERY HOLD DOWNS

Hold-down bolt

Battery case

J-bolts

Courtesy of Chrysler LLC

A battery needs to be held securely in place to prevent it from tipping over. If it was to tip over it

could _____.

This is why battery clamps are used. They need to be checked regularly for security.

BATTERY-CHARGING

A battery is a series of chemical cells. Each cell of a lead-acid battery is capable of producing 2V. The size and number of plates in the cell determine its capacity or output. It is of a secondary cell type.

What does 'secondary cell' mean?

_____

_____

What is a primary cell? Give an example.

_____

_____

The positive and negative plate active materials of the secondary cell have a different chemical composition and when submerged in a suitable electrolyte produce an electrical pressure difference which, when connected to a circuit, allows a current to flow. The chemical reaction then discharges the cell until the plates become chemically similar.

Why is the cell then capable of being recharged?

_____

_____

 When lead-acid batteries are being charged they can produce explosive mixtures of hydrogen and oxygen gases.

Remove or loosen the battery caps. This is to prevent a build-up of gas in the battery which, if left unattended, could cause the battery to explode. Always remember to charge lead-acid batteries in a well ventilated area away from any form of ignition. It is recommended that all batteries are charged in a dedicated area of the workshop. Ideally, a safety shower and an eyewash station should be in the battery-charging area.

When the battery requires charging, the supply, be it from the vehicle's charging system or the mains supply, must be a direct current (DC). Why must only DC be used?

_____

Complete the table to show how the plate materials and electrolyte are affected during the charge and discharge cycle.

Process	Positive plate	Electrolyte	Negative plate
Fully charged	_____	_____	_____
Discharged	_____	_____	_____

During the charge process the chemical reaction causes the plates to give off bubbles. Name the gas given off at the:

positive plates _____ negative plates _____

When batteries are charged on a vehicle by the alternator, they are initially (after engine starting) subjected to a high rate of charge and then controlled to a much lower charge rate.

When removed from the vehicle and externally charged, fast or slow chargers may be used.

It is preferable to use a _____ charger.

What major precautions should be taken if an external charger is used to charge a battery still connected to the vehicle?

_____

_____

_____

Picture kindly provided by permission of Snap-On Industrial

Battery charger

Show on the figure opposite how three 12V batteries are connected to a constant current charger.

How are the cells electrically connected?

_____

When connecting a battery to an external charging system, what way (electrically) are the charger-to-battery connections made?

_____

_____

Constant current battery charger

List the precautions required when charging batteries:

_____

_____

BATTERY STATES OF CHARGE AND CAPACITY

The performance of a battery is affected by many factors; these include its state of charge, temperature, condition and age.

Complete the figure below to show the variation of internal electrical resistance throughout the charge cycle. When is the resistance at its lowest and what would cause it to increase?

_____

_____

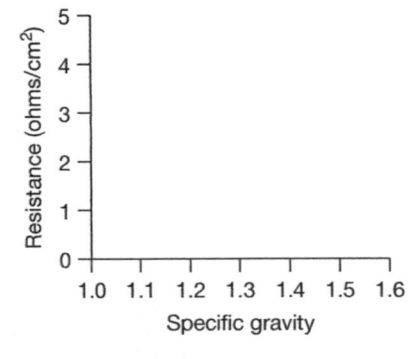

Cold cranking performance

Batteries are given a rating to indicate their ability to crank the engine at –18°C. At this temperature, or lower, the capacity of the battery is greatly reduced.

Complete the figure below to show how the capacities of a fully charged and 80 per cent charged battery are reduced by a fall in temperature to –18°C.

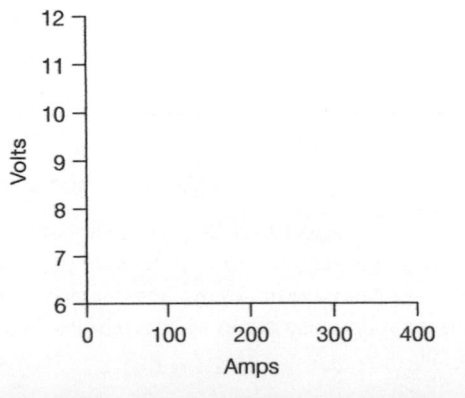

The values on the previous graph show the amount of current being drawn from the battery. Describe the effects of low temperatures on the battery.

_____

_____

_____

Why should batteries never be completely discharged?

_____

_____

_____

Show by means of a graph the effect of the charging cycle on relative density (specific gravity) and cell voltage.

_____

State the effects caused by excessive overcharging.

_____

_____

_____

State the effects caused by a lack of charging or a battery standing idle.

_____

_____

What is the capacity of a battery primarily dependent on?

_____

_____

What can reduce the capacity of a battery?

1 _____

2 _____

3 _____

4 _____

⚡ Disconnect the battery negative terminal first, followed by the positive terminal. This is to prevent the possibility of shorting the spanner out on the vehicle chassis. When connecting the battery attach the positive terminal first, followed by the negative terminal.

↻ **Connecting batteries in series or parallel**
You will need two fully charged batteries for this activity. Connect them together in series and parallel. Complete the diagrams opposite and record your findings.

_____

_____

_____

⚡ Remember to carry out all safety precautions and wear correct PPE whilst being supervised by your lecturer/teacher.

The capacity or output voltage of a system can be changed by connecting batteries together. Show how pairs of batteries are connected in series and parallel on the figures below.

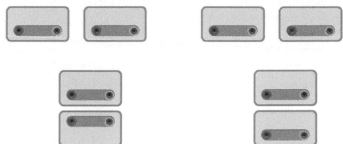

What will be the effect on voltage output and capacity if batteries are connected in:

1 Series?

_____

_____

2 Parallel?

_____

Battery maintenance

What maintenance checks are indicated on the drawing opposite?

1 _____

2 _____

3 _____

4 _____

5 _____

_____

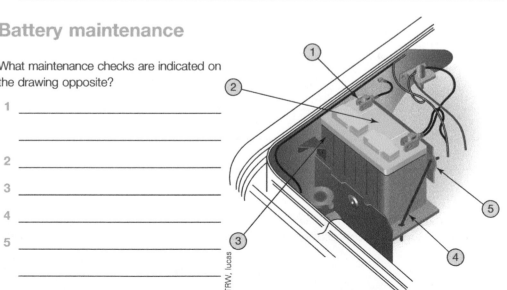

TRW, lucas

What would not require checking on a modern 'low maintenance' or 'maintenance-free' battery? Give reasons.

_____

_____

If a battery is suspected of being faulty, a systematic check of its condition should be carried out. State the basic visual and manual checks that should be made:

1 _____

2 _____

3 _____

4 _____

5 _____

When testing the battery, equipment such as a voltmeter, hydrometer, refractometer, high rate discharge tester and battery charger may be used. List typical faults that testing may reveal:

1 _____

2 _____

3 _____

4 _____

5 _____

6 _____

BATTERY TESTING

Carry out tests 1, 2a, 2b and 3 on a battery and state the function or make comments on the six tests mentioned.

1 Open circuit voltage test

Switch on headlamps for 30 seconds, then switch off all electrical loads. Connect a voltmeter across the battery terminals to determine the stabilized open circuit voltage. Expected voltage readings are:

_____

_____

_____

Reading obtained_____ Comment_____

This test would replace the hydrometer test on sealed batteries.

2a Hydrometer test

How does this test indicate the condition and state of charge of the battery?

Show specific gravity readings for the values indicated on the float

Discharged _____

70% Charged _____

Charged _____

TRW, lucas

Note: Readings should indicate over 70 per cent charged before carrying out further tests.

Show specific gravity readings for the values indicated on the float.

Readings obtained						
Cell number	*1*	*2*	*3*	*4*	*5*	*6*
Specific gravity value						

Comment _____

A few batteries have built-in hydrometers. They have an indicator in the top of the battery case. The state of charge can be checked by this display as it will change colour accordingly.

Built-in hydrometers

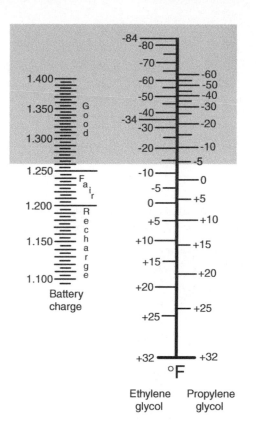

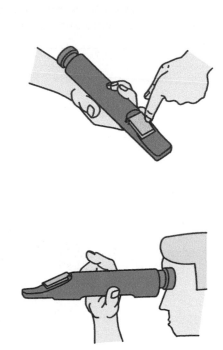

2b Refractometer

This can be used instead of a battery hydrometer. It can be the same refractometer used to test engine coolant and screen wash, which can also be used to test the specific gravity of electrolyte. These tools can feature automatic temperature compensation.

Describe how a refractometer is used to take a reading of the specific gravity electrolyte:

● _____
_____

● _____
_____

3 High-rate discharge test (battery load test)

This is made to ensure that each cell will supply the heavy currents required for starting.

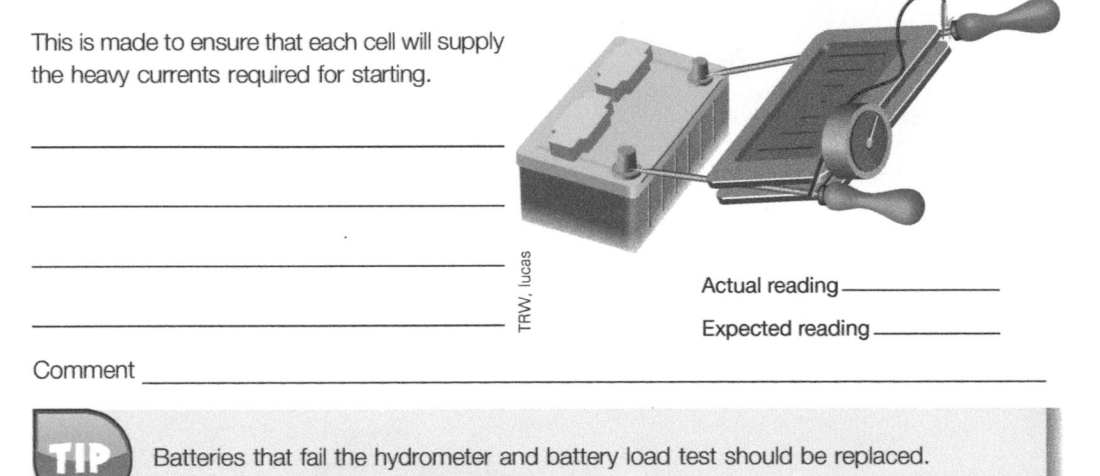

TRW, lucas

Actual reading _____

Expected reading _____

Comment _____

TIP   Batteries that fail the hydrometer and battery load test should be replaced.

4 Cycling test

_____

_____

5 Capacity tests

Reserve capacity: _____

_____

_____

10/20 hour rate: _____

_____

6 Cold start performance

_____

_____

_____

MAINTENANCE-FREE BATTERY

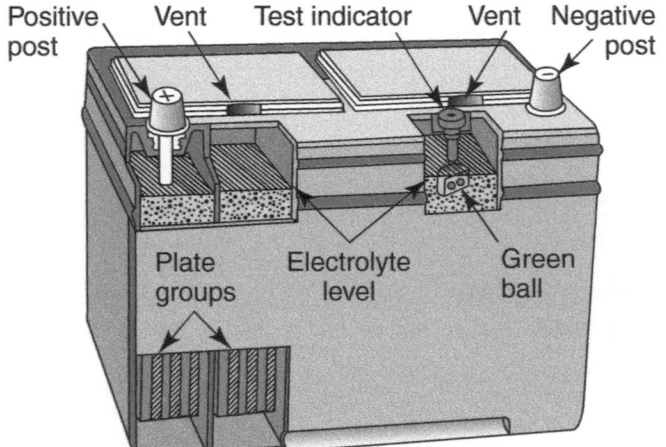

These batteries are completely sealed so a hydrometer state of charge cannot be made. How then is the state of charge determined?

_____

_____

_____

_____

This type of battery is fitted with lead-calcium plate grids to reduce gassing to a minimum. The gassing that occurs when it is fully charged is passed into a gas reservoir in the lid and returned as it forms water droplets.

How is gassing controlled during the charge process?

_____

_____

_____

DIAGNOSTICS: BATTERY – SYMPTOMS, FAULTS AND CAUSES

State a likely cause for each symptom/system fault listed below. Each cause will suggest any corrective action required

Symptom	System fault	Likely cause
Electrical systems are completely dead or lamps are very dim and starter fails to operate	No or low terminal voltage	
Starter will not crank engine first thing in the morning, but is OK after starting	Loss of capacity and failure to hold charge	
Starter cranks engine slowly, alternator seems to be charging correctly	Incorrect relative density of electrolyte or acid loss	
Starter will not crank engine – there is a wet pool on the floor	Physical damage	
Slow cranking of engine. Battery needs continual topping up	Undercharging/ overcharging	

Symptom	System fault	Likely cause
Batteries do not become charged in the expected time – battery gasses excessively	Faulty operation of multi-battery layouts	

BATTERY JUMP STARTING

Indicate how to connect a slave battery to a vehicle using jump leads by placing the numbers 1 to 4 in the circles on the diagram to show the correct order.

State the connecting procedure using a slave battery.

_____

_____

_____

_____

Battery jump starters

It is much more convenient and safe to use a portable rechargeable battery pack to start a vehicle with a discharged battery, rather than use jump leads and a spare battery or a battery still fixed to another vehicle.

These battery packs provide a 12V power source fitted in a sturdy carrying case. The lead-acid battery used is a sealed (gel-electrolyte) type designed for cyclic use and can operate at a maximum boost of up to 400A.

The unit's basic use is to act as a 'booster' to start a vehicle on a cold morning. What other features may these units have?

_____

_____

_____

How should a jump starter unit be kept in good condition?

_____

SURGE PROTECTORS

These are designed to protect the vehicle's sensitive electronics from damaging voltage spikes and surges when either jump starting or welding on a vehicle.

How are surge protectors fitted? _____

_____

What vehicle systems can be protected by a surge protector? _____

_____

_____

Some vehicle jump start cables can contain surge protection devices.

VEHICLE MEMORY SAVER

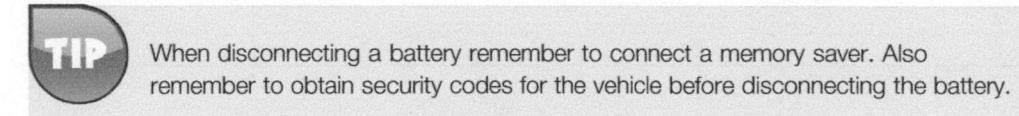

TIP When disconnecting a battery remember to connect a memory saver. Also remember to obtain security codes for the vehicle before disconnecting the battery.

On modern vehicles there are many memory controlled devices which would need to be reset should the battery be disconnected. To prevent this inconvenience a memory saver may be used.

What vehicle units may suffer from loss of memory due to battery disconnection?

- _____
- _____
- _____
- _____
- _____

Picture kindly provided by permission of Snap-On Industrial

The vehicle memory saver is a unit which provides a 12V supply to the electrical system when the battery is disconnected. The supply may be obtained from the mains via a transformer if in the workshop or by a 12V battery if on a breakdown. Where is the memory saver connected to the vehicle's electrical system?

_____

SAFE WORKING PRACTICE

State general rules and precautions to be observed before working on batteries and charging systems.

Handling batteries: _____

_____

Charging batteries: _____

_____

_____

Connecting batteries or alternators: _____

_____

_____

_____

Nickel-Metal Hydride (NiMH) batteries

These batteries are commonly used in Hybrid vehicles. They are environmentally friendly and can be recycled. The electrodes in the cells are made of metal hydride and nickel hydroxide. Potassium hydroxide is used as the electrolyte. The batteries are housed in sealed containers which transfer the heat reasonably well. They have a very high cycle life, being safe and durable in design. Other advantages are that they are lightweight, a compact size and have a long life.

The figure shows a Honda Civic IMA NiMH battery. It consists of 20 modules which are connected in series, and each of these modules contains six cells. Each cell has a voltage of 1.2V, thus giving a total battery-module voltage of 144V, with a capacity of 6.0 Ah.

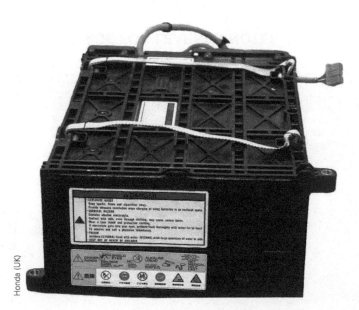

Honda (UK)

Multiple choice questions

Choose the correct answer from a), b) or c) and place a tick [✓] after your answer.

1 **When charging a battery, what gas is given off by the positive plates?**

   a) hydrogen [ ]

   b) oxygen [ ]

   c) nitrogen. [ ]

2 **What is SG an abbreviation of?**

   a) specific gravity [ ]

   b) special gravity [ ]

   c) specific ground. [ ]

3 **What is the specific gravity of a fully charged lead-acid battery cell?**

   a) 1.270 – 1.290 [ ]

   b) 1.110 – 1.130 [ ]

   c) 1.230 – 1.250. [ ]

4 **In a 12V battery, how are the cells linked together?**

   a) parallel [ ]

   b) series [ ]

   c) manually. [ ]

5 **When disconnecting a battery from a vehicle always disconnect the positive terminal first.**

   a) True [ ]

   b) False. [ ]

SECTION 4

Starting systems

USE THIS SPACE FOR LEARNER NOTES

_____

_____

_____

_____

_____

_____

_____

_____

Learning objectives

After studying this section you should be able to:

● Describe the layout, construction and operation of inertia and pre-engaged starter motors.
● Explain the function and operation of a one-way clutch (pre-engaged starter motor).
● Describe the operation of the starter gear reduction.
● Explain the function and operation of the starter ring gear.
● Explain the function and operation of the starter solenoid.
● Identify common starter motor faults.

Key terms

Pinion   A small gear on the end of the starter motor which meshes with the ring gear.
Inertia starter motor   The pinion is 'thrown' into engagement due to its inertia.
Pre-engaged starter motor   The pinion is moved into mesh before the motor starts to turn.
Overrun clutch   Releases the pinion drive when the engine is turning faster than the starter.
Flywheel ring gear   The starter pinion engages in this gear which is mounted on the rim of the flywheel.
Solenoid   In the case of the starter motor, a heavy-duty electromagnetic switch.

WWW   http://www.phys.unsw.edu.au/hsc/hsc/electric_motors.html
http://electronics.howstuffworks.com/motor2.htm
http://www.animations.physics.unsw.edu.au/jw/electricmotors.html#DCmotors

TIP   Before fitting a starter motor remember to check the starter ring gear for condition and security. It is important to check all of the ring gear teeth. To do this mark one tooth then slowly rotate the engine, visually inspecting all of the other teeth until the marked tooth reappears.

MAGNETISM

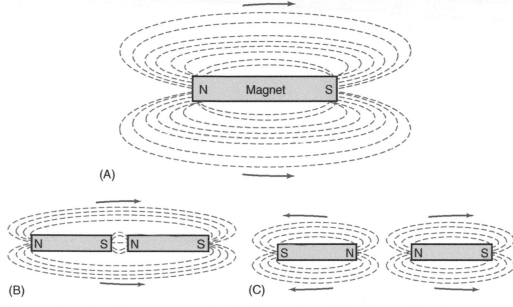

(A)

(B)   (C)

A magnet has the ability to attract steel, nickel, iron and cobalt (ferrous materials). There are two points, one at each end of a magnet, which are maximum points of attraction. These are known as the North and South poles. When opposite poles (North and South) are brought together they attract. When like poles (North and North or South and South) are brought together they repel each other (push away).

Every magnet has an imaginary field of force around it, known as the flux field.

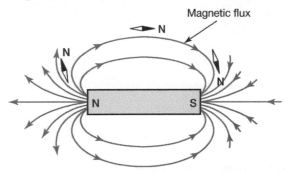

Magnetic flux

Electromagnetism

When an electric current is passed through a wire conductor a magnetic field is created around the wire. Increasing the current will increase the strength of the magnetic field. By adding more wires the magnetic field can be increased.

To concentrate the lines of flux and increase the magnetic field, wires can be coiled. By adding a soft iron core in the centre of the wire coil the magnetic strength increases considerably.

The basic electrical principles of all starter motors are the same and can be demonstrated by passing a current through a coil of wire positioned between the poles of a permanent magnet and noting the effects.

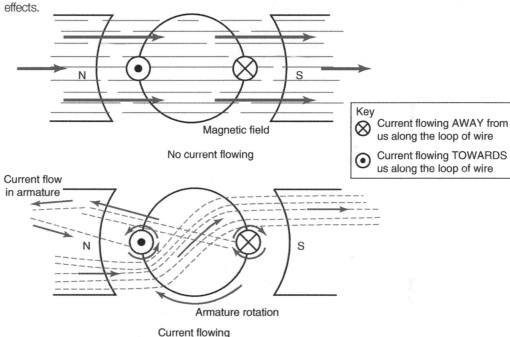

Magnetic field

No current flowing

Current flow in armature

Armature rotation

Current flowing

Key

⊗ Current flowing AWAY from us along the loop of wire

⊙ Current flowing TOWARDS us along the loop of wire

The top diagram in the figure above shows an end view of the armature windings and magnets. No electrical current is flowing through the armature winding in this diagram.

Briefly explain what is happening in the lower diagram:

_____

_____

_____

MOTOR OPERATION

Describe, with the aid of the figure below, the basic operation of an electric motor (such as a starter motor):

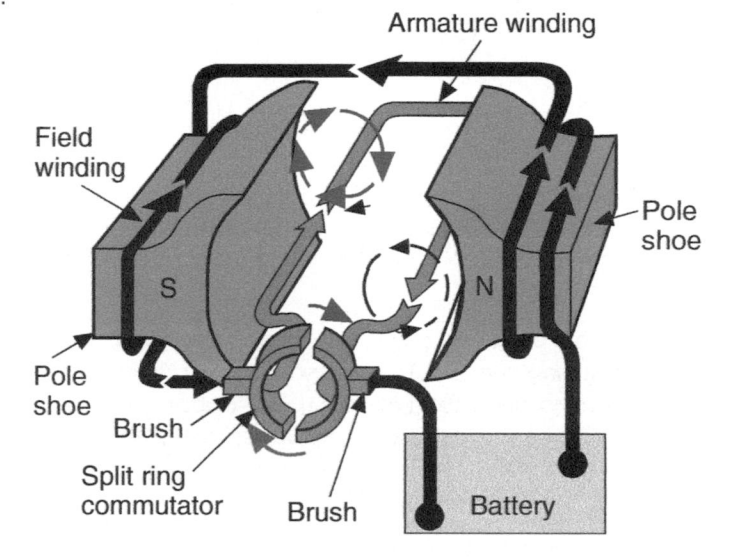

Armature winding

Field winding

Pole shoe

S

N

Pole shoe

Brush

Split ring commutator

Brush

Battery

_____

_____

_____

Commutator

Insert the missing words into the following paragraph. Note: there are two extra distracter words.

reversing	winding	armature	current	split
switch	ends	copper	half	steel
torque	steady	magnetic	one	

The commutator prevents the _____ on the motor from _____ every time the armature _____ moves through one _____ field and enters the next. It acts like a _____ which reverses the _____ flow in the _____ windings every time the commutator makes _____ a turn, so that a _____ rotating force can be produced. The commutator is a _____ ring device made from _____ which connects to the _____ of the armature windings.

How can the rotation of the armature be maintained?

_____

_____

Examine the armature of a starter motor. Count the number of segments and the number of loops. Use a digital multimeter set to show resistance and check to find the beginning and end of each loop.

STARTER MOTORS

Remove rings, wrist watches and other jewellery that might come into contact with live electrical cables. Wear safety glasses and personal protective clothing.

Passenger car starters can be classified by the following criteria:

Type of power transfer: _____

Type of magnetic field generation: _____

Type of engagement: _____

A starter motor has one basic functional requirement which is:

_____

_____

It achieves this by converting electrical energy into _____

What is the required rotational speed of a petrol engine in order for it to start?

_____

Cutaway of a starter motor

Pre-engaged permanently excited starters with gear reduction are the most commonly used systems in cars today.

The different sizes and types of engines make it necessary to require different designs of starter motor.

What characteristics are desirable for a starter motor?

1 _____

2 _____

3 _____

4 _____

5 _____

6 _____

Starter motor construction

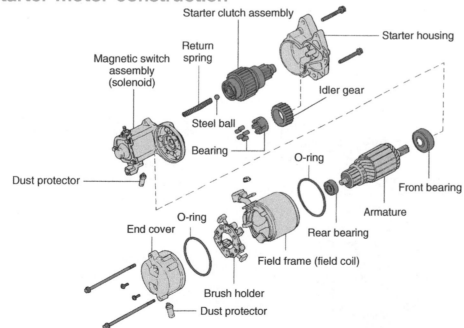

A typical starter motor assembly

- Starter clutch assembly
- Return spring
- Magnetic switch assembly (solenoid)
- Starter housing
- Idler gear
- Steel ball
- Bearing
- Dust protector
- O-ring
- O-ring
- End cover
- Front bearing
- Armature
- Rear bearing
- Field frame (field coil)
- Brush holder
- Dust protector

Armature

This part consists of a series of conductor loops. It is mounted in the starter motor by bearings at either end, which allow it to rotate. The conductor loops are connected at their ends on to the commutator assembly which is made up of heavy copper segments.

Brushes are connected to one terminal of the battery supply. These brushes pass the current to the conductor loops.

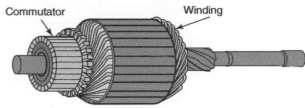

The armature of a starter motor

- Commutator
- Winding

Field coils

Insert the missing words into the following paragraph. Note: there are three extra distracter words.

current	iron	carbon	produced	armature
magnetic	strong	coil	weak	terminal
rotate	resistance	field	insulated	aluminium

_____ coils are used to provide the _____ field which causes the _____ to _____. Field coils consist of _____ copper wires, wrapped around soft _____ cores. They are connected in opposite pairs, to provide North and South poles. Each _____ pair has one end connected to a supply _____. The other end of each coil pair is connected to a _____ brush. When _____ is allowed to flow through the coils, a _____ magnetic field is _____.

- Field windings
- Pole shoe

Example of a field coil and pole shoe

Starter motor electrical configurations

Most starter motors use four pole shoes surrounded by field coils and so produce double the magnetic effect.

Correctly label the figure opposite with the following terms:

Field coils in series
Field coils in parallel
Armature in series with field coils

Starter motor windings can be configured to provide low, medium and high torque. As already mentioned, the most common arrangement uses four field coils, which will provide medium torque. To achieve this, two pairs of field coils are wired in series, then both series pairs are wired in parallel.

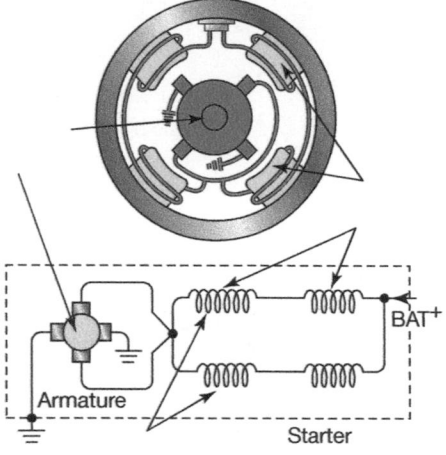

How much current is required and when does a starter motor achieve its maximum torque?

_____

_____

_____

Electric motor wiring

Field windings can be wired in a number of ways. Complete the labelled diagrams in the figure below to show these basic theoretical methods.

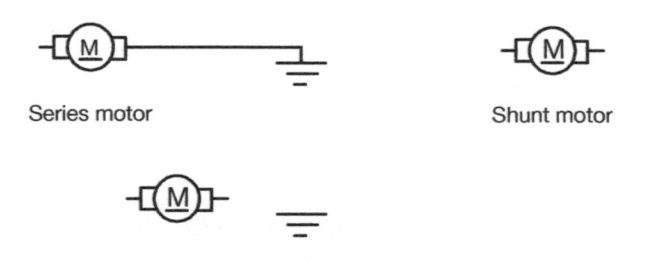

Series motor

Shunt motor

Compound motor

Which two major factors determine the type of starter motor required for any particular application?

1 _____

2 _____

State the characteristics possessed by:

1 A series wound starter motor: _____

_____

_____

2 A parallel (shunt) wound general purpose motor: _____

_____

_____

3 A compound motor: _____

_____

_____

Permanent magnet motors

A recent change in the design and construction of starter motors is the use of permanent magnets instead of electromagnets as field coils.

State three advantages of using permanent magnet motors:

1 _____

2 _____

3 _____

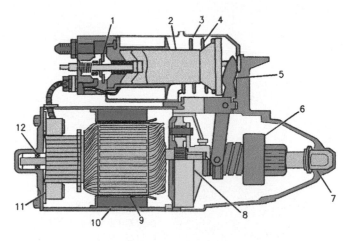

Permanent magnet type starter assembly

Correctly name the numbered parts on the above diagram:

1 _____

2 _____

3 _____

4 _____

5 _____

6 _____

7 _____

8 _____

9 _____

10 _____

11 _____

12 _____

TYPES OF STARTER MOTOR

Inertia starter motor

This type of starter motor may still be found on older types of vehicles.

What does the word 'inertia' mean?

_____

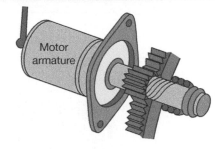

Motor armature

This type of starter motor uses the inertia of the pinion to engage the flywheel.

As shown in the figure above, the drive consists of a pinion mounted on the armature shaft.

Describe its operation when the starter turns.

_____

_____

_____

_____

Pre-engaged starter motor

The most commonly used starter motor on motor vehicles is the pre-engaged starter motor.

What are the advantages of the pre-engaged starter motor when compared with the inertia type?

_____

Pinion

The ends of the starter pinion gear are chamfered to help the teeth mesh more easily without damaging the flywheel ring gear teeth.

A pinion gear on the starter motor engages with the teeth on the flywheel ring gear.

Chamfered pinion gear

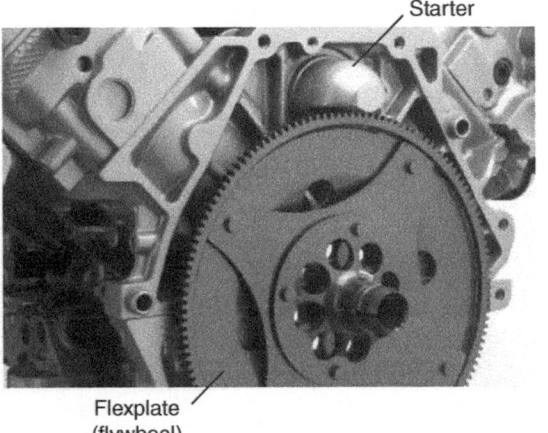

Starter

Flexplate
(flywheel)

Starter drive pinion gear

Flywheel ring gear

The starter ring is generally 'shrink' fitted to the outside of the flywheel. The teeth will normally have a chamfered edge on the side facing the starter pinion.

The figure opposite shows the relationship in size between the starter pinion and the starter ring gear. The ratio between the pinion and ring gear is typically 10:1 to 20:1.

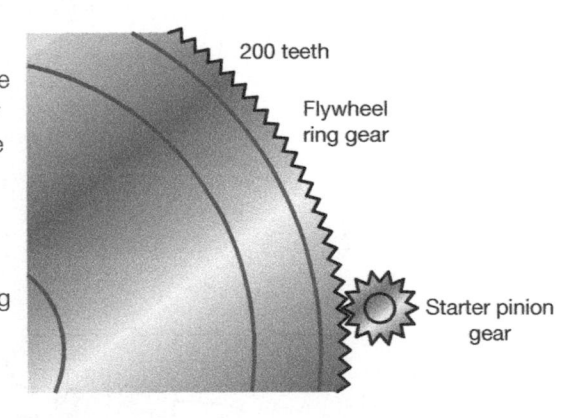

200 teeth

Flywheel ring gear

Starter pinion gear

STARTER SOLENOID

A solenoid is an electromagnetic switch. Why is a solenoid switch necessary in a starter circuit?

_____

_____

The figure below shows a typical basic solenoid. Correctly label the diagram from the list below and describe the basic operation of a solenoid switch.

Heavy duty terminals for starter leads   **Push-off spring**   **Field windings**
Plunger   **Bridge contacts**

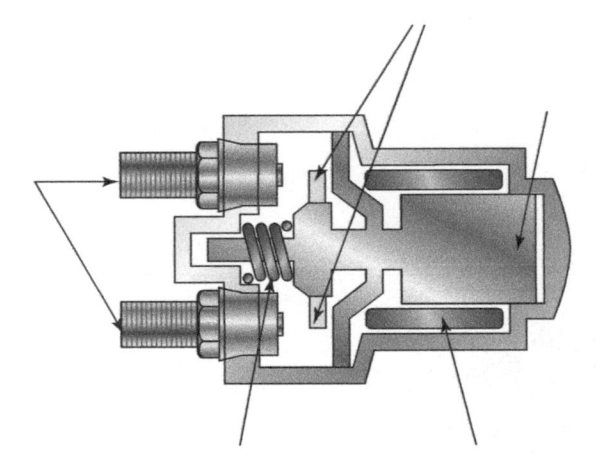

_____

_____

_____

The solenoid used on the modern type of starter motor, such as the pre-engaged type, is mounted on the starter motor. It contains two separate, but connected, electromagnetic windings.

ADDITIONAL STARTER RELAY

Give reasons why some starter systems using a pre-engaged type drive use an additional relay as shown in the figure below.

_____

In pairs discuss and describe how the additional starter relay operates.

Name and label the main parts in the figure below.

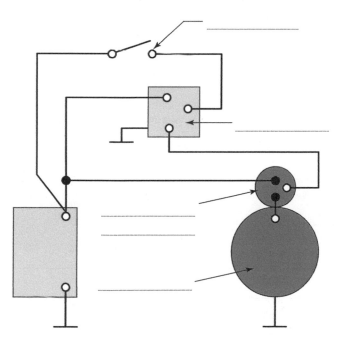

The figure below shows a typical pre-engaged starter motor with the solenoid mounted directly on it. Label the diagram below using the following terms:

contacts	shift lever	ignition switch
pinion	contact bridge	neutral safety switch
pull-in winding	flywheel ring gear	fusible link
hold-in winding	plunger	to battery positive (+)

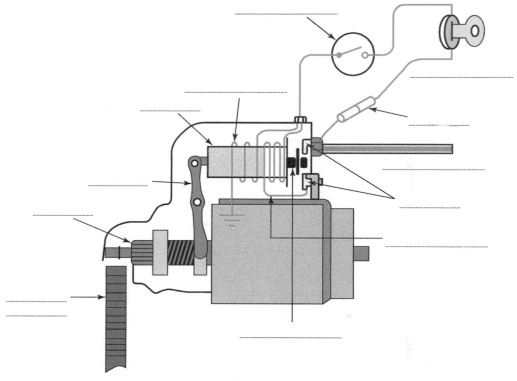

What is the purpose of the neutral safety switch, as shown in the figure above?

_____

_____

_____

Another common name for this type of switch is _____

Movement of the solenoid plunger results in a similar movement of the pinion gear in the opposite direction. This allows the gear to engage the flywheel before the armature rotates.

What is the purpose of the engagement spring on the pinion?

_____

_____

_____

TIP A repeated clicking sound from a solenoid can be an indication that the hold-in winding is defective.

Insert the missing words into the following paragraph. Note: there are three extra distracter words.

start	shift	parallel	magnetic	energized
solenoid	ignition	mesh	spring	larger
move	off	battery	hold	plunger
smaller	pinion	rotate	current	
series	pull	turn	neutralizes	

The starter _____ has a number of functions. When the _____ switch is turned to the _____ position the _____-in windings and hold-in windings are _____. These windings are connected in _____. The plunger is pulled axially into the solenoid due to _____ force. This plunger movement is relayed to the _____ by means of the _____ lever causing the pinion to be pulled into _____ with the ring gear. It is at this point, when the contact bridge touches the contacts, that the battery _____ is conducted to the starter motor and electrically _____ the pull-in winding. This is when reduced current only passes through the hold-in windings, allowing a _____ magnetic field, to hold the solenoid plunger in place. Current flows from the _____ through the contact bridge and contacts into the starter motor, which begins to _____ causing the pinion and ring gear to _____. When the engine starts, the ignition switch is then released. The _____-in windings are de-energized, causing the solenoid _____ to return to its original position under the return _____ pressure. The movement of the solenoid plunger causes the pinion to _____ out of mesh from the ring gear.

GEAR REDUCTION STARTER DRIVE

The armature of this starter motor, commonly fitted to petrol engines, rotates at a higher speed than the conventional type. It incorporates an epicyclic reduction gear assembly which increases torque output by approximately 3:1.

Name four advantages of using a reduction type of starter motor:

1 _____

2 _____

3 _____

4 _____

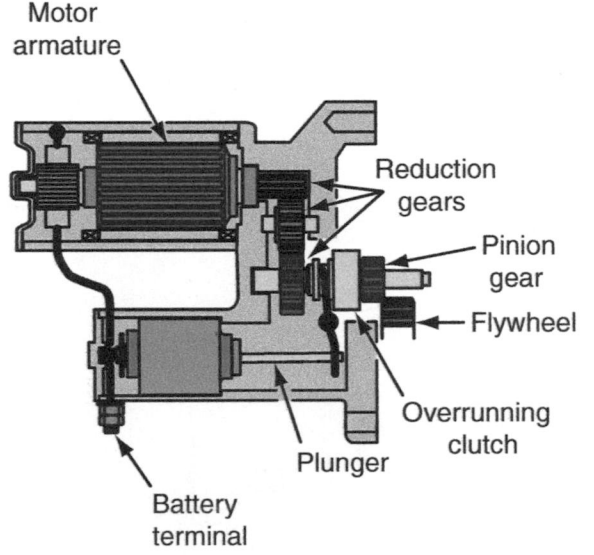

A gear reduction drive starter

How does the solenoid act in the reduction type of starter motor?

_____

Reduction gear

What are the advantages of using a planetary reduction gear to drive the pinion?

_____

_____

Name the gear drive parts and indicate the rotational movement of the gears.

Describe the operation of the epicyclic gear drive:

_____

_____

_____

_____

_____

Overrun (one-way) clutch

In addition to the epicyclical gearing, this type of starter drive incorporates an overrun clutch, which is also commonly used in other types of pre-engaged starter motors. What is its purpose?

_____

_____

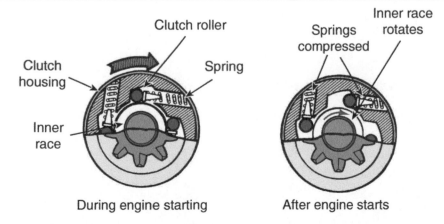

Overrun clutch

As shown in the figure above the rollers are spring-loaded and according to direction of drive are either free running or wedge locked.

If the starter motor is considered suspect, for example it is cranking the engine slowly, a systematic check should be carried out to determine if there is an excess voltage loss (high resistance) in the circuit. Typically, the current for a four-cylinder engine should not exceed 140A when cranking evenly at a speed of no lower than 180 rpm. However, in low temperature conditions the current flow will be substantially greater.

Inspection

Before tests are carried out, what are the preliminary checks that must be made?

_____

In all six starter tests shown on the following page, the engine must be cranked without starting. How is this achieve on a:

SI engine: _____

CI engine: _____

Carry out a series of voltage checks to determine the condition of a starter circuit using a 0–20V voltmeter. State the function of each check and show the position of the voltmeter for each test on the diagrams on page 292.

State the expected and actual readings.

Test	1	2	3	4	5	6
Expected voltage						
Actual voltage						

1. Battey voltage on load _____

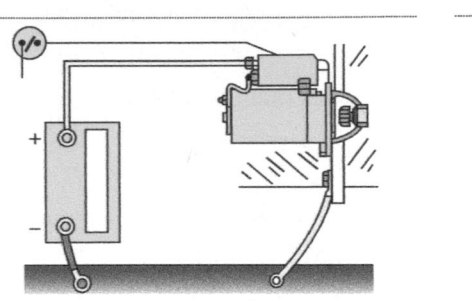

4. Voltage drop insulated link _____

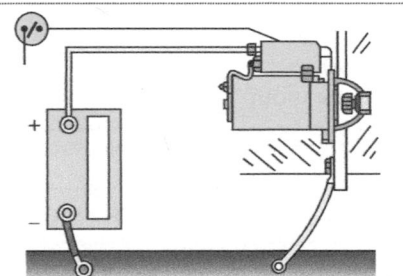

2. Voltage at solenoid operating terminal _____

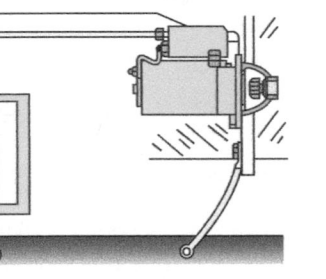

5. Voltage drop solenoid contacts _____

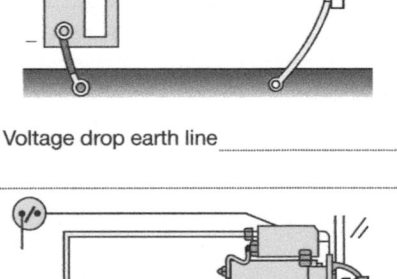

3. Voltage at starter on load _____

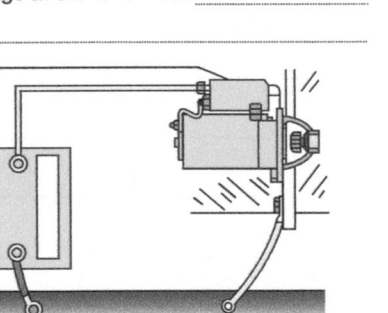

6. Voltage drop earth line _____

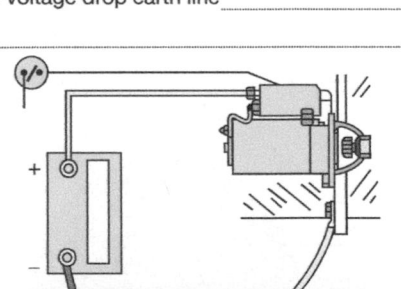

TIP When checking cranking voltage, always refer to the manufacturer's technical data. Most literature states that cranking voltage should be at least 9.6V but some engines will have difficulty starting at voltages lower than 10.2V.

It is necessary to supply a very large amount of current to the starter motor to enable it to turn the engine from a stationary position. This causes a considerable voltage drop in the circuit. The starter motor is wired to produce maximum torque immediately on turning (locked torque). This maximum torque is required to overcome the resistance to movement of the engine.

It is possible to carry out a lock torque test by clamping the starter motor in a suitable test rig.

Inspection for faults

If an external fault has been determined by testing then the starter motor must be dismantled for examination to identify the possible faults.

With the aid of workshop manuals describe the main points to be observed when removing a starter motor from a vehicle.

Vehicle make _____ Model _____

1 _____

2 _____

3 _____

4 _____

5 _____

6 _____

Make sure the handbrake is applied and the vehicle's transmission is in neutral before carrying out this test; otherwise the starter motor might move the vehicle and possibly cause injury to yourself and others.

DIAGNOSTICS: STARTER MOTOR SYSTEM – SYMPTOMS, FAULTS AND CAUSES

State a likely cause for each symptom/system fault listed below. Each cause will suggest any corrective action required.

Symptom/System faults	Likely cause
Starter cranks engine very slowly	
There is an abnormal clonking noise when starter is operated	
There Is a grating sound as pinion engages and disengages flywheel	
When starter key is turned nothing happens, but clicking sound is heard	
Starter cranks engine slowly and its torque output is low	
Starter cranks engine slowly and its current consumption is low	
Starter cranks engine slowly and its current consumption is high	
There is a very heavy clonking noise when engine is started	

Describe how the following should be protected from hazards during normal use or repair.

Cables: _____

Terminals (starter battery): _____

_____

Multiple choice questions

Choose the correct answer from a), b) or c) and place a tick [✓] after your answer.

1   **The purpose of an overrun clutch is to:**

   a) allow the armature to turn in both directions [ ]

   b) prevent the engine from turning the starter [ ]

   c) help the starter pinion to smoothly mesh. [ ]

2   **Which starter components commonly use permanent magnets?**

   a) armature [ ]

   b) solenoid [ ]

   c) field coils. [ ]

3   **Which component contains a hold-in winding and a pull-in winding?**

   a) field coil [ ]

   b) starter solenoid [ ]

   c) ignition switch. [ ]

4   **Which component allows a small current, using light gauge wiring, to switch a high current which uses heavy gauge wiring?**

   a) ignition switch [ ]

   b) solenoid [ ]

   c) brushes. [ ]

5   **Which one of the following starter motors is likely to be found on a modern motor car?**

   a) inertia [ ]

   b) coaxial [ ]

   c) pre-engaged. [ ]

6   **Permanent magnet motors are lighter, more compact and simpler in design and construction as compared to electromagnetic magnet coil wound motors.**

   a) True [ ]

   b) False. [ ]

SECTION 5

Charging systems

USE THIS SPACE FOR LEARNER NOTES

_____

_____

_____

_____

_____

_____

_____

_____

Learning objectives

After studying this section you should be able to:

● **Identify charging system drive assemblies.**

● **Identify charging system components.**

● **Describe the construction and operation of charging system components.**

● **Describe how to remove and replace charging system units and components.**

● **Compare charging system components and assemblies against alternatives.**

● **State common terms used in conjunction with charging systems.**

Key terms

Stator   The static part of the alternator which houses the three windings.
Rotor   Electrically magnetized North and South poles. Rotates inside the stator.
Slip ring   Carbon brushes run on these which are made of copper.
Three phase   Three evenly spaced alternating currents.
Rectifier   Changes alternating current (AC) to direct current (DC).
Diode   An electronic component which allows current to flow in one direction only.
Regulator   Controls the maximum alternator output.
Transistor   An electronic component operated by low voltage to switch higher voltages.

WWW   **http://www.animations.physics.unsw.edu.au/jw/electricmotors.html**
Has some very good animations on alternators and motors.

http://www.alternatorparts.com/understanding_alternators.htm
Describes how alternators work.

   Never run an engine with the alternator electrics disconnected and the drive belt still attached. This can cause damage to the internal circuitry.

CHARGING SYSTEMS

The type of charging system found on all modern vehicles consists of an alternator and storage battery. State the purpose and functional requirements of (a) the charging system and (b) the battery:

a   _____

    _____

b   _____

    _____

    _____

Complete the figure below to show a typical layout of a battery-charging system. Name the parts and describe the electrical flow when operated.

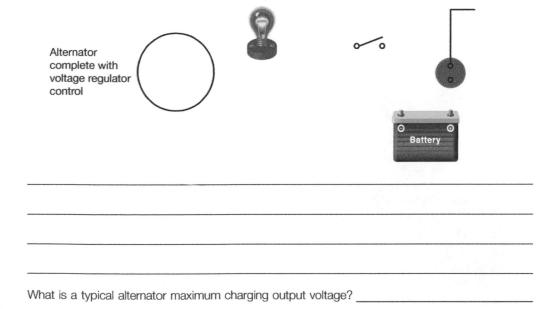

Alternator complete with voltage regulator control

Battery

_____

_____

_____

_____

What is a typical alternator maximum charging output voltage? _____

Alternator drives

The alternator uses mechanical power to produce electricity. Alternators are normally driven by the engine by a V-belt or, as more commonly fitted to modern vehicles, a multi-ribbed type belt (also known as a poly belt or serpentine belt). The alternator is driven from the crankshaft pulley. Because of the design of multi-ribbed belts they can operate under a greater tension without stretching, and on smaller radius pulleys. They are adjusted by automatic adjusters. The ribbed side of the belt can drive such components as the alternator or sometimes the rear smooth side is used.

> ⚡ When the engine is running keep fingers and loose clothing away from moving components, e.g. alternator drive belts.

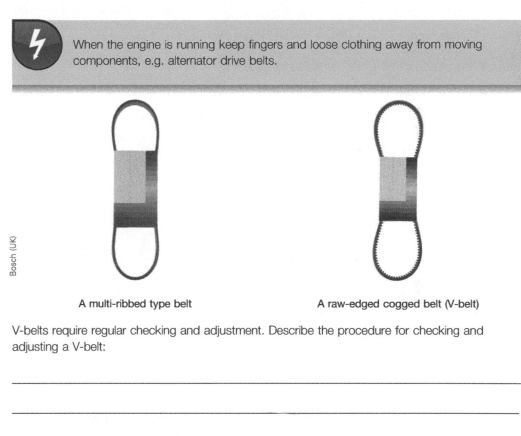

Bosch (UK)

A multi-ribbed type belt   A raw-edged cogged belt (V-belt)

V-belts require regular checking and adjustment. Describe the procedure for checking and adjusting a V-belt:

_____

_____

_____

_____

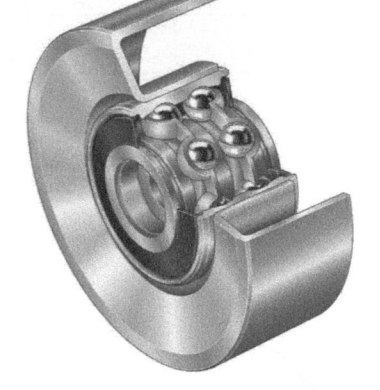

Poly-ribbed belt tension and idler pulleys

Hydraulic automatic poly-ribbed belt tensioner

Schaeffler.com

TIP When servicing a vehicle always remember to visually inspect the drive belts for condition, contamination, security and damage. It takes a short period of time to do this and is good preventive maintenance.

How is a poly-ribbed type belt checked and adjusted?

_____

_____

The automatic tensioner also dampens out the oscillations.

Alternator drive pulley

This is retained on the front end of the alternator on to the rotor by a locking nut.

What checks need to be carried out on the pulley?

_____

_____

Compact alternators

Correctly label the exploded diagram of a compact type of alternator using the terms listed below:

Voltage regulator and brushes	**Rectifier assembly**	**Bearings**
Rotor	**Slip rings**	**Cooling fans**
End cover	**Drive pulley**	**Stator**

Bosch (UK)

Look at an alternator. Either strip one down or use one that is already disassembled. Identify the main alternator components and compare them with those in the diagram above.

Most modern vehicles are now fitted with compact alternators. The main differences between this and a conventional alternator are that the compact alternator has two internally mounted fans, one at either end of the armature. This helps to improve cooling. It also has smaller collector rings and the rectifier is located outside the collector ring end shield.

Courtesy of Robert Bosch GmbH, www.bosch-presse.de

Rotor

Correctly label the parts indicated on the
rotor with the following terms:

Field windings
Slip rings
Bearing
Magnets (finger poles)

The rotor rotates inside the stator and is mounted at each end in bearings. It consists of field
windings, wound over an iron core and both ends are attached to the slip rings. An iron claw, with
finger poles, is at either end of the rotor core surrounding the field windings.

Slip rings

These are two _____ rings, insulated from each other and the rotor on which they are mounted.
The carbon brushes press against each of these slip rings.

What are these slip rings connected to?

_____

Carbon brushes

These are made from soft carbon
and are pressed against the slip
rings by springs. The springs
automatically compensate for any
wear in the carbon brushes. The
photograph to the right shows the
brushes running against the slip
rings.

Describe what happens when voltage passes through the carbon brushes:

_____

_____

_____

The alternator output voltage depends upon the speed of the rotor and its magnetic field strength.

Stator

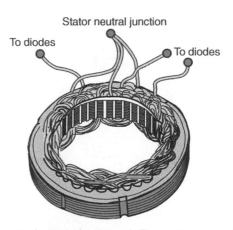

As the name suggests the stator is static (stationary). The stator consists of three evenly spaced
insulated enamel copper wire windings, held in a frame of soft iron laminations. The windings are
connected in either star or delta configurations. Light vehicles mainly use the star configuration.
The stator shown above is connected as a star configuration.

The voltage and current output is different for each of the configurations for the same magnetic
field strength and rotor speed.

Which configuration is commonly used in modern alternators and why?

_____

Sketch a star and delta configuration below:

Star configuration   Delta configuration

So that an AC voltage can be produced the windings are interlaced and set 120° apart.

What are these windings connected to? _____

The rotor rotates inside the stator and as it rotates it induces an EMF in the stator windings and current begins to flow.

What does EMF stands for? _____

Cooling fans

These provide air to cool the alternator's internal workings and electronics.

Older types of alternators have an external fan at the front of the alternator, mounted behind the drive pulley. Where are the cooling fans located in a modern compact type alternator?

_____

Function of the charge warning lamp

The ignition charging warning light is part of the rotor field excitation circuit. When the ignition is switched on it functions as part of the circuit to magnetize the rotor field windings, until the engine is running and self-excitation takes place. It functions as a resistor and determines the size of the pre-excitation current.

What is the typical rating for the charging warning light bulb?

_____

What causes the lamp to go out?

_____

_____

VOLTAGE GENERATION

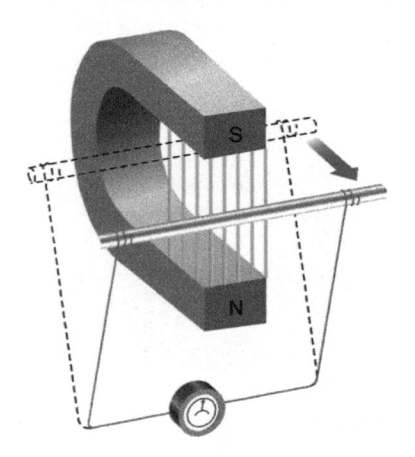

An alternator is a mechanically operated device which generates electrical current. It is important to have a basic understanding of how voltage and current flow are generated. The figure above shows a permanent horse-shoe magnet and a length of copper wire connected to an ammeter. The wire is being passed between the magnetic field of the North and South poles of the magnet.

What happens when a wire is moved in a magnetic field as shown in the diagram opposite?

_____

Describe what happens when the North and South poles of the magnet are reversed.

_____

Single phase output

When a permanent _____ rotates in a _____ it induces a _____ and _____.

The figure opposite shows the North and South poles of the permanent magnet (rotor) rotating inside a coil (stator). The arrows show the direction of current flow.

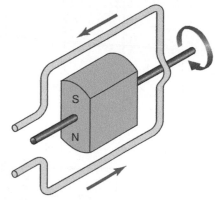

As the rotor gets closer to the winding at 90° (see the figure below) the voltage gets stronger generating a positive maximum voltage. As the magnet moves away from the winding the voltage gets weaker at 180° (zero volts). When it has gone through 270°, maximum negative voltage is reached. As it continues to 360° zero voltage is reached. This is known as single phase.

The figure opposite also shows a single phase output. This output is known as an _____. This is because it changes from positive to negative voltage.

What name is given to the wave pattern shown? _____

Single phase output generation

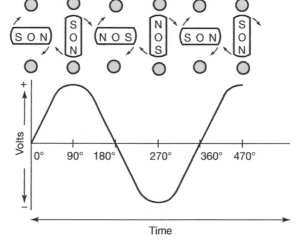

Single phase output generation

Three phase output

To make better use of an alternator three windings are used. The result is a three phase current output.

How is a three phase output produced (see the diagram of a rotor in a three phased stator opposite)?

_____

_____

The figure below shows the three sine waves of a three phase output. Each sine wave is generated by each of the windings which are evenly spaced.

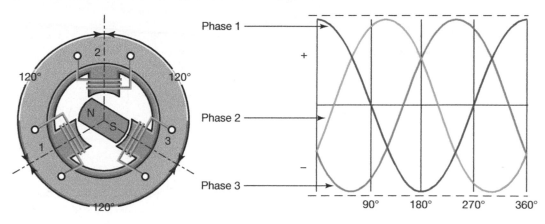

Rotor in a three phased stator

Three phase waveform

The windings in a three phase alternator can be connected in either a star or delta configuration.

Diode rectification pack

Insert the missing words into the following paragraph. Note: there are two extra distracter words.

rectified	voltage	diodes	alternating	resistance
transistors	direct	alternator	battery	

An _____ produces alternating current (AC). The vehicle _____ and electrics need direct current (DC). The _____ current flow produced by the alternator must be _____ to _____ current. This will then produce a steady useable _____. This is achieved by using a number of static rectifiers called _____.

Show a diode's electrical symbol and indicate the direction of current flow.

_____ __ _____

The function of a diode is to: _____

It is made from: _____

A

B

The four diodes shown above make up a single full wave bridge rectification system. Show, using arrows, how diodes rectify the supply of current induced in a single coil of wire.

A   The diodes passing current are numbers: _____

B   The diodes passing current are numbers: _____

Correctly label the components on the diagram below.

What is being shown in the diagram?

_____

_____

EMF AC

Sketch the waveform which the single full wave bridge rectifier will produce.

EMF 0

0   90   180   270   360

EMF supply to battery rectified to DC

Describe what effect the single full wave bridge rectifier has had on the AC voltage:

_____

_____

_____

Three phase rectification

Describe the function of a rectifier:

_____

_____

How many diodes are used in a bridge rectifier?

_____

How many diodes are used for each stator winding?

_____

Using arrows, indicate the rectified flow path through the system on each diagram below. The first diagram shows the North pole passing a stator coil, the second diagram shows the South pole passing the same stator coil.

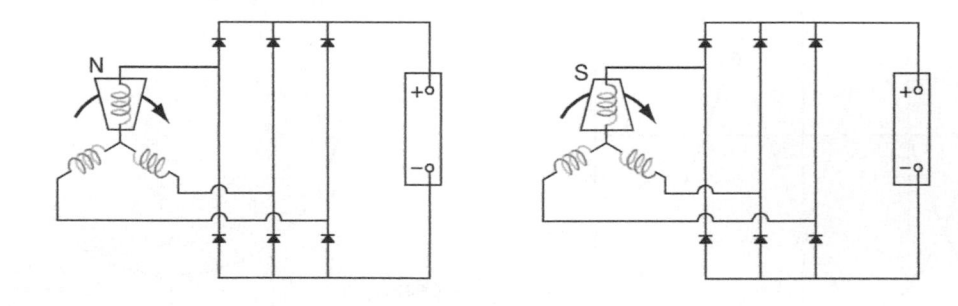

Select some diodes and bridge rectifiers. Use a digital multimeter to check the flow through the diodes.

Draw a three phased rectified voltage on the chart below. Clearly show the small DC voltage ripple.

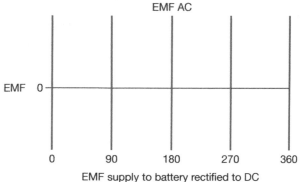

EMF AC

EMF   0

| 0 | 90 | 180 | 270 | 360 |

EMF supply to battery rectified to DC

Field diodes

Describe the function of the field diodes which are shown in red on the figure opposite:

_____

_____

_____

_____

_____

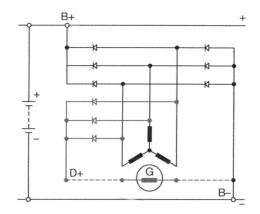

Field diodes

The maximum output of an alternator needs to be regulated. Most alternators fitted on UK vehicles use electronic regulators fitted at the rear end of the alternator to control the maximum alternator output. This makes the unit a completely self-contained electrical power source.

Since these units cannot be adjusted it is only necessary to understand how they switch on and then control the field supply current.

An integrated voltage regulator

Examine the machine sensing regulator drawing on the next page.

When the ignition is switched on, current flows through the warning light to the IND connection on the alternator, then to + on regulator, through resistor to switch T1 on. This allows flow through the alternator field coil to F. T2 switches T3 on and the current flows to earth unrestricted.

Alternators may be 'machine-sensed' or 'battery-sensed'.

The term 'sensing' relates to where the regulator picks up the supply current flowing through the regulator to the zener diode.

If machine-sensed the supply is via the three field diodes, through the field winding, to the regulator.

If battery-sensed, the regulator sensing circuit is connected directly by a separate cable to the battery. With this arrangement the alternator is more sensitive to the charging loads placed upon the battery.

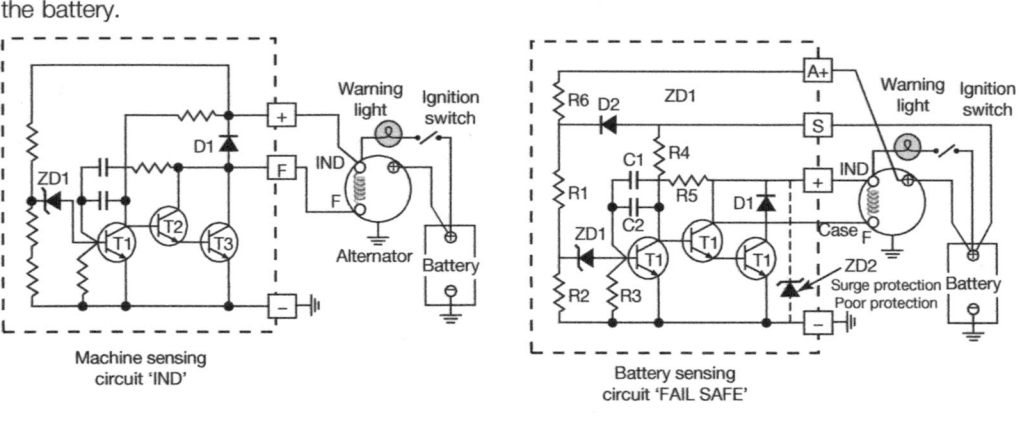

Machine sensing circuit 'IND'

Battery sensing circuit 'FAIL SAFE'

The regulator makes use of transistors. Explain the basic function of a transistor:

_____
_____

Transistor

Correctly label the drawing of a transistor.

What is the function of the surge protection diode (zener diode) which is an integral part of the voltage regulator shown in the figure at the top of the page?

_____
_____
_____

Sketch a surge protection diode symbol on to the figure below.

—— ——

On earlier types of alternators the surge protection diode is positioned on the alternator's end plate next to the regulator. The connections electrically are in the same position, between IND and earth.

What other methods offer protection to the alternator?

1 _____

2 _____

3 _____

4 _____

A TYPICAL ALTERNATOR CIRCUIT

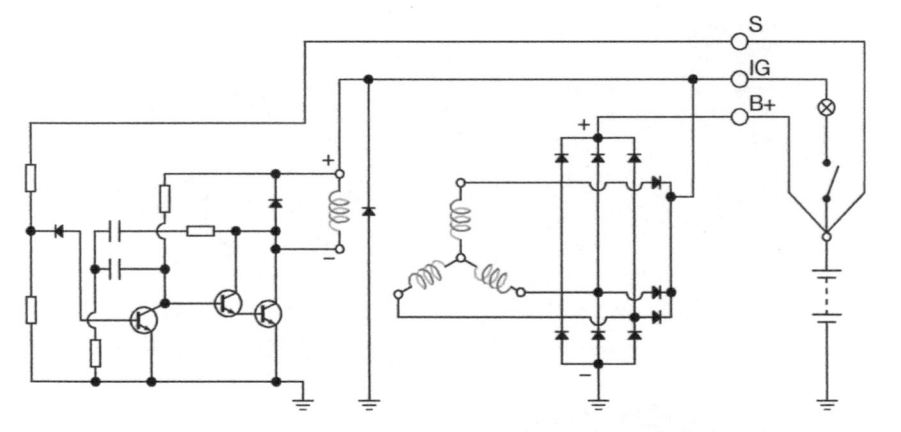

The figure above shows a typical alternator circuit. Identify the following components on the diagram by labeling their corresponding number:

1 **Battery**   4 **Stator windings**
2 **Transistors** 5 **Rotor winding**
3 **Diodes**   6 Charging lamp

Try and use different coloured pens/pencils and circle the following on the diagram:

Regulator   **Rectifier**

Alternator terminals

There are a number of electrical terminals on the rear of an alternator. State the purpose of the following terminals:

'B+' terminal: _____

The 'B+' is live, even with the ignition switched off. Avoid shorting this terminal to earth.

For this reason it is important to always disconnect the battery before removing the alternator.

The connector contains:

'IG' terminal: _____

'S' terminal: _____

'L' terminal: _____

For specific applications there may be an additional terminal. Describe its function:

'W' terminal: _____

_____

DIAGNOSTICS: ALTERNATOR SYSTEM – SYMPTOMS, FAULTS AND CAUSES

State a likely cause for each symptom/system fault listed below. Each cause will suggest any corrective action required.

Symptom	System fault	Likely cause
Battery cranks engine slowly, it is in a low state of charge or electrolyte is low	No low or high current output	
Alternator is making an unusual noise and is too hot to touch	Overheating	

Symptom	System fault	Likely cause
A squealing noise when engine is revved. A whining noise	Abnormal noises	
Light remains on or is intermittent – light not operating before starting	Incorrect warning lamp operation	

Diagnostic test equipment

List the test equipment used for testing batteries and alternators:

1 _____

2 _____

3 _____

4 _____

Carry out checks on a vehicle's alternator and charging system. Also remove and refit an alternator in accordance with health and safety procedures.

Alternator performance check

If an alternator is suspected of being faulty it should be tested on the vehicle before being removed for repair.

After completing the simple preliminary checks, carry out the basic alternator tests shown on page 304.

List the preliminary checks that should be made before the alternator is tested:

- _____
- _____
- _____
- _____

TIP Before using electrical test equipment always check it for condition and calibration to ensure it will give accurate readings.

Describe the procedures and show where the meters and leads should be fitted for tests 1, 2 and 3.

Test 1. Alternator output

Expected output _____

Actual reading _____

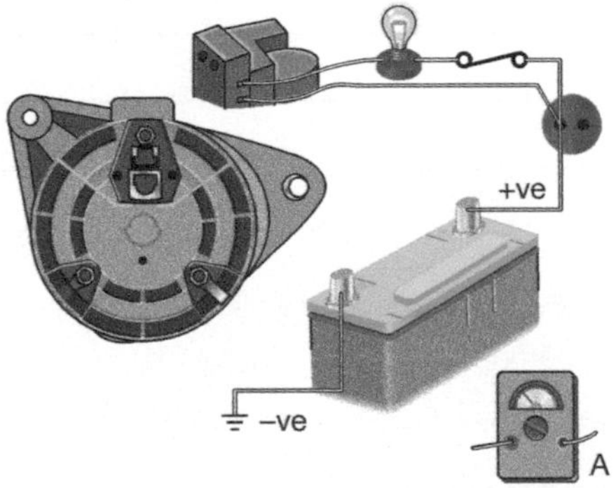

_____

_____

_____

_____

_____

Test 2. Charging circuit

V_1 actual reading _____

V_2 actual reading _____

1 _____

_____

_____

_____

2 _____

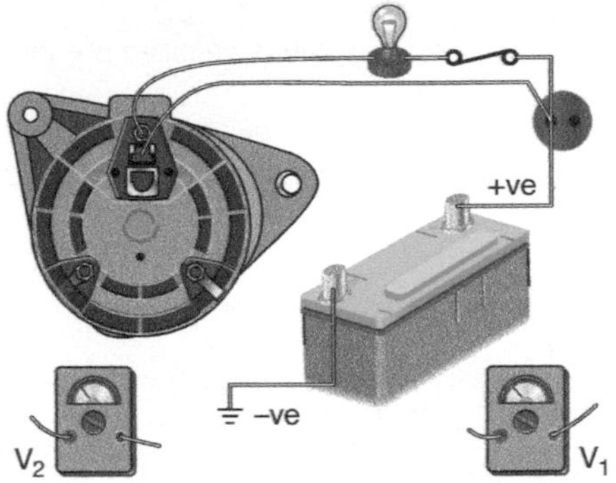

_____

Test 3. Check alternator control

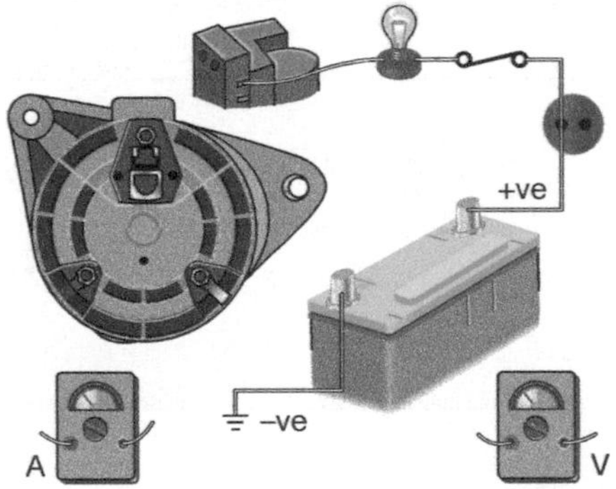

Actual reading _____

_____

_____

ALTERNATOR REMOVAL AND REPAIR

Describe the procedure to remove an alternator from the following types of engine:

V-belt type: _____

_____

_____

Multi-ribbed belt type: _____

_____

_____

LIQUID COOLED ALTERNATOR

This is a recent development which uses the engine coolant to circulate around passageways in the alternator casing. This system efficiently cools the diodes keeping them at optimum operating temperatures.

What are some other advantages of liquid cooled alternators?

1 _____

2 _____

3 _____

_____

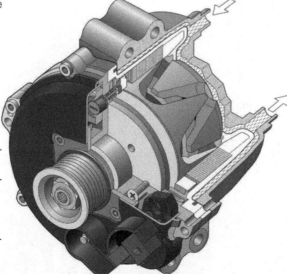

Multiple choice questions

Choose the correct answer from a), b) or c) and place a tick [✓] after your answer.

1 The alternator component which has the three interlaced windings is the:

  a) rotor [ ]

  b) stator [ ]

  c) rectifier. [ ]

2 The purpose of a zener diode is to allow current to flow:

  a) in one direction [ ]

  b) in two directions [ ]

  c) very slowly. [ ]

3 Which one of the following controls the maximum alternator output?

  a) regulator [ ]

  b) rectifier [ ]

  c) stator. [ ]

4 How many diodes are used in a bridge rectifier?

  a) 5 [ ]

  b) 3 [ ]

  c) 6. [ ]

5 Which one of the following has a base, emitter and a collector?

  a) stator [ ]

  b) diode [ ]

  c) transistor. [ ]

6 When checking an alternator V-belt adjustment, it should have a deflection of no more than:

  a) 1mm [ ]

  b) 40mm [ ]

  c) 10mm. [ ]

7 To check a voltage regulator connect voltmeter across the battery. Run the engine at 3000 rev/min, until ammeter indicates less than 10A. The voltmeter should give a steady reading of between 11V and 17.2V.

  a) True [ ]

  b) False. [ ]

8 The carbon brushes run on the:

  a) slip rings [ ]

  b) bearings [ ]

  c) stator. [ ]

SECTION 6

Auxiliary electrical systems

USE THIS SPACE FOR LEARNER NOTES

_____

_____

_____

_____

_____

_____

Learning objectives

After studying this section you should be able to:

● Identify components and explain the operation of lighting systems.

● State checks and adjustments made to the lighting system.

● Describe the operation of direction indicators.

● Describe the construction and operation of windscreen wipers, electric windows, central door locking, instrumentation, horns and heater circuits.

● Describe how to interpret simple light vehicle wiring diagrams used in auxiliary circuits.

Key terms

High intensity discharge (HID) A modern method of lighting.

Construction and Use Regulations Legal regulations that manufacturers must adhere to in the UK. They cover all aspects of vehicles including weights and dimensions, safety items and environmental standards.

Negative temperature coefficient (NTC) Material where its electrical resistance decreases when temperature increases.

Bimetallic Two metals joined with different rates of expansion.

Voltage stabilizer A device that maintains a steady voltage usually about 2V below nominal battery voltage.

Integrated circuit A miniaturized electronic circuit consisting of components such as resistors, capacitors and transistors.

Capacitor An electrical component which can store a charge for a period of time.

Resistor A device that lowers current and voltage.

VEHICLE LIGHTING SYSTEMS

The lighting system of a car may be split into two basic circuits. These are:

1 _____

2 _____

List the basic components that make up a lighting circuit and draw their electrical symbol:

When more than one lamp is used the circuit is usually wired in: _____

The sidelight circuit consists of two lamps at the front and at least two at the rear plus a lamp to illuminate the vehicle's rear registration plate.

State the legal requirements for these lamps regarding:

Colour: _____
_____

Rating: _____

The headlight circuit consists of two circuits. List the two types of circuits and when they are used:

1 _____

2 _____
_____

TIP The Construction and Use Regulations stipulate requirements for vehicles relating to lighting and other safety aspects. Go to **http://www.unece.org** and search vehicle regulations or visit **http://www.legislation.gov.uk** and search lighting regulations.

Draw a basic sidelight circuit using the following components and the appropriate circuit symbols with an earth return system.

**Battery   Fuse   Switch   Two rear tail bulbs   One number plate bulb
Two front sidelight bulbs   Three instrument illumination bulbs**

Complete the following diagram of the basic headlight circuit using the earth return system.

What component is used to enable the light switch and dip switch current to be reduced?

_____

Dip switch

Lighting switch

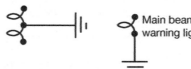
Main beam warning light

List four other circuits which are sometimes considered part of the basic lighting system, but use lights for signalling purposes or for better visibility:

1 _____   3 _____

2 _____   4 _____

Types of bulb

Examine light bulbs of different types.

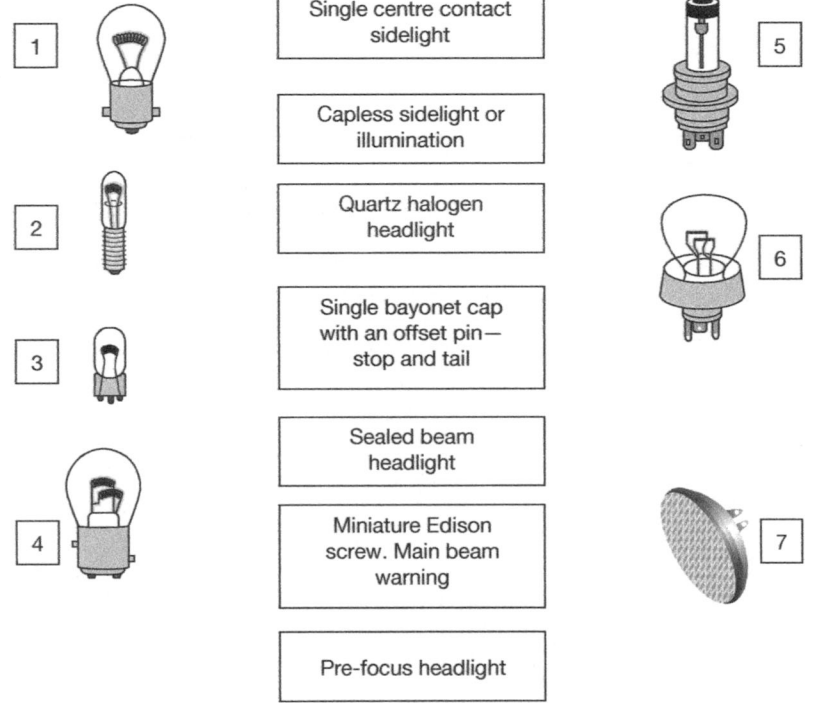

1	Single centre contact sidelight
2	Capless sidelight or illumination
3	Quartz halogen headlight
4	Single bayonet cap with an offset pin— stop and tail
	Sealed beam headlight
	Miniature Edison screw. Main beam warning
	Pre-focus headlight
5	
6	
7	

Draw a line to the description of the bulb.

Note: These sketches are not to scale.

In what important way does bulb 3 differ from the other small bulbs?

_____

Why does bulb 4 have a staggered pin fitment?

_____

Bulbs 5 and 6 are known as pre-focus bulbs, where the base filament is in the form of a flat edge and has to position the main beam filament at the focal point of the reflector.

How is the correct location of the beam achieved?

_____

What does unit 7 include?

_____

For the same wattage rating headlamp bulb 5 will give off a much brighter light than bulb 6. How is this improvement achieved?

_____

_____

The inert gas used inside the bulb is usually: _____

The resistance wire used for the filament is: _____

The temperature reached by this wire is approximately: _____

A composite headlamp with a replacement halogen bulb

High intensity discharge (HID) or xenon headlamps

How are these lights recognized? _____

This light uses two electrodes which require more than 15000V to jump a gap between the electrodes. This excites xenon gas inside the headlamp which vaporizes metallic salts that maintain the arc and emit a powerful light.

To provide a high voltage to start the arc to jump the gap a voltage booster is required. Once the high voltage bridges the gap, only about 80V is needed to keep current flowing. It takes approximately 15 seconds for the lights to reach maximum intensity, therefore these lights are commonly used for dip beam and additional conventional lights are used for main beam.

Xenon bulb

METHODS OF DIPPING THE BEAM

Two-headlamp system

When a vehicle's lights are on main beam the rays illuminate the road as far ahead as possible. What occurs when the lights are dipped?

_____

_____

Show the position of the dipped beams.

Four-headlamp system

With some systems, when on main beam the outer lamps are on permanent dip while the inner lamps throw out a brighter longer range beam along the road.

What occurs when the lights are dipped?

_____

Show the position of the light beams when on main beam.

Lenses

The lens in, for example, a headlamp distributes the light rays to provide correct illumination of the road ahead during main beam and dip operation. State the purpose of the area (X) on the lens shown below:

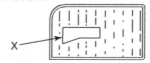

_____

_____

Modern headlamps

Headlamps now use a plain front cover to enable increased light output. The headlamp shape can be designed to fit the body styling to improve looks and aerodynamics.

The lamp opposite has a cylindrical bulb housing which concentrates the beam of light without too much scattering.

HEADLAMP ALIGNMENT

All vehicle headlamps in the UK must comply with the Department of Transport Road Vehicle's Headlamp Regulations which state the position, or angle, of the dipped beams.

TIP Lighting systems in the UK must comply with Construction and Use Regulations.

Manufacturers must adhere strictly to these regulations which also apply to many other aspects of the vehicle.

The specialist equipment used to check the alignment of headlamps measures the angle of dipped beam and the beams' positions relative to one another when both are on main and dipped beam. Show a sketch of such equipment in its testing position on the diagram below. Draw a sketch of the equipment in your workshop or search for a suitable headlamp aligner.

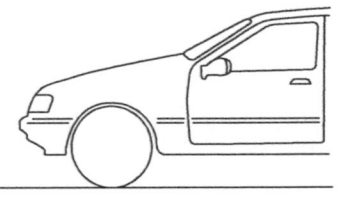

Which organization does the equipment have to be approved by when used for MOT purposes?

_____

State three checks to be carried out on the headlight assembly during a routine service:

● _____
● _____
● _____

Give four pre-checks that are necessary to ensure that headlamp alignment is accurately carried out:

1 _____
2 _____
3 _____
4 _____

Check the headlamp alignment of a vehicle, using available equipment, and describe the major points of the alignment procedure.

Make and model of equipment_____

● _____
● _____
● _____
● _____
● _____

Where would you find the headlamp settings for the vehicle you are checking?

_____
_____

TIP The headlamps are adjusted using mechanical adjusters on the back of the headlamp unit.

OBLIGATORY LIGHTING

The direction indicator, stop light, reversing light and rear fog light are all obligatory lighting/warning systems on a modern car.

Direction indicator/hazard warning

Amber coloured flashing lights are used to signal a hazard or the intention to change the direction of a vehicle to other road users. In the indicator circuit a flasher unit is used to make and break the electrical supply to the lamps causing them to flash.

State the flash rate for direction indicators: _____

Complete the simple wiring diagram of an indicator circuit below. Include the flasher unit and panel indicator lamp.

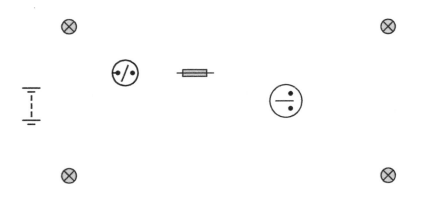

Electronic flasher unit

As is the case with many other electrical components, the use of electronic circuitry has improved the performance of the flasher unit.

The system comprises an integrated circuit, capacitor and resistors working in conjunction with a relay. Intermittent current pulses operate the relay which controls the current to the lamps.

Stop lamps

Complete the diagram above to show how stop lamps are operated.

_____

WINDSCREEN WIPER SYSTEM

Statutory regulations require that a road vehicle should be equipped with one or more windscreen wipers and washer to give the driver a good view of the road ahead in all weather and driving conditions. Most windscreen wiper mechanisms are electrically operated. A relatively powerful motor is required to drive the wipers and it is desirable that the motor and mechanism are quiet in operation.

Describe the operational features of a windscreen wiper/wash system:

- _____
- _____
- _____
- _____

Windscreen wipers

The wipers are operated by a mechanism which is given a to-and-fro action by a suitably designed linkage connected to the armature shaft of a small electric motor.

The electric motor may be of a single, twin or variable-speed type. Almost all types have a limit switch device incorporated into the drive assembly.

The figure below shows the wiring diagram of a single-speed windscreen wiper motor with the wiper control switch on the earth side.

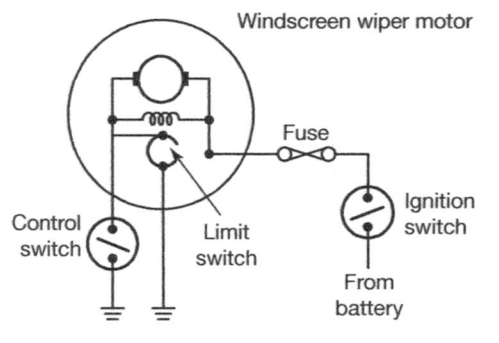

Windscreen wiper motor

Explain the operation of the limit switch (park switch) when the main wiper control switch is opened.

_____

_____

Single and two-speed wiper motors

The electric motors used for wiper operation are normally a two-speed permanent magnet type.

Low speed operation

During low speed operation, current flows through the commutator using a large number of windings due to the brushes being on opposite sides of the commutator. This causes a large amount of magnetism, resulting in a back electromotive force (back EMF) which lowers the current and therefore the motor speed.

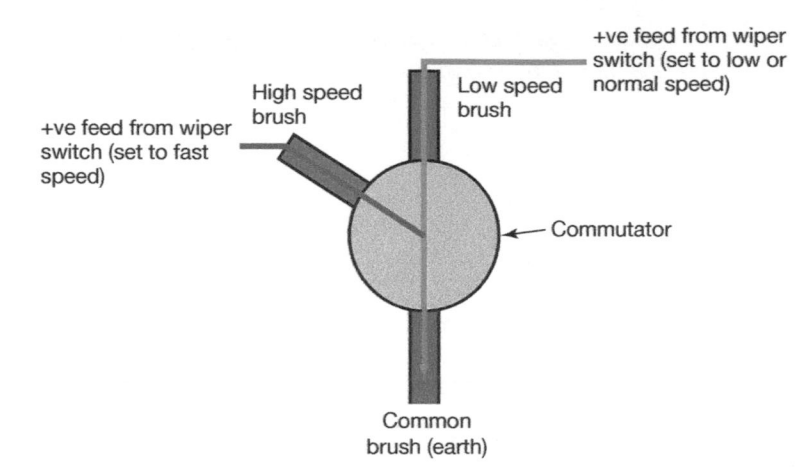

High speed operation

On this setting current passes through the offset brush, which is closer to the common brush, to earth. This results in fewer windings which reduces the back EMF, causing the current flow to be higher and therefore a faster motor speed.

How many brushes would a single speed motor require? _____

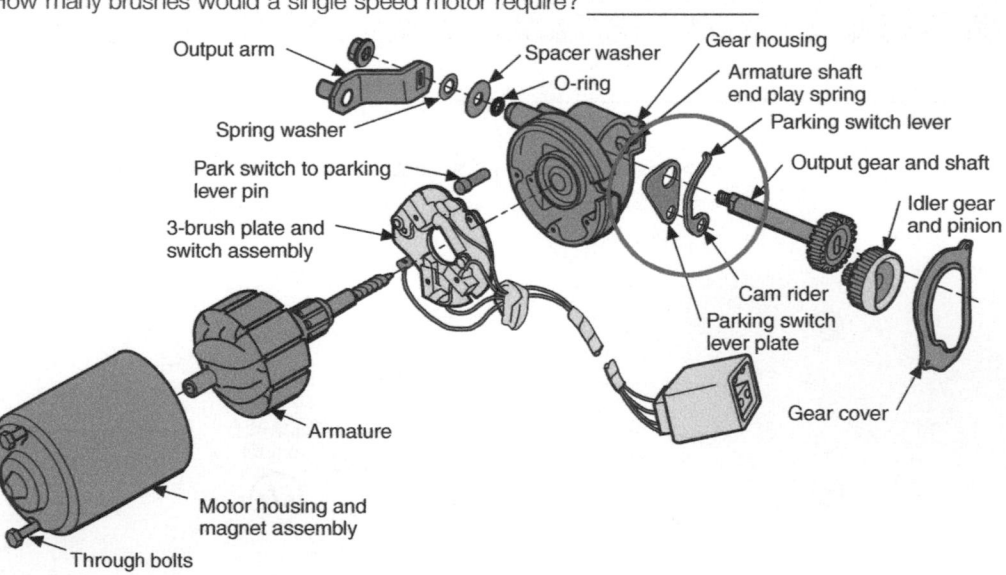

Exploded view of a wiper motor

HORNS

A requirement of the Construction and Use Regulations is that all vehicles must be equipped with an audible warning device.

Two types of horn used for this purpose are:

1 _____

2 _____

The figure opposite includes the horn circuit. Identify the components circled:

1 _____

2 _____

3 _____

4 _____

Explain why the component labelled 2 is needed in this circuit?

_____

Describe the operation of the horn circuit using the circuit diagram:

● _____

_____

● _____

_____

Windtone horns shown are often used in pairs on a vehicle. Why is this?

_____

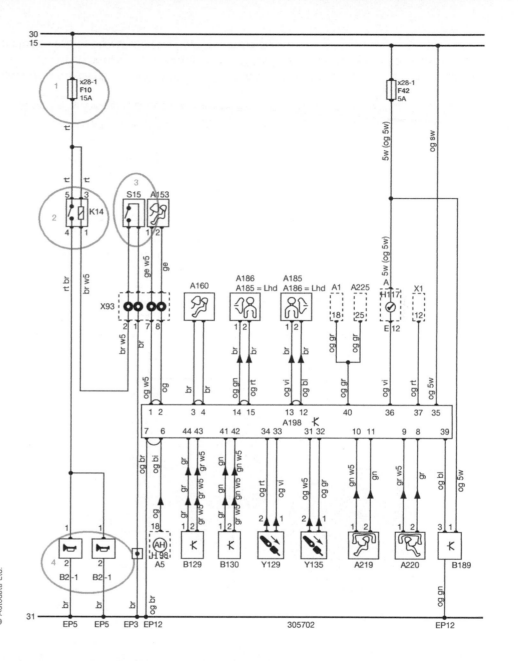

© Autodata Ltd.

TEMPERATURE AND FUEL GAUGES

The current supply to both fuel and temperature gauges is controlled by a voltage stabilizer. The sender units in both the fuel tank and cooling system have resistors that vary depending on either the amount of fuel in the tank or the engine temperature.

Voltage stabilizer

Why is a voltage stabilizer required in a gauge circuit?

_____

_____

Balancing coil temperature gauge

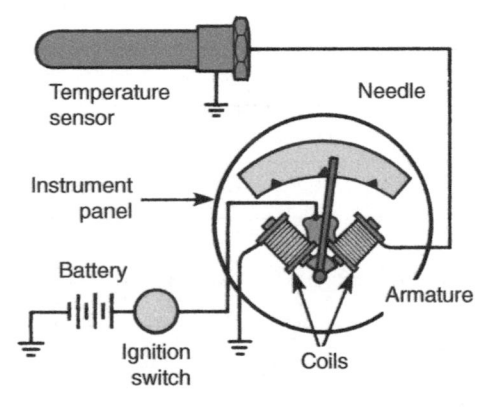

This gauge has a moving ferrous needle which is attracted to magnetic forces. Two coils are used, both sharing a positive feed. The left coil is earthed to the body, while the right coil earths through a temperature sensor which alters its resistance as coolant temperature changes. Current flowing through the coil causes a magnetic field, which attracts the ferrous needle. The coolant sensor is a negative temperature coefficient thermistor, which lowers its resistance as temperature rises.

Explain why the gauge moves to the right.

_____

Thermal or bimetal fuel gauge

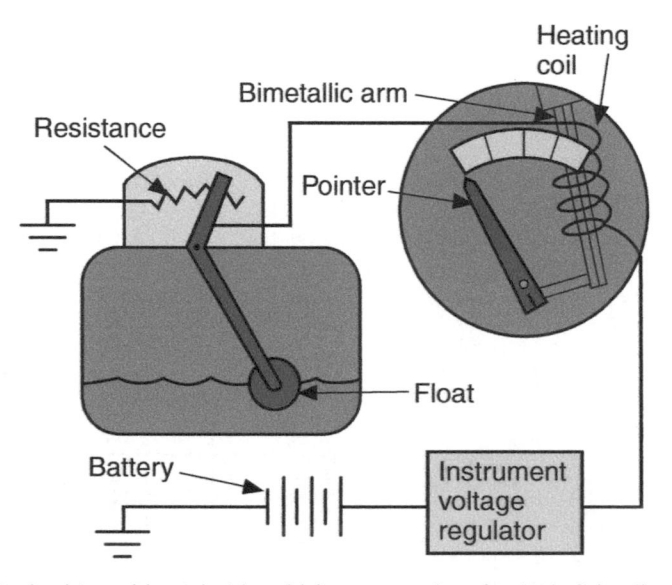

The needle is attached to a bimetal strip which moves when heated. A heating coil is wrapped around the bimetal strip which is connected to the voltage stabilizer and earths through the gauge sender in the fuel tank. When the fuel level is low the resistance is high and reduces as the fuel level rises.

Describe the operation of the gauge.

_____

_____

_____

ELECTRICALLY OPERATED WINDOWS

Side windows can be operated by electric motors connected to a geared linkage which moves the windows up or down, depending on the direction of motor rotation. A DC permanent magnet motor is used to power each window drive mechanism.

Using the diagram below, describe the operation of switches to move the right-hand window motor down. Complete the statement using the words in the word bank below. Note: there are three distractor words.

terminal 1	positive	closed	contact	relay
open	negative	earth	fuse	

The _____ switch contacts at 3 receive a _____ feed from the window switch connection at 4 from _____ F2, located in the internal fuse box.

The positive feed from terminal 3 enables a positive feed to _____ of the right-hand window lift motor (M15), the motor earths through the closed switch _____ at 2, completing the circuit to the main _____ point.

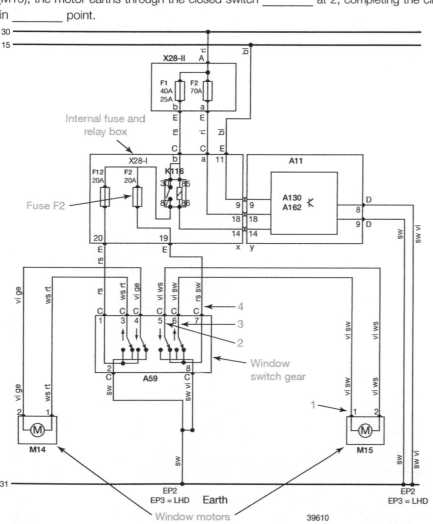

CENTRAL DOOR LOCKING

A central door locking system enables all doors, boot or tailgate to be locked or unlocked simultaneously when the key is turned in the door lock or the remote control operated. Electric motors or solenoids in each door and boot or tailgate are activated to operate the door locking mechanism.

The figure on the next page shows a four-door central door locking system. Match the numbers to the following components.

Number:

3 – Right-hand rear door lock motor.

_ – Left-hand front door lock motor and switch assembly.

_ – Right-hand front door lock motor and switch assembly.

_ – Central door locking module.

_ – Left-hand rear door lock motor.

_ – Motor link connector 1.

_ – Motor link connector 2.

Starting at the central door locking module draw coloured lines from terminals 10 and 11 to all four central door lock motors.

How is direction of motor rotation changed?

_____

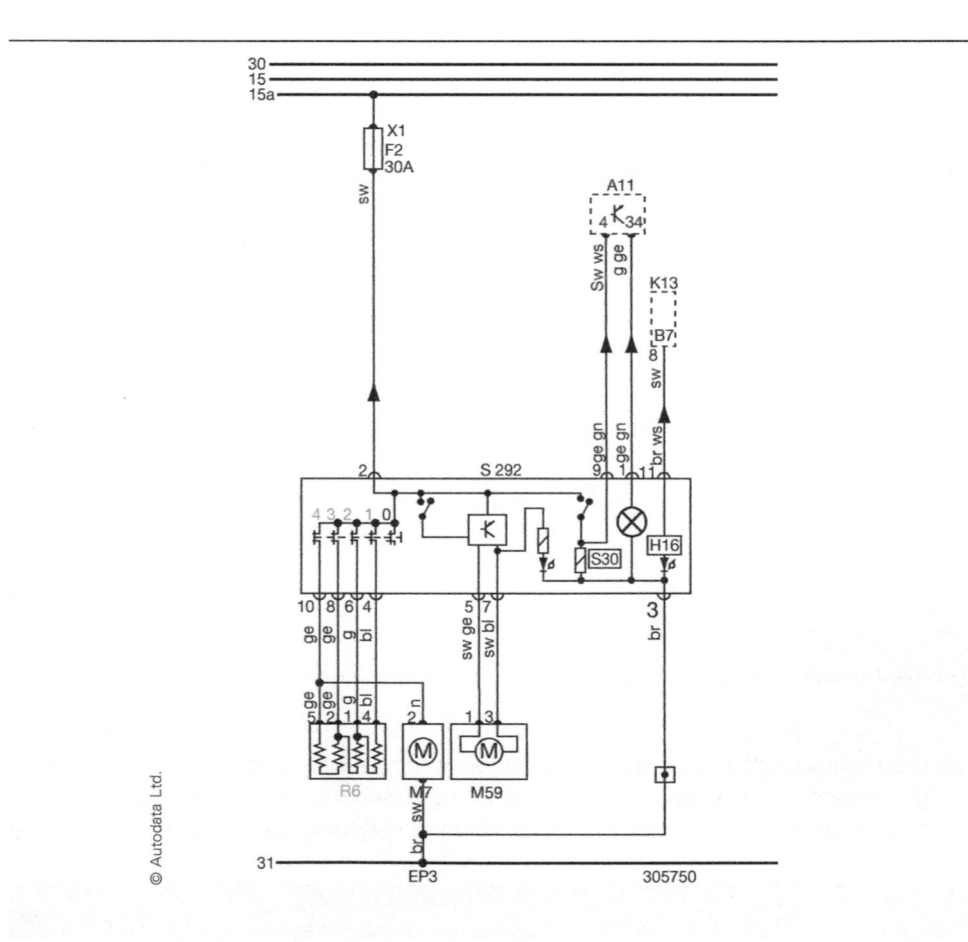

COMFORT AND CONVENIENCE

The heater motor is situated in a box where it can blow air over the heater matrix, drawing air from outside the cabin or recirculating air from inside the cabin.

To suit different heating or cooling conditions several speeds are used.

How is this achieved in most heater motor circuits?

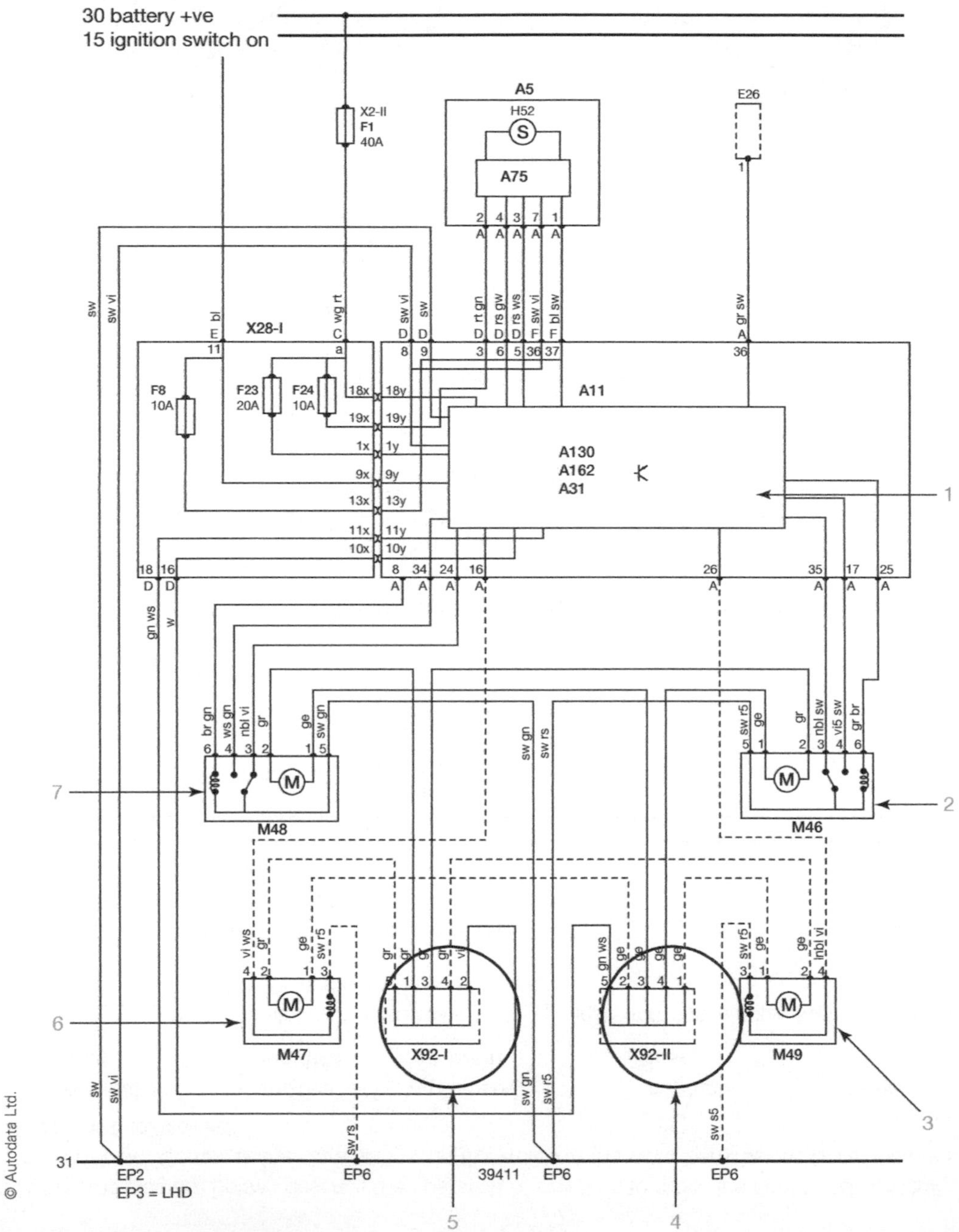

Using the wiring diagram on the previous page determine the current flow in the heater motor circuit if component R6 has all resistors with a value of 2Ω, the circuit voltage is 12V and motor resistance is 4Ω:

Position 1 – _____

_____

Position 2 – _____

_____

Position 3 – _____

_____

Position 4 – _____

_____

Multiple choice questions

Choose the correct answer from a), b) or c) and place a tick [✓] after your answer.

1 **How often should direction indicators flash?**

a) 60–120 times per minute [ ]

b) 30 times per minute [ ]

c) every 3 seconds. [ ]

2 **How many brushes are fitted to a two-speed wiper motor?**

a) 2 [ ]

b) 3 [ ]

c) 4. [ ]

3 **What is the purpose of the relay in the headlamp circuit?**

a) to reduce the current to the headlight switch [ ]

b) to allow more current to flow through the headlight switch [ ]

c) to reduce the circuit voltage to the headlamp switch. [ ]

4 **What component is used to keep fuel and temperature gauges accurate?**

a) voltage regulator [ ]

b) current regulator [ ]

c) voltage stabilizer. [ ]

5 **How is speed controlled in a heater motor circuit?**

a) using three brushes on the motor [ ]

b) using different resistors [ ]

c) using diodes. [ ]

SECTION 7

Diagnosis and Testing

USE THIS SPACE FOR LEARNER NOTES

_____

_____

_____

_____

_____

_____

_____

Learning objectives

After studying this section you should be able to:

- Explain systematic test procedures.
- Identify electrical test equipment.
- Describe how to use electrical test equipment.
- Describe the meaning of electrical short, high resistance and open circuit and parasitic drain.
- Explain basic systematic electrical testing procedures.

Key terms

Logic probe Electrical test equipment.
Open circuit No current flows in the circuit, e.g. due to a broken wire.
Short circuit Electrical shorts to earth before its dedicated earth, e.g. chaffed wire.
High resistance Current flow is restricted in the electrical circuit, e.g. corroded terminal.
Digital multimeter (DMM) Electrical meter to test a number of electrical functions, e.g. current, resistance, voltage, frequency.

ELECTRICAL DIAGNOSIS AND TESTING

General fault finding procedures

Increased electrics and electronics on vehicles has meant that effective diagnosis and repair are two of the main skills for a vehicle technician.

Electrical diagnostics requires a step-by-step approach to solving the problem. Knowledge of the vehicle systems is essential as well as access to manufacturer test data and wiring diagrams. The ability to diagnose and rectify electrical faults quickly and accurately are essential skills to have when working on modern vehicles which have lots of electrical and electronic systems.

The following is a recommended step-by-step systematic approach to finding and effectively repairing electrical/electronic faults.

Symptoms

These are the results of the fault. Remember – the symptom may not be directly related to the fault.

What are the symptoms? How would you find this out?

_____

Confirmation

Confirm the fault by checking the vehicle yourself. Carry out a full system check.

This is to _____

_____

Evaluation

Evaluate the evidence.

Take time to look at the evidence. Check _____

Further tests

Carry out further tests. This is to _____

Rectification

Repair or replace the faulty component. Rectify the problem and carry out _____

Check all systems

After completing the repair you should thoroughly check all of the vehicle systems, e.g. lights, wipers. Why is this?

_____

_____

 TIP   Where possible use manufacturer's specific test equipment and testing procedures.

VOLTAGE

In order to diagnose electrical faults accurately and quickly, it is important to have a good understanding of voltage and how to correctly use a multimeter.

Read through the following statements and fill in the missing words from the word bank below.

current	reading	difference	resistor	value
two	circuit	resistance	current	
voltage	greater	size	voltage	

It is important to understand the following rules:

- _____ is electrical pressure (this pushes the _____ around the circuit).
- Voltage drops across a resistor (e.g. bulb, motor).
- Voltage drop only happens when _____ is flowing.

In a circuit it is the _____ of the resistor _____ which determines the _____ drop across the resistor. When compared to other resistors in the _____, the _____ the resistor's value the more of the available voltage will be used by it. There will be a _____ of zero volts after the last _____ in a circuit, as long as there is current flowing (the circuit is complete and switched on).

When a voltmeter is connected either side of the _____ it will read the _____ in voltage between these _____ points.

Automotive multimeter

A good quality multimeter is one of the most essential tools in a technician's tool kit. The multimeter shown measures voltages, current and ohms. It also has additional functions, some of which are specific for automotive applications.

Some functions are self-explanatory such as engine RPM, which may be connected using an inductive clamp and temperature measurement to check cooling and air conditioning systems, and exhaust catalysts.

The table shows some of the other functions which a good automotive multimeter should have. State the type of sensor or item each function checks.

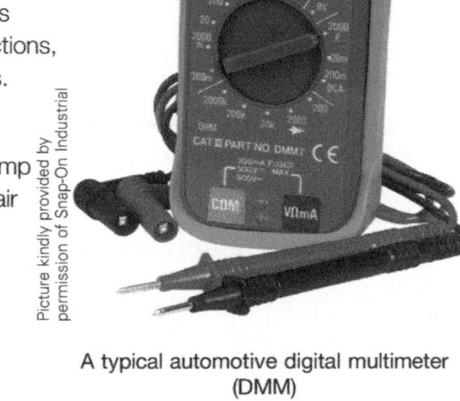

Picture kindly provided by permission of Snap-On Industrial

A typical automotive digital multimeter (DMM)

Multimeter function	Sensor or item checked
Duty cycle/Dwell This is a measurement of the time a component is switched on expressed as a percentage (dwell is a duty cycle expressed in degrees)	_____ _____ _____
Pulse width	_____ _____ _____
Low AC voltage measurement	_____ _____
Bar graph display Shows movement similar to an analogue display	_____
Data hold function Will hold minimum and maximum values to be recalled later	_____ _____
Capacitance.	_____ _____ _____
Hz (Frequency)	_____
USB output	_____

TIP Before using a multimeter check the casing and leads are not damaged. Switch the scale to ohms, plug the leads into the meter and touch the probes together; this should show continuity and not fluctuate when wiggling the leads. Make sure the internal battery is in good condition.

WWW Visit **www.multimeterwarehouse.com/techcenter.htm** or **www.doctronics.co.uk/meter.htm** for information on how to use a multimeter.

TIP Some meters can have adapter cables and amp clamps, that plug into the 'A' socket on the multimeter, which are designed to increase the amount of current the meter can check (the black lead is plugged into the COM port).

When using a digital multimeter it is important to connect the test leads in the correct plug sockets on the meter. Look at the digital multimeter in the figure opposite and explain what each of the plug sockets are used for and how to connect them when carrying out different tests.

A _____

_____

mA TEMP _____

_____

COM _____

VΩ Hz % _____

_____

Examine the type of DMM used in the workshop and sketch its face in the block below, naming the important features.

GENERAL ELECTRICAL TESTING EQUIPMENT

Circuit test lamp

This circuit tester is for use on 6 and 12V systems. It can be used to detect power, earth and find shorts and breaks. It has an insulated alligator clip and a coiled extendable cord.

Picture kindly provided by permission of Snap-On Industrial

Circuit tester (test lamp)

Briefly describe how to use a test lamp: _____

_____

_____

_____

TIP   Where possible, follow manufacturer's test plans and safety procedures. Use dedicated testing equipment.

Logic probe

The logic probe has a high internal resistance that does not affect the circuit under test (suitable for digital components).

This piece of equipment has two large crocodile clip connectors (one red, one black) and a smaller crocodile clip connector (black). The main unit has a probe and a rocker switch. There is a light-emitting diode (LED) which illuminates either red or green.

How is this equipment connected to check for a live feed or an earth in an electrical circuit?

Sealey Group

Logic probe

_____

_____

_____

What happens to the LEDs when testing an electrical circuit?

_____

_____

An electric window motor is suspected to be faulty. Briefly describe how a logic probe would be connected and used to check its operation.

_____

_____

_____

Use a logic probe to check a headlamp circuit for a live feed and an earth.

Operate the light switches. Use the probe and see what happens to the LEDs on the probe.

Amp clamp

How is an amp clamp connected to check for current flow in a circuit?

_____

_____

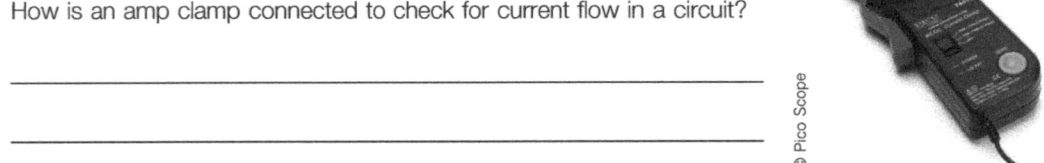

© Pico Scope

Amp clamp

Scanner

Modern vehicles have a lot of electronic systems. This piece of equipment is designed to plug into the 16 pin, data link connector (DLC) using a European On Board Diagnostics (EOBD) dedicated socket. All petrol cars sold within Europe since 1 January 2001, and diesel cars manufactured from 2003, must have on-board diagnostic systems to monitor engine emissions. This piece of equipment reads any fault

Picture kindly provided by permission of Snap-On Industrial

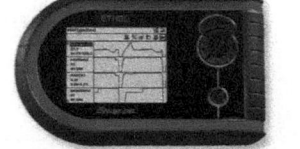

Ethos scanner

code and shows live data as well as other functions. It is an essential piece of equipment for diagnosing and rectifying engine management faults.

Lab scope

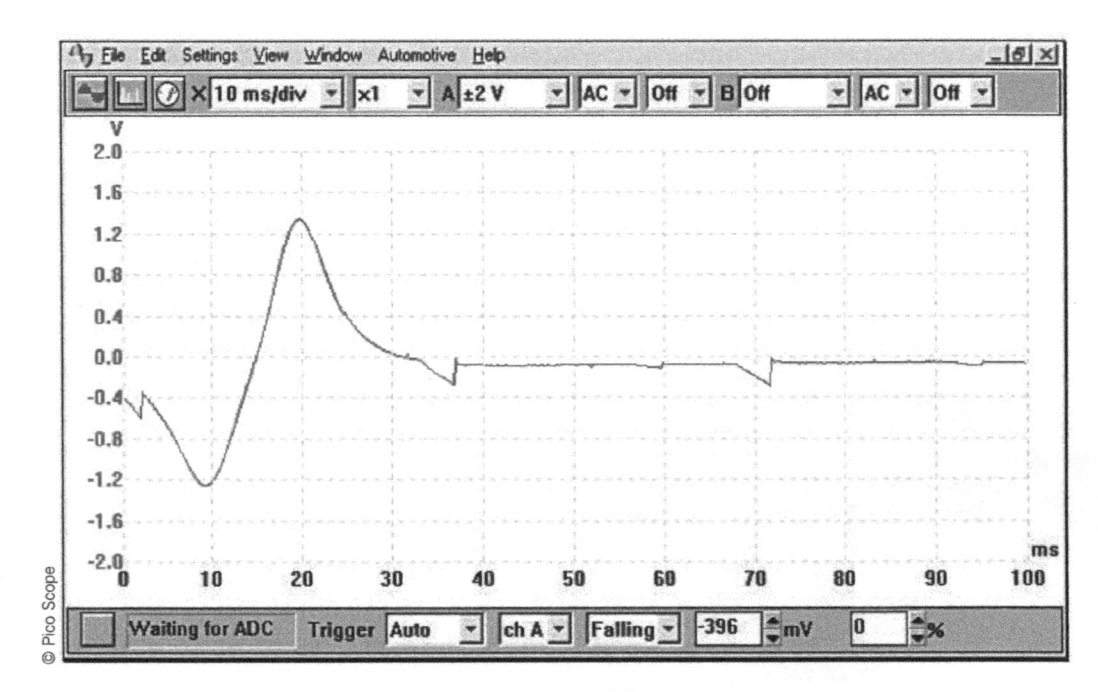

© Pico Scope

An inductive cam sensor lab scope trace

Due to the electronics on modern vehicles an essential piece of test equipment is a lab scope. It is used to check output signals of different sensors and other electronic signals. These can be compared to a database of images for the signal being checked in order to confirm that it is correct and any faults can be identified and rectified.

www.picotech.com This website contains information on using lab scopes. It also shows how to test components and what wave patterns to expect.

Manufacturer's dedicated test equipment

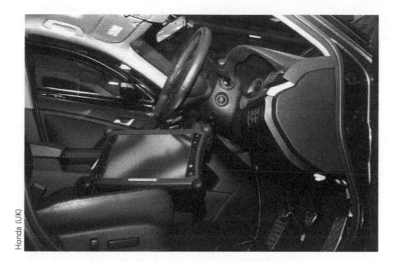

Honda (UK)

Honda dedicated diagnostic system

Manufacturers have their own dedicated diagnostic test equipment. They normally connect to the data link connector (DLC), which is normally close to the steering wheel.

What types of vehicle systems can these test?

_____

Manufacturer dedicated diagnostic systems have the capabilities to read and erase fault codes, set vehicle system parameters, read live data, re-programme and code control units.

TESTING CIRCUITS

WWW Visit **www.ladyada.net/learn/multimeter/** for information on how to use a multimeter to check current, resistance and voltage.

State the type of tests being carried out by the multimeter shown in the figure below.

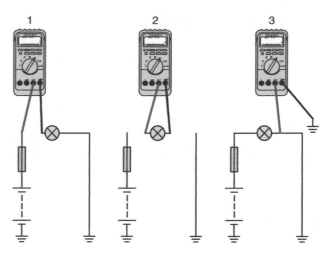

1 _____

2 _____

3 _____

How is a multimeter connected into a circuit to check for current flow?

_____

The unit of measurement for current is: _____

How is a multimeter connected into a circuit to check for resistance?

_____

_____

The unit of measurement for resistance is: _____

What is an earth switched circuit?

_____

In an earth switched circuit if the switch was open the voltage after resistance should be:

_____

If the switch is closed in an earth switched circuit, the voltage after the last resistance should be:

_____

TYPES OF CIRCUIT FAULTS

Two simple light circuits are shown in the figure below, one is an open circuit, the other has a short circuit.

Identify each circuit, giving reasons for choice:

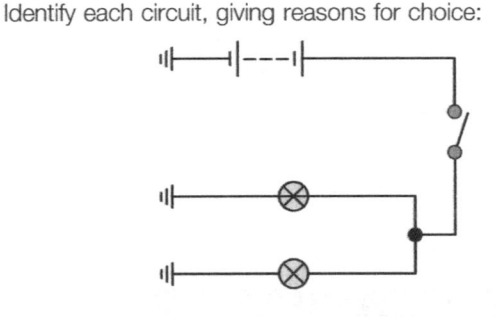 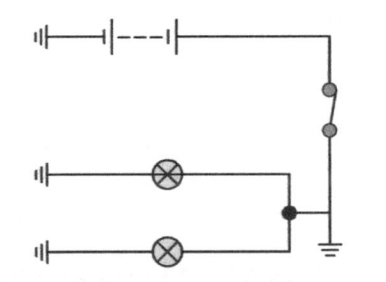

The circuit _____

_____

_____

The circuit _____

_____

_____

What could be a probable cause and effect of the short circuit?

_____

_____

_____

What is used in an electrical circuit to protect the electrics against a short circuit to earth?

_____

_____

Open circuit fault finding

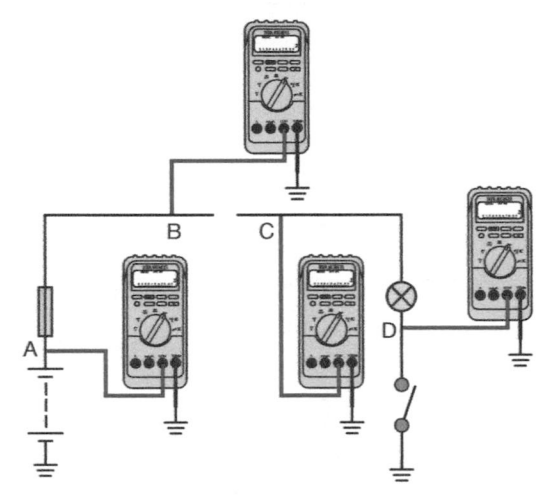

The diagram shows a simple 12V light circuit. The circuit is switched on. The light fails to illuminate. A voltmeter is connected at points 'A', 'B', 'C' and then 'D'. What would be the expected voltage in each of the voltmeters if there was an open circuit between points 'B' and 'C'.

_____

_____

When checking voltage on a motor vehicle the multimeter should be set to _____

SHORT CIRCUITS

There are three areas where a short can occur in a circuit. Each of these areas will have a different effect on the circuit.

Before the resistance

The short to ground (earth) reduces the circuit resistance to virtually zero. Current will still continue to be supplied, even though there is no consumer (e.g. bulb, motor) to use the current. This will result in a serious over-current situation in the circuit, causing the fuse to blow or lead to a serious fire.

Draw a simple electrical circuit to show an electrical short before the resistance. Include a bulb fuse, battery (or motor) and switch in the circuit. Remember this is an earth switched circuit.

After the resistance but before the switch (earth switched circuit)

When the short is before the switch and after the resistance (e.g. bulb, motor) there will be a permanent supply to the consumer, so the driver is unable to switch off the affected circuit. This is because the circuit now has a permanent earth. If the fault affects a circuit such as the glove box light, where the driver cannot see the light is constantly on, then over a period of time the battery can go flat. This is known as a 'parasitic drain' as the faulty circuit will drain the battery flat (like a parasite which benefits at the expense of the host, in this case the battery).

Draw a simple electrical circuit to show an electrical short after the resistance and before the switch. Include a bulb fuse, battery and switch in the circuit. Remember this is an earth switched circuit.

After the resistance and after the switch (earth switched circuit)

If this happens there will probably be no apparent symptoms as it is in the earth side of the circuit after the switch.

HIGH RESISTANCE FAULTS

A common fault of electrical circuits is unwanted series resistance. What is meant by this term and how does the resistance occur?

_____

_____

_____

_____

_____

Describe how a voltmeter should be used to check the continuity of a circuit for high resistance.

_____

_____

_____

When checking for continuity of a non-live circuit, the multimeter needs to be set to read:

_____

When using an ohmmeter to check an electrical circuit for resistance, it is important to isolate the component/circuit from the power supply.

The figure opposite shows a simple 12V light circuit. When the circuit is switched on the lamp is dim. There is a high resistance between points 'C' and 'D' which is causing the fault. Correctly connect the multimeter into the circuit at points 'A', 'B', 'C' and 'D'.

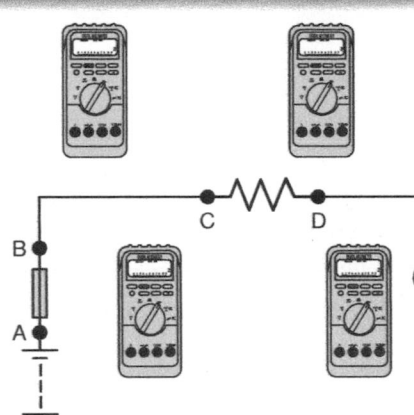

Write the expected readings for the multimeter at each point, choosing the correct four values from the following list:
5V 12V 12V 12V 12A 5A 5Ω 12Ω 12A.

Describe how the resistance of a component should be tested using an ohmmeter or multimeter.

1 _____

2 _____

3 _____

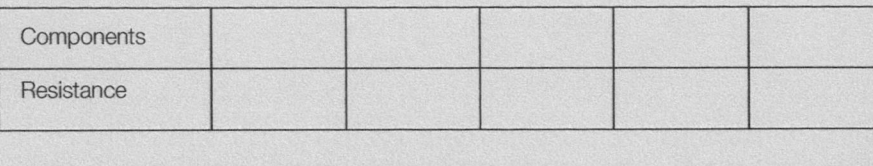

Select FIVE electrical/electronic components and measure their resistance. Write the results in the table.

Components					
Resistance					

Checking current flow

Correctly connect the circuit in the figure above, showing how the multimeter is connected to check current flow.

How is current flow through the circuit being tested?

_____

_____

When checking current flow in a circuit the meter is connected in _____

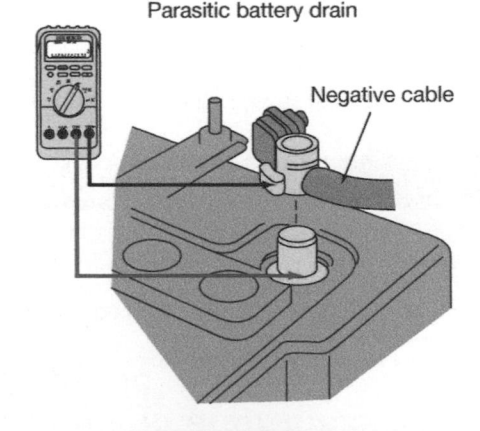

Parasitic battery drain

Negative cable

Measuring current drain

Read through the following paragraph and fill in the missing words from the following word bank (note there are four extra distracter words):

series   parallel   10A   0.5A   negative   positive   current   off   on   drain

If a vehicle is left to stand for a while and the battery goes flat this could be caused by a parasitic _____. To check for this the battery _____ terminal needs to be disconnected and the multimeter set to measure _____ flow (amperage) is connected in _____, between the negative battery lead and the negative battery terminal. With the key removed from the ignition, all doors shut (so that the interior lights are off), and all consumers switched _____, there should be a reading of no more than _____.

Is a current drain of up to 0.5A acceptable?

_____

PRECAUTIONS WHEN TESTING OR REPLACING COMPONENTS

Describe any special precautions to be observed when:

1 Disconnecting electrical components.

_____

_____

2 Replacing components/units.

_____

_____

_____

3 Testing components.

_____

_____

_____

What faults are especially dangerous to electronic components?

1 _____

2 _____

_____

Multiple choice questions

Choose the correct answer from a), b) or c) and place a tick [✓] after your answer.

1 **When checking a circuit for an unwanted resistance, what is the maximum allowed voltage drop?**

   a) 12V [ ]

   b) 0.5V [ ]

   c) 0V. [ ]

2 **An unwanted high resistance in a light circuit would cause the:**

   a) light to be dim [ ]

   b) fuse to blow [ ]

   c) light to be bright. [ ]

3 **When checking a circuit for current flow, the multimeter needs to be set to:**

   a) volts (AC) [ ]

   b) amps (A) [ ]

   c) ohms (Ω). [ ]

4 **It is the size of the resistor value in a circuit that determines the voltage drop across the resistor.**

   a) true [ ]

   b) false. [ ]

5 **Which one of the following is correct?**

   a) a parasitic drain can cause the battery to go flat [ ]

   b) high resistances will cause a parasitic drain [ ]

   c) parasitic drains are the result of reduced current flow. [ ]

GLOSSARY

Absolute pressure True pressure (gauge pressure + atmospheric pressure).

Accident An unplanned event that results in injury, ill-health or damage.

Ackerman linkage Form of steering arranged to give true rolling motion round corners.

Active restraint A safety restraint system which the vehicle's occupants must make a manual effort to use.

After sales The section of a business which deals with the repair, maintenance and fitment of auxiliary components once a new vehicle has been sold to the customer.

Air gap The gap between the electrodes of a spark plug or reluctor and pick-up.

Airbag A bag which inflates on impact and saves the driver and passengers from injury.

All-wheel drive (AWD) A term associated with vehicles that have permanent four-wheel drive.

Amplifier A device used to increase the electrical signal in an electronic ignition system.

Antifreeze Is added to water in the correct percentage to reduce freezing and act as a corrosion inhibitor.

Approved repairer A company/business which has been approved to undertake repairs on behalf of specific manufacturers.

Aspect ratio The ratio between the height and width of a tyre (expressed as a percentage).

Battery capacity How well a battery will perform under certain conditions (e.g., cold cranking and reserve capacity).

Baulk ring A component that connects with the gear cone to match speeds.

Beam axle Rigid axle.

Bias belted A design of tyre which is a hybrid, a cross between a ply and radial design.

Bimetallic Two metals joined with different rates of expansion.

Boundary A term applied where the film of lubricant is applied by splash and mist.

Brake fade Loss of friction in the brakes due to overheating.

Bulkhead Panels at the rear of the engine compartment and also separating the passenger compartment from the boot, which span the full width of the body.

Bump Upward movement of suspension.

Caliper Housing for the piston(s) which operates the brake pads.

Camshaft Rotates at half engine speed and opens the valves at the correct time and duration.

Capacitor An electrical component which can store a charge for a period of time.

Carburettor A mechanical device for mixing fuel with air.

Carcinogenic Cancer forming.

Catalytic converter A honeycomb construction which converts harmful gases into harmless gases.

Cell Contains a positive and negative plate surrounded by an electrolyte in a container. Six linked cells make a 12 volt battery.

Centre bearing A bearing that supports long propshafts to reduce vibration and whip.

Closed loop An electronic control system using feedback to maintain correct ignition timing/mixture strength.

Clutch drag A fault where the clutch does not release fully.

Clutch judder A fault causing unwanted vibration.

Clutch plate/centre plate/drive plate Turns the gearbox input shaft.

Clutch release bearing Thrust bearing that pushes on the diaphragm spring fingers.

Clutch slip A fault where torque transmission is reduced.

CO Carbon monoxide.

Coefficient of friction Ratio of the amount of force needed to cause an object to start to slide.

Compliance Slight controlled movement within the suspension bushes on the vehicle.

Composite Vehicle construction which utilizes a separate chassis and body.

Compression ignition A term used when an engine compresses the air mixture to generate ignition.

Compression ratio (CR) The ratio between the maximum cylinder volume and the minimum volume it is compressed into.

Conductor A material that has freely moving electrons, which allow current to flow.

Constant velocity When speed does not vary.

Construction and Use Regulations Legal regulations that manufacturers must adhere to in the UK. They cover all aspects of vehicles including weights and dimensions, safety items and environmental standards.

Contact breaker A mechanical switching device used in older ignition systems.

Corrosive Substance that can destroy tissue, usually strong acid or alkali.

COSHH Control of Substances Hazardous to Health.

Crankshaft Rotates in the cylinder block. Converts reciprocating motion to rotational motion.

Cross ply A tyre in which the plies are placed diagonally across each other at an angle of approximately 30 to 40 degrees.

Cross-flow radiator Coolant flows horizontally and transfers heat from the engine to the surrounding air.

Crown wheel Large gear driven by the pinion.

Crumple zones Zones of controlled deformation under impact which are built into the front and rear of modern vehicles.

Current Flow of electrons.

Damper A device which dampens the oscillations or vibrations of the road springs.

Dead axle Non-driven axle (usually on the rear).

Detent A plunger or ball used to positively locate the selector shaft in gear.

Detergent Chemical used for cleaning, usually diluted with water.

Diaphragm spring A dish type spring used in pressure plates.

Differential Series of gears to allow road wheels to rotate at different speeds whilst cornering.

Digital multimeter (DMM) Electrical meter to test a number of electrical functions, e.g. current, resistance, voltage, frequency.

Diode An electronic component which allows current to flow in one direction only.

DIS Distributor-less ignition system.

Disc brake A brake in which friction pads grip a rotating disc in order to slow the vehicle down.

Distributor A mechanical component, which distributes high voltage to the spark plugs.

Dog teeth Teeth on baulk rings and syncrohub assemblies.

Drag link Connects drop arm to first steering arm.

Drive shaft A shaft designed to take drive from the final drive to the driven wheels.

Drop arm Connects steering box to drag link of steering system.

Drum brake A brake in which curved shoes press on the inside of a metal drum to produce friction.

Dry sump Oil is supplied and returned to a separate oil tank away from the engine.

Ductility Ability for a cold metal to be stretched and formed without breaking.

Dynamic imbalance Lack of balance of a rotating part such as a wheel in motion which can cause vibration or judder.

Earth return The conductive part of the body is connected to the negative side of the battery.

ECU (ECM) Electronic control unit (module).

Electrolyte Active fluid in a battery cell.

Electron Negatively charged part of an atom.

EPA Environmental Protection Act.

Exhaust gas recirculation (EGR) A system used to reduce NO_x and the levels of unburnt hydrocarbon.

Exhaust manifold Provides a means for the burnt gases to be directed to the exhaust system.

Ferrous A metal which contains iron, making it magnetic.

Firing order The sequence in a multi-cylinder engine in which ignition takes place in each of the cylinders.

Flammable/inflammable Substance likely to catch fire.

Flywheel ring gear The starter pinion engages in this gear which is mounted on the rim of the flywheel.

Four-stroke The number of cycles which take place during full combustion.

Four-wheel drive (4WD or 4×4) This term is usually used where vehicles have selectable four-wheel drive.

Franchised dealer A firm selling and servicing a particular make of vehicle, appointed by the manufacturer.

Full-fluid film The film of lubricant is thick, where no metal-to-metal contact takes place.

Fuse Weak link in an electrical circuit to protect from current overloads.

Half shaft A shaft enclosed in an axle casing to transmit drive from the final drive to the wheel.

Harmful Substance that can cause ill-health or injury.

Hazard Something likely to cause harm or loss – a source of danger.

HC Hydrocarbon.

Helical gear Gear teeth cut at an angle.

High intensity discharge (HID) A modern method of lighting.

High resistance Current flow is restricted in the electrical circuit, e.g. corroded terminal.

Hybrid A vehicle powered by at least two power sources, e.g. petrol engine and electric motor.

Hydrodynamic lubrication Using the natural movement of the oil 'wedge' to separate the surfaces of highly loaded bearings when shafts rotate.

Hydrometer Tests equipment to check the specific gravity of a liquid.

Hygroscopic Capable of absorbing moisture from the atmosphere.

Hypoid gears Where the pinion is offset from the centre line where it meshes with the crown wheel.

Idler arm Similar to the drop arm but having a guiding function only.

IFS Independent front suspension.

Impeller A water pump which circulates coolant around the engine.

Inertia starter motor The pinion is 'thrown' into engagement due to its inertia.

Inlet manifold Allows the air/fuel mixture to be distributed from a single duct to the many required branches.

Insulator Material offering considerable resistance to current flow.

Integral Vehicle construction with no separate frame, using panels joined together to give overall strength. Can be called monocoque.

Integrated circuit A miniaturized electronic circuit consisting of components such as resistors, capacitors and transistors.

Interlock Means of preventing two gears being engaged at the same time.

IRS Independent rear suspension.

Job description States the duties and responsibilities of a particular job role.

Lambda 1 The point at which the lowest values for CO, NO_x and HC are achieved.

Laminated glass Mainly used in windscreens. Can break and puncture but will not shatter.

Lead dioxide Composition of a positive plate in a fully charged cell.

Lead sulphate Composition of a positive plate in a fully discharged cell.

Leading shoe Shoe in a drum brake system which pivots outwards into the drum.

Live axle Driven axle.

Logic probe Electrical test equipment.

Longitudinal (in-line) engine The engine is positioned in the centre line of the vehicle.

Maintenance-free battery A sealed battery which requires limited maintenance.

Malleability Ability of a metal to be shaped by compression loads.

Master cylinder A cylinder in the hydraulic circuit which pressurizes fluid.

Memory saver A device connected via the cigarette lighter socket, which provides voltage to essential consumers when the vehicle battery is disconnected.

Mission statement An explanation of the main objectives of the company, its aims and why it is trading.

Multi-point An injection system in which each cylinder has its own injector. Only air enters the inlet manifold and injectors are situated in the inlet manifold close to the valve ports.

Multigrade Oil which meets the viscosity requirements of several different single-grade oils.

Negative temperature coefficient (NTC) Material where its electrical resistance decreases when temperature increases.

NO_x Oxides of nitrogen.

Open circuit No current flows in the circuit, e.g. due to a broken wire.

Organizational chart A chart with vertical lines of authority and horizontal lines linking people with equal status.

Organizational structure Who does what in a company.

Overrun clutch Releases the pinion drive when the engine is turning faster than the starter.

Panhard rod A rod mounted between the body or chassis and the axle, to control the lateral (sideways) movement of the axle.

Parallel Two or more paths for current to follow.

Passive restraint A safety restraint system which operates automatically.

Pinion A small gear on the end of the starter motor which meshes with the ring gear.

Pitch Forward and backward rocking motion of the vehicle.

Plates A positive and negative plate are in a single cell.

Polarity conscious A device must be connected to the positive and negative terminals correctly.

Positive crankcase ventilation (PCV) A system which reduces hydrocarbon pollution by drawing crankcase vapours into the engine to be burnt and form water and carbon dioxide.

Potentiometer A variable resistor.

Power Electrical energy.

Pre-engaged starter motor The pinion is moved into mesh before the motor starts to turn.

Pre-tensioner Tensions seat belts under impact to prevent occupants from sliding under the seat belts on impact.

Pressure cap Maintains the correct operating pressure for the cooling system.

Pressure plate Bolted to the flywheel and presses the clutch plate to the flywheel face.

Propeller shaft (propshaft) Transmits torque from the gearbox to the final drive.

Rack Toothed bar.

Radial A tyre in which the plies are placed at right angles to the rim.

Rebound Downward movement of suspension.

Rectifier Changes alternating current (AC) to direct current (DC).

Regulator Controls the maximum alternator output.

Relay Electrically operated switch.

Release mechanism Means of the driver temporarily stopping clutch plate rotation.

Reluctance When the magnetic field is impaired.

Resistance Slows the flow of electrons.

Resistor A device that lowers current and voltage.

Risk Likelihood or chance of harm being caused.

Roll Sideways sway or 'heel-over' of a vehicle on corners.

Rotor Electrically magnetized North and South poles. Rotates inside the stator.

Safety cage The reinforced central section of the car body which acts as the passenger compartment.

Self-servo action Self-energizing effect which helps to apply a brake shoe to a drum.

Series Single path for current to follow.

Shock absorber A suspension device designed to absorb road shocks, i.e. a spring.

Short circuit Electrical shorts to earth before its dedicated earth, e.g. chaffed wire.

Single point A single injector system which sprays fuel for all cylinders into the air at one place, usually by the throttle body in the inlet manifold.

Slave cylinder Hydraulic cylinder moving the clutch arm.

Slip ring Carbon brushes run on these which are made of copper.

Solenoid In the case of the starter motor, a heavy-duty electromagnetic switch.

Solvents Chemicals used to clean and remove oil or grease that are often highly flammable.

Space-frame A strong chassis construction used for sports cars.

Spark ignition A term used when an engine uses a spark to ignite the fuel and air mixture.

Spongy lead Composition of a negative plate in a fully charged cell.

Spur gear 'Straight cut' teeth cut at 90° to the gear.

Static and dynamic timing Ignition timing set when the engine is stationary and when running.

Static balancing Checking a wheel's balance by seeing if it stops in the same position when rotated. If it does the wheel is imbalanced so a small weight is attached to the rim opposite the heavy spot to counter the imbalance.

Stator The static part of the alternator which houses the three windings.

Steering box Changes rotary movement into linear movement.

Stoichiometric Chemically correct ratio of fuel and air for complete combustion.

Sub-frame Detachable assembly that is mounted to the underbody of the car to support the engine, transmission, suspension, etc.

Sun gear Side gear in the differential, which is splined to the axle shaft.

Synchromesh Matching gear speeds to allow easy engagement.

Syncrohub Splined to the mainshaft and links to a selected gear.

Thermostat A temperature controlled valve which assists engine warm-up.

Three phase Three evenly spaced alternating currents.

Tie rod Connecting rod or bar, usually under tension.

Toe in or toe out Inward or outward inclination of the leading edge of the front wheels.

Torque A turning force measured in Newton metres (Nm).

Toxic Poisonous, likely to cause injury or death (often chemical).

Track rod Bar connecting the steering arms.

Trailing shoe Shoe in a drum brake system which is forced away from the drum by its rotation.

Transistor A semi-conductor which can be used to switch electronic circuits and also amplify voltage.

Transverse engine The engine is fitted across the vehicle.

Un-sprung weight The weight of those parts of the car which are not carried by the suspension.

Universal joint (UJ) Allows small angular changes on shafts.

Viscosity The resistance to flow or 'thickness' of a liquid.

Viscosity index A number which indicates how the viscosity of a liquid changes with temperature.

Volatile Evaporates readily – can cause fire or explosion, e.g. petrol.

Voltage Electrical pressure.

Voltage drop Loss of voltage occurring across any part of a circuit which is using current (consuming power).

Voltage stabilizer A device that maintains a steady voltage usually about 2V below nominal battery voltage.

VPE Vehicle Protective Equipment.

Well base rim A rim with a centre channel which enables easy removal and re-fitting of the tyre.

Wet sump Oil is returned from the engine by gravity and carried in a sump below the engine.

Working relationships The interaction between colleagues when working together towards a common goal.

Yield strength or yield point The stress at which a material begins to deform.